THE
UNIVERSAL FORCE
VOLUME 1

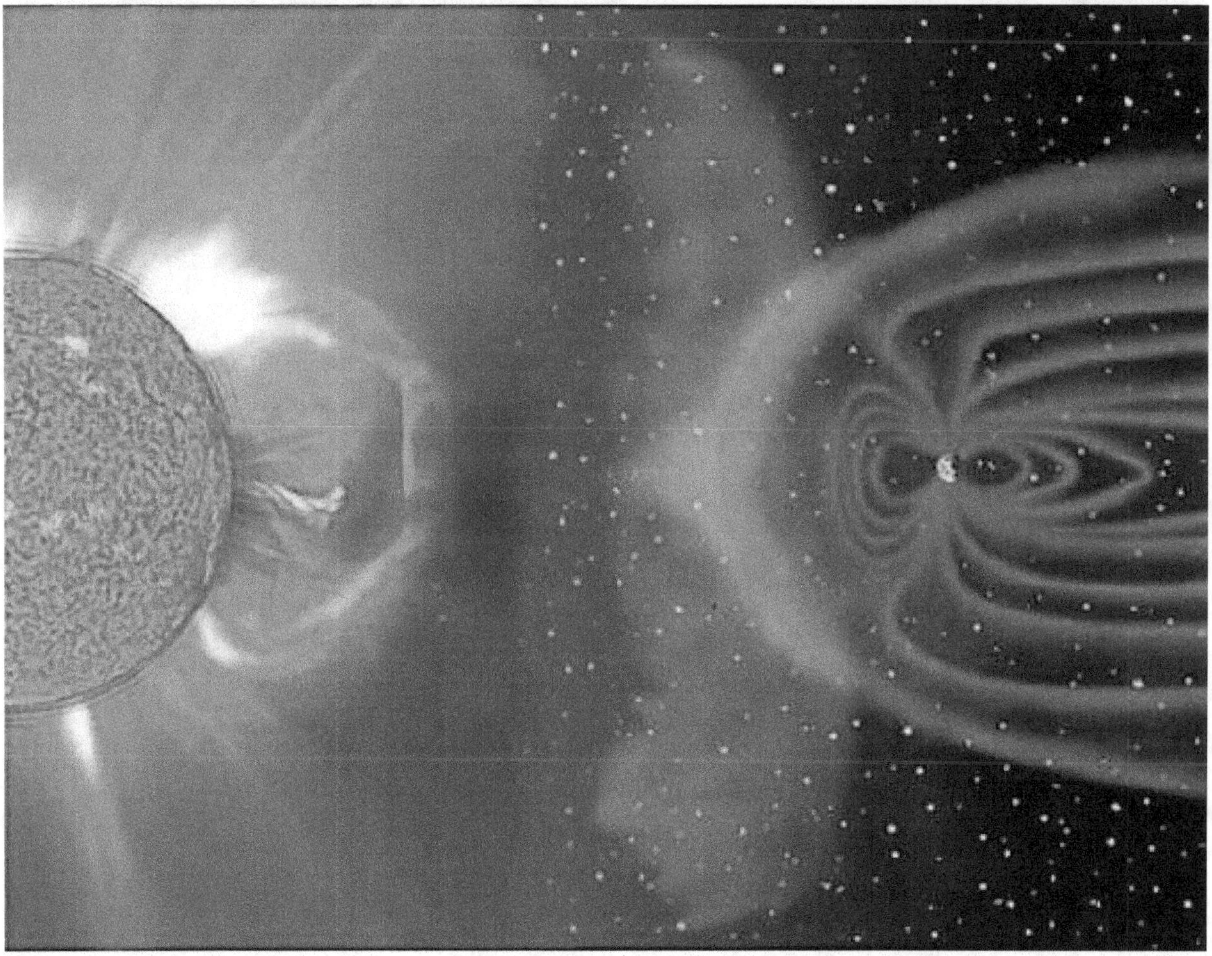

Derived From A More Perfect Union of the

Axiomatic and Empirical Scientific Methods

CHARLES W. LUCAS, JR.

Version History

1.0   May 5, 2013 203 pages

2.0   July 2013 Added Figures in Chapter 1 for Axiomatic, Empirical, and Existential Scientific Methods 207 pages

3.0   October 2013 Added Chapter 7 Hierarchy of Electromagnetic Interactions, Appendix E Poincare's Views on Relativity, Gravity, Lorentz Transformation, and Electrodynamics, and a new cover 223 pages

4.0   December 2013 Added version history, information at the beginning of Chapter 1 on the original assumptions underlying the pursuit of science which are denied by modern science, and a paragraph on the origin of quantization for gravitational orbits of planets and moons in Chapter 8   227 pages

5.0   March 2014 Added explanation of inertial mass and the anomalous Pioneer 10 and 11 acceleration to Chapter 11 on Mach's Principle. Updated Figures in Chapter 1 for Axiomatic, Empirical, Existential and Post-Modern, and Structural Scientific Methods, added sections 6.7 and 6.8 and improved formatting of the chapters 242 pages

6.0   July 2014 Added reference to the mass defects in nuclei indicating that mass is not a fundamental property of elementary particles. Forward updated. Added a new Chapter 7 Logical Arguments from Metatheory 250 pages

7.0   January 2015 Added additional Chapter 11 showing from the very precise NIST atomic and nuclear data that there is no evidence requiring the existence of neutrons in the nucleus, the strong interaction force in the nucleus, or the weak interaction force in the nucleus 263 pages

Dedication

This book is dedicated to all lovers of truth and especially the following:

Euclid and the ancient Greeks that developed geometry and the axiomatic method to "prove" or derive theories of natural philosophy in a systematic and logical way

Sir Isaac Newton who developed the empirical scientific method to measure and mathematically define the minimal set of force equations to explain nature

James Clerk Maxwell who showed how to combine four of the empirical laws of electrodynamics to develop his wave equations for electrodynamics which allowed the separate electric and magnetic force laws to be combined into a single electrodynamic force. He explained the wave nature of light which became the foundation of optics. He followed Michael Faraday and Andrè-Marie Ampère in emphasizing the role of fields in extending the electrodynamic force to great distances to replace Weber's action-at-a-distance electrodynamic force.

William J. Hooper, President and Director of Research at Electrodynamic Gravity, Inc. who published the book **New Horizons in Electric, Magnetic, and Gravitational Field Theory** in which he documented some experiments that showed that there are three types of electric and magnetic fields with different properties and these fields remain attached to the source charges to great distances making the electrodynamic force a local contact force.

Thomas L. Barnes, professor of Physics at the University of Texas at El Paso, who showed the way to eliminate Einstein's Special Relativity Theory from electrodynamics by taking into account the electrical feedback effects on finite-size charged particles.

Andre Koch Torres Assis, professor of physics at the University of Campinas – UNICAMP in Brazil, who showed the way to explain gravity as a fourth order electrodynamic effect between vibrating neutral electric dipoles using Weber's electrodynamic force.

Alice Pittard Lucas my faithful and loving wife who encouraged and supported my research that resulted in this series of books.

TABLE OF CONTENTS

Foreward 1

Preface 9

   References 11

Acknowledgments 15

    References 16

Chapter 1 The Development of Classical Science 17

  1.1   The Axiomatic Method 17

  1.2   Limitations of the Axiomatic Method 18

  1.3   Newton's Empirical Method of Axiom Discovery 19

  1.4   Limitations of the Empirical Method in Mechanics 25

  1.5   The Existential Method in Electrodynamics 26

  1.6   Limitations of the Existential Method in Electrodynamics 30

  1.7   References 31

Chapter 2 The Development of Modern Science 33

  2.1   Special Relativity Theory 33

    2.1.1   Michelson-Morley Experiment of 1886 39

    2.1.2   Photoelectric Effect 41

    2.1.3   Fields of Moving Charges 44

    2.1.4   Michelson-Morley Experiment of 1887 45

    2.1.5   Sagnac Experiment 46

  2.2   Quantum Mechanics 48

    2.2.1   Blackbody Radiation 50

    2.2.2   Atomic Emission Spectra 52

  2.3   Summary 60

  2.4   References 61

Chapter 3 Structuralism – Key to Reality and Meaning in Science 67

  3.1   Foundations of Natural Philosophy 67

  3.2   Axiomatic Philosophy 67

  3.3   Empirical Philosophy 68

  3.4   Existential Philosophy 68

  3.5   Poincaré's Philosophy of Structural Realism in Science 69

    3.5.1   Goal of Science is Mathematical Relations 70

    3.5.2   Scientific Theories Are Structures Describing Empirical Data 70

    3.5.3   Arithmetic Based on a Priori Intuition 71

    3.5.4   Geometry Based on a Priori Intuition 71

    3.5.5   Role of Generalizations in Empirical Science 72

    3.5.6   Three Ways Generalizations Are Based on a Priori Intuition 72

     3.5.7   Changes in the Size of Things with Technology 74

   3.6   **Structural Philosophy** 75

   3.7   **Postmodern Philosophy** 76

   3.8   **Universal Force Approach Based on Structural Philosophy** 77

   3.9   **References** 78

Chapter 4 Derivation of an Improved Electrodynamic Force Law for Constant Velocity .. 81

   4.1   **Proper Axioms of Electrodynamics** 81

   4.2   **Derivation of Electrodynamic Force Law** 83

   4.3   **References** 93

Chapter 5 Extension of Improved Electrodynamic Force Law to Include Acceleration 95

   5.1   **Generalized Electromagnetic Potential U(r, v)** 95

   5.2   **Acceleration Fields and Radiation** 100

   5.3   **Radiation from Real Finite-Size Particles** 101

   5.4   **References** 102

Chapter 6 Extension of Improved Force Law to Include Radiation Reaction da/dt 103

   6.1   **Radiation Reaction and Self-fields** 103

   6.2   **Outstanding Problems Using Self fields** 103

   6.3   **Derivation of Non-Relativistic Radiation Reaction Force** 104

   6.4   **Derivation of Relativistic Radiation Reaction Force** 105

   6.5   **Significance of the Radiation Reaction Boundary Condition** 106

   6.6   **The Demise of Inertial Reference Frames** 107

   6.7   **The Demise of Relativity Theory and Covariance** 107

   6.8   **References** 108

Chapter 7 Logical Arguments from Metatheory 109

   7.1   **Metatheory** 109

   7.2   **Argument that Special Relativity is of Electrodynamic Origin** 109

   7.3   **Argument that Gravity is of Electrodynamic Origin** 110

   7.4   **Argument Based on the Superposition Principle** 110

   7.5   **Conclusions from Metatheory** 111

   7.6   **References** 111

Chapter 8 Hierarchy of Electrodynamic Interactions 113

   8.1   **Charge to Charge – Coulomb Force** 113

   8.2   **Charge to Vibrating Electric Dipole – Force of Inertia** 113

   8.3   **Vibrating Electric Dipole to Vibrating Electric Dipole – Force of Gravity** 114

   8.4   **Charge to Vibrating Electric Quadrupole – Force of Inertia** 114

   8.4   **Vibrating Electric Dipole to Vibrating Electric Quadrupole – Force of Gravity** .. 114

   8.5   **Vibrating Electric Quadrupole to Vibrating Electric Quadrupole – Force of
Gravity** 115

8.6   Vibrating Charge Structures in Atoms 115

8.7   References 118

Chapter 9 Electrodynamic Origin of Gravitational Forces 119

9.1   **Introduction** 119

9.2   **Origin of Gravitational Forces** 121

9.3   **Computation of Radial Force Term** 123

9.4   **Corroborating Evidence for Radiative Decay of Gravity** 129

9.5   **Decay of the Force of Gravity** 131

9.6   **Computation of Non-Radial Gravitational Force Term** 137

9.7   **Corroborating Evidence for Spiraling Orbits** 139

9.8   **Origin of Hubble's Law Due to Gravitational Redshifts** 140

9.9   **Significance of Quantized Red Shifts** 142

9.10 **MOND vs. Dark Matter** 145

9.11 **Gravitational Bending of Starlight** 148

9.12 **Summary** 149

9.13 **Summary of the Electrodynamic Approach to Gravity** 150

9.14 **References** 151

Chapter 10 Electrodynamic Origin of Inertial Forces 155

10.1 **Introduction** 155

10.2 **Derivation of Force of Inertia from Universal Force Law** 156

10.3 **Derivation of Newton's 2nd Law from 1st Acceleration Term** 159

10.4 **Additions to Newton's 2nd Law from 2nd Acceleration Term** 161

10.5 **Summary** 164

10.6 **References** 165

Chapter 11 Evidence for Strong and Weak Forces in Nuclei? 167

11.1 **Does Nuclear Data Require the Existence of Strong or Weak Forces in Nuclei?.** 167

11.2 **An Analysis of Isoelectronic Data for One Electron Atoms** 167

11.3 **Explanation of Z^2 Dependence in Atomic Ionization Energy** 169

11.4 **Analysis of Nuclear Isotopic Masses** 170

11.5 **1st Proof that Neutrons and Weak Force Do Not Exist in Nuclei** 171

11.6 **2nd Proof That Neutrons and Weak Force Do Not Exist in Nuclei** 172

11.7 **Proof that the Strong Interaction Does Not Exist Within Nuclei** 172

11.8 **Conclusions** 177

11.9 **References** 177

Chapter 12 Structure and Symmetry of the Universe 179

12.1 **Importance of the Structure and Symmetry of the Universe** 179

12.2 **Structure Is From Symmetry of Universal Force** 180

12.3 **Symmetry of Structure of Elementary Particles** 182

12.4   Symmetry of Structure of Nuclei 183

12.5   Symmetry of Structure of Atoms 184

12.6   Symmetry of Structure of Molecules 184

12.7   Symmetry of Structure of Crystals 186

12.8   Symmetry of Structure of Plant Leaf Patterns 187

12.9   Symmetry of Structure of Plant Flower Petal Patterns 189

12.10 Symmetry of Structure of Plant Seed Head Patterns 190

12.11 Symmetry of Man and Animals 190

12.12 Symmetry of Structure of Orbits in Solar System 190

12.13 Symmetry of Structure of Milky Way 192

12.14 Symmetry of Structure of Whole Universe 193

12.15 Conclusions 194

12.16 Universal Force Law Symmetry Is Source of All Beauty 194

12.17 References 194

Chapter 13 Mach's Principle and the Concept of Mass.................................... 195

13.1   Inertial Mass 198

13.2   Gravitational Mass 198

13.3   Conclusions 202

13.4   References 204

Chapter 14 Conclusions.. 205

14.1   References 209

Chapter 15 Epilogue 211

15.1   Hooper-Monstein Experiment 211

15.2   Aharonov-Bohm Effect 212

15.2.1 Potentials vs. Fields 213

15.2.2 Global Action vs. Local Forces 214

15.2.3 Universal wave function vs. Fields 214

15.3   Unipolar Induction 215

15.4   Marinov Motor 216

15.5   References 218

About the Author.. 219

Education – degrees and awards 219

Overview of Scientific Work 219

Scientific Publications and Professional Talks 222

TV Network Shows 228

Postlude 229

Additional Books in the Series... 230

The Universal Force Volume 2 – An Electrodynamic Model of Elementary Particles 230

The Universal Force Volume 3 – An Electrodynamic Model of the Atom and the Nucleus 230

The Universal Force Volume 4 – An Electrodynamic Model of Molecules and the Origin of Life 230

Appendix A: Derivation of the Biot-Savart and Grassmann Form of Ampere's Law ... 231

    A: References 235

Appendix B: Helmholtz Decomposition Theorem 237

    B: References 238

Appendix C: Derivation of Blackbody Radiation Formula 239

    C: References 242

Appendix D: Velocity Dependent Generalized Potentials 243

    D: References 245

Appendix E: Poincaré's Views on Relativity, Gravity, Lorentz Transformation and Electrodynamics 247

    E: References 248

Index 237

TABLE OF FIGURES

Figure 1-1 Axiomatic Scientific Method..18

Figure 1-2 Newton's Empirical Scientific Method...20

Figure 1-3 Existential & Post-Modern Scientific Method 27

Figure 2-1 Scattering Data Showing Proton and Neutron Internal Structure and Finite-Size..........36

Figure 2-2 Planetary and Moon Data Supporting the Modern Version of Bode's Law 37

Figure 2-3 Red Shift Quantization Viewed From (a) Center of the Universe (b) the Earth 38

Figure 2-4 Modified Fizeau Experiment Measuring the Velocity of Light through Moving Water 40

Figure 2-5 Extinction Effect Changes Results of Michelson-Morley Experiment 41

Figure 2-6 Photoelectric Effect Apparatus and Typical Data 42

Figure 2-7 Bergman & Wesley Spinning Ring Model of Electron 43

Figure 2-8 Absorption of Light on Optical Antenna Arrays 44

Figure 2-9 Static and Moving Electric Fields of a Charge 44

Figure 2-10 Iron Filings Showing Pattern of Magnetic Fields Attached to a Magnet..........45

Figure 2-11 Michelson-Morley Experiment of 1887 45

Figure 2-12 Sagnac Interferometer and Path of Light..47

Figure 2-13 Blackbody Radiation Intensity I vs. Frequency 50

Figure 2-14 Atomic Emission Spectra for Hydrogen Atoms 52

Figure 2-15 Parson's Model of Atom for Neon 55

Figure 2-16 Lewis Dot Diagrams 55

Figure 2-17 Bostick Plasmoid Helical Spring Fiber on Toroidal Surface............................57

Figure 2-18 Bostick E and H Fields of Torus 57

Figure 2-19 Extreme Ultraviolet Spectrum for Helium and Hydrogen 59

Figure 3-1 Scattering Data Showing Proton and Neutron Internal Structure and Finite-size 74

Figure 4-1 Induction Field $B_i(r', t)$ Due to Charge q Moving with Velocity V 83

Figure 4-2 Repulsion of Like Charges 85

Figure 4-3 Attraction of Opposite Charges 85

Figure 4-4 Definition of Fields for Faraday's Law...86

Figure 4-5 Geometry of Gauss's Law...91

Figure 5-1 Curling of Corkscrew Motion of Charge...99

Figure 5-2 Searchlight Effect in Radiation Pattern for Charge in Motion $\beta \approx 1$ 102

Figure 6-1 Scattering Data Showing Proton and Neutron Internal Structure and Finite –Size.......106

Figure 8-1 Electric Charge to Electric Charge Force..113

Figure 8-2 Electric Charge to Electric Dipole Force 113

Figure 8-3 Electric Dipole to Electric Dipole Force 114

Figure 8-4 Electric Charge to Electric Quadrupole Force 114

Figure 8-5 Electric Dipole to Electric Quadrupole Force 114

Figure 8-6 Electric Quadrupole to Electric Quadrupole Force 115

Figure 9-1 Eskimo Nebula and Cat's Eye Nebula 119

Figure 9-2 Oscillations of Electrons in Neutral Dipoles 122

Figure 9-3 Cosmic Background Radiation from NASA's COBE1 Satellite 130

Figure 9-4 Spatial Distribution of Cosmic Background Radiation COBE2 NASA Skymap 130

Figure 9-5 Stretch Marks of Earth's Expansion 132

Figure 9-6 Close-up of Earth's Expansion Stretch Marks 133

Figure 9-7 US Office of Naval Research World Ocean Floor Map 1977 134

Figure 9-8 Movement of Tectonic Plates 134

Figure 9-9 Parallel Magnetic Ocean Floor Stripes 135

Figure 9-10 Expansion Cracks in Jupiter's Moon Ganymede (NASA) 136

Figure 9-11 Mares or Seas of the Earth's Moon Showing Where It Has Expanded 136

Figure 9-12 Radar Image Showing Expansion of Venus 136

Figure 9-13 Expansion Crack in Surface of Planet Mars 137

Figure 9-14 Tilt of Planetary Orbits with Respect to Equatorial Plane of Sun 139

Figure 9-15 Spiral Orbits of Jupiter's Moons 140

Figure 9-16 Red Shift of Hydrogen Absorption Lines for Near, Medium, Far Distance Stars 140

Figure 9-17 Hubble's Law Red Shifts vs. Distance or Brightness 141

Figure 9-18 Quasar Mark 205 Bound to NGC4319 Galaxy 142

Figure 9-19 Planetary Data Supporting Modern Bode's Law 143

Figure 9-20 Uranus Moon Data Supporting Bode's Law 143

Figure 9-21 Tifft's Quantized Red Shifts Support Bode's Law on Universal Scale 143

Figure 9-22 Effect of Observing Red Shifts Away From the Center of Universe 145

Figure 9-23 NGC 2403 Spiral Galaxy Graph of Rotational Velocity vs. Distance 146

Figure 9-24 Relationship of Rotational Velocity to Galactic Luminosity 147

Figure 9-25 Analysis of 900 Optical Rotation Curves of Spiral Galaxies 147

Figure 9-26 Quantization of Galaxy Luminosity and Size 148

Figure 9-27 General Relativity Predicted Bending of Starlight by the Sun 149

Figure 9-28 Observed Bending of Starlight by the Sun 149

Figure 10-1 Oscillations of Electron in Vibrating Neutral Electric Dipole 158

Figure 11-1 Ionization Energies for H Isoelectronic Sequence 168

Figure 11-2 Nuclear Isotopic Masses for Odd Atomic Weight 181 174

Figure 11-3 Isotopic Masses of Isotopes of Odd Atomic Weight 181 174

Figure 11-4 Isotopic Weights of Isotopes of Odd Atomic Weight 104 175

Figure 11-5 Mass of Isotopes of Even Atomic Weight 104 and Even Nuclear Charge 176

Figure 11-6 Mass of Isotopes of Even Atomic Weight 104 and Odd Nuclear Charge.................... 176

Figure 12-1 Spherical and Chiral Symmetry of the Electron and Toroidal Rings 183

Figure 12-2 Spherical and Chiral Symmetry of the Oxygen-16 Nucleus 183

Figure 12-3 Spherical and Chiral Symmetry of Neon-20 Atom 184

Figure 12-4 Chiral Symmetry of Simple Molecules 185

Figure 12-5 Chiral Symmetry of Complex Organic Starch and Protein Molecules 185

Figure 12-6 Chiral Symmetry of Complex Organic DNA Molecule 186

Figure 12-7 Spherical and Chiral Symmetry of Snowflakes 187

Figure 12-8 Spherical and Chiral Symmetry of Plant Leaf Patterns 188

Figure 12-9 Chiral Symmetry of Leaf Shapes 188

Figure 12-10 Spherical and Chiral Symmetry of Flower Petal Patterns 189

Figure 12-11 Spherical and Chiral Symmetry of Seed Head Patterns 190

Figure 12-12 Chiral Symmetry of Human Body 190

Figure 12-13 Spherical and Chiral Symmetry of Planetary Orbits 191

Figure 12-14 Spherical and Chiral Symmetry of the Orbits of Jupiter and Its Moons 191

Figure 12-15 Spherical and Chiral Symmetry of the Milky Way Galaxy 192

Figure 12-16 Spherical and Chiral Symmetry of Hoag's Ring Galaxy 193

Figure 12-17 Spherical and Chiral Symmetry of the Universe from Red Shift Data 193

Figure 13-1 Schiff Spin-Spin Precession and Gravity Probe B 197

Figure 13-2 NGC 6946 Spiral Galaxy Graph of Rotational Velocity vs. Distance R from Center.... 200

Figure 13-3 Relationship of Rotational Velocity to Galactic Luminosity 201

Figure 13-4 Anomalous Acceleration on Pioneer 10 and 11 201

Figure 15-1 Hooper-Monstein Experiment 211

Figure 15-2 Aharonov-Bohm Effect... 212

Figure 15-3 Magnetic Field of a Solenoid 214

Figure 15-4 Faraday Disc Generator and Schematic... 215

Figure 15-5 Marinov Motor 216

Figure C-1 Bergman Toroidal Ring Electron 232

Figure C-2 COBE Blackbody Spectral Distribution 235

Figure D-1 Dirac Strings and Magnetic Monopoles in Spin Ice Dy2Ti207 237

Figure E-1 Ptolemy's Epicycle and Deferent Orbits 240

Foreword

Dr. Lucas was an honor student as an undergraduate at the College of William and Mary in Williamsburg, VA where he received his B.S. and Ph.D. in physics. He was a member of the ΣΠΣ Physics National Honor Society and the ΦBK liberal arts National Honor Society. Historically the members of ΦBK at the College of William and Mary were instrumental in leading the American Revolution. Following in that liberal arts tradition, Dr. Lucas is attempting to lead a revolution in modern science based on universal truth.

He notes in this book that the Postmodern Philosophy of Science currently dominates the college and university approach to science. In the Postmodern Philosophy each field of scientific study has its own internally defined notions of truth or reality. As a consequence, the truth or validity of different fields of study are no longer based on logical notions of truth shared across all fields as in the liberal arts approach of the "uni-versity = one-version or truth". Each field is supervised by a group of experts in order to police the borders of that field with criteria for inclusion and exclusion. Each field of study defines its own unique criteria for truth and meaning. Each field is now a somewhat independent silo of knowledge and truth. Dr. Lucas wants to change that and return to the ancient Greek approach to natural philosophy based on the logical approach of the axiomatic method. He wants the university to return to a "unified" logical approach to knowledge and truth instead of independent silos of knowledge and truth.

According to Dr. Lucas the ancient Greeks developed natural philosophy by defining the appropriate axioms or postulates of natural philosophy which were constructed from a few terms taken as primitives. They were most successful in geometry and mathematics where they could intuitively guess reasonable primitive terms and axioms. Euclid showed how to develop geometry from a minimal set of these primitive terms and axioms using the axiomatic method of logical "proofs" of theorems. These terms and axioms needed to be defined and constructed by some method which could warrant their truth. To the extent that the postulates and primitive terms used were valid, the logically developed theorems would be valid.

The ancient Greeks academies of Plato and Aristotle were so impressed with geometry that they felt that all educated men should be knowledgeable of geometry. They believed, as does Dr. Lucas, that geometry holds the key to understanding and explaining the structure of the universe on all size scales. The axiomatic method of geometry is very important not only as a method of logical organization and presentation of the field, but also geometry is intimately involved in the very core or essence of all fields of natural philosophy through the symmetry of structures. This book develops that theme.

Despite all the successes of the ancient Greek axiomatic method for developing natural philosophy, there were some problems. In physics and other areas of natural philosophy, the ancient Greeks were unable to **discover** the appropriate axioms and primitive terms. Thus little progress was made for thousands of years in some areas. This was due to the fact that the axiomatic method is a method of logical organization of proofs of theorems or theories, but not a method of discovery.

To the rescue comes Isaac Newton. When he published his **Principia** or (**Mathematical Principles of Natural Philosophy**), he stated that he intended to illustrate a new way of doing natural philosophy that overcomes some of the limitations of the axiomatic method. Newton claimed that in the past natural philosophers tried to understand nature in vain, because they did

1

not use an empirical approach based on experimentation. The goal of Newton's method was to find empirically the complete set of the forces of nature and to express them in mathematical form in terms of primitives.

Newton successfully developed the force of gravity and the force of inertia. However, he found that the empirical approach does not lead to all truth at once. Even though Newton could successfully define mathematical equations for the force of gravity and inertia, he did not understand the causes and nature of gravity and inertia. Also some of the primitive terms were not well understood. These included the gravitational mass and the inertial mass. When the experimental inertial and gravitational masses were found to be proportional to one another for the same bodies, Newton realized that instead of the force of inertia and the force of gravity being different fundamental forces, they might have a common cause. Furthermore, Newton did not know or understand how the force of gravity was conveyed from the earth to the moon or from the earth to the sun through empty space.

Later empirical scientists Ampere, Faraday, Gauss, Lorentz, and Lenz developed a set of six empirical laws for electric and magnetic phenomena in terms of the primitives of charge, distance, velocity and fields. Maxwell took four of the six known empirical laws and developed his wave equations for electrodynamics. Maxwell's wave equations allowed the separate electric and magnetic force laws to be combined into the electrodynamic force law. The wave nature of light was explained and served as the foundation of optics. In order to get his wave equations into an acceptable form, Maxwell had made the point particle approximation, assumed no contributions from non-linear field effects, assumed static and induced fields obeyed the Superposition Principle, and defined an idealized displacement current within capacitors.

At the time of Faraday, Ampere, Gauss, Weber and Maxwell, the concept of conservation of energy was just being developed by many different scientists. No two scientists involved defined energy conservation quite the same way. Rather the development of the concept was rapid and disorderly being driven by the needs of the industrial revolution. Faraday never accepted these developing concepts of energy conservation for religious reasons. As a result Maxwell's electrodynamics was developed by his followers in terms of the vector potential of vector calculus based on the properties of vectors instead of the potential energy for energy conservation as had been done by Weber a few years earlier. The vector potential approach was specifically based on Gauss's Law for magnetism that stated that there were no magnetic monopoles.

As time passed a number of experiments were discovered that did not agree with Maxwell's electrodynamics. These included the Michelson-Morley experiment for the velocity of light in moving media, the blackbody radiation experiments, and the photoelectric experiments. Einstein's Theories of Special Relativity and the Copenhagen version of Quantum Mechanics were invented to supplement Maxwell's electrodynamics. They were incorporated into electrodynamics via the vector potential. Then in 2009 magnetic monopoles were discovered bringing into question the whole vector potential approach. (See **Appendix D**)

At this point in time this book begins by implementing a more complete or perfect union between the axiomatic and empirical scientific methods in electrodynamics as specified by Isaac Newton. This is done by (1) using the complete set of the six empirical equations for electric and magnetic fields, (2) not using the point particle approximation or idealization, (3) not assuming only a linear superposition of fields, and (4) not using the idealized displacement current in electrodynamics but using a more general form of Ampere's Law known as the Biot-Savart Law. The resulting version of electrodynamics appears to be able to explain all the electrodynamic experiments for which Einstein invented Special Relativity Theory, but without needing to use any part of Special Relativity Theory.

Dr. Lucas makes the claim that the electrodynamic force is a local contact force based on the theoretical work of Cullwick and the experiments of Hooper. The experiments of Hooper show that the electric and magnetic fields of a charged particle, such as an electron, always remain attached to it. **Thus the particle of light, called the photon, does not exist separately from the charged particle! Only waves of electromagnetic radiation in the fields of charged particles exist!** More recent versions of the Michelson-Morley experiment, the photoelectric effect on amorphous metals and optical circuits, and blackbody radiation are consistent with this view. **Thus the outstanding problem of classical and modern science to explain how forces are delivered has been explained! No aether or quantum vacuum is needed to convey the force!**

Dr. Lucas then goes on to extend the electrodynamic force law to include acceleration and radiation by using conservation of energy to define the electrodynamic potential energy. From the potential energy Dr. Lucas derives an extended electrodynamic force law including acceleration terms. He then shows that these terms in the electrodynamic force describe both the non-relativistic and the relativistic empirical laws for radiation.

Dr. Lucas extends the electrodynamic force a second time using conservation of energy to include the relativistic da/dt terms for radiation reaction or recoil. This is done by integrating the total power radiated formula by parts assuming the electrodynamic force is a continuous and regular function mathematically. The experimentally observed radiation reaction force is obtained under the unique boundary condition that charged elementary particles consist of finite-size closed loops of moving charge. This boundary condition then becomes the basis for the electrodynamic structure of all elementary particles using combinatorial geometry which is the subject of **The Universal Force Volume 2 - An Electrodynamic Model for Elementary Particles**.

Once the electrodynamic force is more completely defined as a function of the relative variables **r, v, a,** and **da/dt** instead of just **r** and **v**, idealized inertial reference frames are no longer appropriate for electrodynamics. One can now use the true reference frame that takes into account all types of electromagnetic interactions with all the matter of the universe including absorption and emission of radiation and the radiation reaction or recoil from absorption and emission which are missing in relativistic Maxwellian electrodynamics.

Dr. Lucas notes that all the results of special relativity for electrodynamics were derived by using the complete set of the empirical equations of electrodynamics and removing the point particle idealization. Without the point particle idealization the vector potential approach to electrodynamics cannot support the notion of covariance.

From metatheory, the theory of theories developed by Henri Poincaré, Dr. Lucas presents the argument that no two fundamental theories can use the same fundamental constant such as c. Since electrodynamics uses c in the wave equation, special relativity uses c in the definition of the space-time interval, quantum mechanics uses c in the definition of the energy quantum, and general relativity uses c in the Einstein field equation, the logical argument is made that only one of these four theories could be a fundamental or true theory.

Already Dr. Lucas has shown in his derivation of an improved electrodynamic force law that the results of special relativity are due to electrical feedback effects on finite-size elementary particles. Furthermore he has shown in Appendix C of this book that black body radiation is really due to physical oscillations of charge in the finite-size toroidal ring structure of the electron and not the unphysical point electron oscillations of quantum mechanics which are too large for the electron to stay bound to the atom.

Next Dr. Lucas presents the argument from metatheory that no two fundamental force laws can have the same mathematical form such as $1/R^2$. He notes that the electrodynamic force law and Newton's universal force of gravity both have a $1/R^2$ form. Since Einstein's general theory of relativity involves the fundamental constant c, Dr. Lucas reasons that gravity must also be of electrodynamic origin.

Finally Dr. Lucas uses an argument from metatheory based on the Superposition Principle for coherence and stability. According to the Superposition Principle only linear theories may be combined to coherently describe stable matter in the form of an elementary particle, atom or molecule. Since electrodynamics is a nonlinear theory, special Relativity is a nonlinear theory, quantum mechanics is a linear theory, and General Relativity is a nonlinear theory, no two of these theories may be combined to describe a coherent stable system. Only linear theories may be combined. Only one nonlinear theory is possible. If a nonlinear theory is valid, it has to be the one and only theory, i.e. the universal theory.

Thus Dr. Lucas concludes that for his derived nonlinear version of electrodynamics to be a proper theory, it must be the universal theory. It must explain all the forces in nature, relativistic effects, and quantum effects. This book is the first in a series of four books on the universal force in which Dr. Lucas is attempting to show that his derived version of the electrodynamic force explains all the forces in the universe, the nature of all elementary particles, atoms, nuclei, molecules, crystals, living systems, solar system, Milky Way galaxy, and the overall structure of the universe.

At this point in the book Dr. Lucas begins an examination of the hierarchy of electrodynamic multipole interactions. He starts by deriving the force of gravity as a small fourth order $(v/c)^4$ effect in his electrodynamic force between vibrating neutral electric dipoles consisting primarily of atomic electrons and nuclear protons. In his derivation he uses only one of the many periodic motions of charges in the atom. Later volumes in this series will use all the periodic motions within elementary particles and the atom to predict the measured value of the universal gravitation constant G as well as the masses of all the elementary particles.

In the simplified derivation for only one periodic vibrational motion in the atom, the derived gravitational force was found to have the customary attractive only radial term of Newton's Universal Law of Gravitation ($F = -Gm_1m_2/R^2$) plus a new non-radial term. From the radial term the gravitational mass can be associated with certain electrodynamic parameters such as the charge, the frequency of periodic vibration, and the amplitude of vibration. The non-radial term gave rise to an $(\mathbf{R \cdot V}) \mathbf{R} \times (\mathbf{R} \times \mathbf{V})$ effect which causes the ground state of the orbits of the planets about the sun to spiral around a thick toroidal circle about the sun giving the appearance of an elliptical orbit tilted with respect to the equatorial plane of the sun. This effect has not been explained by previous theories of gravity. Also this term explains the origin of quantization in gravitation known as Stanley Dermott's modern version of Bode's Law. This type of term is also the source of quantization in the electrodynamics of the atom which is the subject of **The Universal Force Volume 3 - An Electrodynamic Model for Atoms and Nuclei**.

The vibrational mechanism causing the gravitational force was found to decay over time giving rise to the cosmic background radiation and Hubble's red shifts versus distance law due to gravitational red shifting. A reasonable range of vibrational amplitude for the electron is able to explain many orders of magnitude change in the observed red shifts of light from distant stars. The vibrational mechanism combined with the $(\mathbf{R \cdot V}) \mathbf{R} \times (\mathbf{R} \times \mathbf{V})$ effect also explains Tifft's quantized red shifts as a type of Bode's Law indicating that there is a geometrical center to the universe. The vibrational mechanism by which gravity decays by radiation over time explained Tifft's measured rapid decay of the magnitude of red shifts over time. The Tulley-Fisher relationship for luminosity of spiral galaxies and Roscoe's observed quantization of the luminosity of 900 spiral galaxies in a manner reminiscent of Bode's Law is explained by the $(\mathbf{R \cdot V}) \mathbf{R} \times (\mathbf{R} \times \mathbf{V})$ term of the electrodynamic theory of gravity. The unexpectedly high velocity of the outer stars of spiral galaxies is explained simply by the decay of mass and conservation of energy without resorting to the use of outlandish ideas such as dark matter and dark energy. The measured quantization of the luminosity and size of spiral galaxies is also explained by the new $(\mathbf{R \cdot V}) \mathbf{R} \times (\mathbf{R} \times \mathbf{V})$ term in the derived gravitational force.

This electrodynamic explanation of gravity indicates that mass is not a fundamental quantity of nature. Thus the notions of mass that are intrinsic to Newton's Universal Law of Gravitation and Einstein's General Relativity Theory appear to be false. This is further emphasized by the unexpectedly large number of diverse astronomical phenomena explained by this electrodynamic approach to gravity.

Next Dr. Lucas goes on to derive the force of inertia from his improved electrodynamic force law assuming it is due to a small secondary electrodynamic force between a unit charge and a vibrating neutral electric dipole. As in the case of gravity, he again obtains two terms. The first term corresponds to Newton's force of inertia. From that term the electrodynamic definition of inertial mass is found to be proportional to the electrodynamic definition of gravitational mass. The second term describes the unusual gyroscope experiments of the famous British engineer Eric Laithwaite that violate Newton's laws.

When Albert Einstein developed his general theory of relativity, he stated without proof the assumption that the correspondence between inertial and gravitational mass is not accidental and that no experiment will ever detect a difference between them. Dr. Lucas points out that in General Relativity Theory the effects of *gravitation* are ascribed to space-time curvature instead of a *force*. Thus in General Relativity Theory gravity is not a force, but a mass energy effect not subject to Newton's third law. **So from the framework of General Relativity Theory the equality of inertial and gravitational mass remains an unexplained mystery!**

Finally this approach to gravity, which is based on the derivation of an improved electrodynamic force law, appears to be more satisfying than all previous approaches. **First**, it confirms the modern version of Bode's Law and the quantization of gravitation due to the $(\mathbf{R \cdot V})$ $\mathbf{R} \times (\mathbf{R} \times \mathbf{V})$ term which requires all physical systems involving motion to be quantized in order to have spatial stability. The motion of the planets spiraling around the sun must return to the same starting point on the spiral or there is no stability. **Second**, this approach is simpler, since it is based on a single electrodynamic force law. **Third**, this force is a local contact force based on the electromagnetic fields of a charge extending the range of the force instead of an action-at-a-distance concept like that used in Newton's Universal Force Law and relativistic Maxwellian electrodynamics. Natural philosophers have known for thousands of years that there is no such thing as an action-at-a-distance force. Some mechanism such as the electromagnetic field is needed to extend forces through space. **Fourth**, this approach explains more gravity-relevant data than all previous theories of gravity combined including the modern version of Bode's law for the quantization of gravity, the tilts of the orbits of the planets with respect to the equatorial plane of the sun, the expansion of the planets and moons of the solar system, the origin of the cosmic background radiation, Hubble's law for red shifts, the quantization of red shifts, the fast decay of red shifts, the quantized Tulley-Fisher relationship for luminosity and size of spiral galaxies, and the unexpectedly high velocity of the outer arms of spiral galaxies according to Newtonian dynamics.

Following Newton's methods, Dr. Lucas declares his derived electrodynamic force law as the universal force law. It has already explained the electric, the magnetic, the gravitational, and the inertial force laws. The only force laws remaining to be explained are the strong interaction force within the nucleus and elementary particles and the weak nuclear force for beta decay. Both of these forces are very short-ranged with a range on the order of the measured size of elementary particles. Dr. Lucas claims that these forces are due to finite-size effects and were invented to compensate for incorporating the point particle idealization in electrodynamics, relativity theory and quantum mechanics.

In order to prove his point about the strong and weak interaction forces being due to just finite-size effects between elementary particles, Dr. Lucas examines the highly accurate NIST atomic and nuclear data to show that the forces within the nucleus are exclusively electromagnetic with no need to invent the idealized strong and weak forces to explain nuclear data. The volumes in this series on elementary particles and the atom and nucleus will explain all the experimental data for elementary particles, the atom, and the nucleus without resort to these imaginary idealized forces.

Mach's Principle is the name given by Einstein to a general principle credited to the physicist and philosopher Ernst Mach. Einstein once said in a letter to Ernst Mach that "inertia originates in a kind of interaction between bodies". He found that certain experiments like the Lense-Thirring effect provided evidence for Mach's Principle. In this book Dr. Lucas claims that the kind of interaction between bodies to which Einstein attributed the force of inertia has been found. It is the electrodynamic force involving vibrating neutral electric dipoles.

The concept of mass upon which Einstein founded his General Theory of Relativity and Newton founded his Universal Force of Inertia and Universal Force of Gravitation is no longer valid. The empirical constant, called mass m_0 in the past, is not really constant. It is decaying over time. The value of the mass changes universally with distance from the center of the universe according to Mach's Principle that "local physical laws are determined by the large scale structure of the universe". Local asymmetries cause mass to change with distance from the center of spiral galaxies, with distance from a star to a planet, and with distance from a planet to a moon. Unlike the Standard Model of Cosmology based on General Relativity Theory and quantum mechanics, the universal electrodynamic force approach does not need to invent dark matter and dark energy to explain the higher than expected velocities of the spiral arms of spiral galaxies and the expansion of the universe. These phenomena, for which idealized dark matter and energy were invented to explain, are explained directly from the universal electrodynamic force law.

In summary it appears that this book presents a very strong case for a major revolution in modern science, where science will essentially go back to a more perfect version of classical science. The previous imperfections of classical science are removed by perfecting the union of the axiomatic and empirical scientific methods and removing idealizations.

Furthermore, the universal electrodynamic force has a combination of spherical and chiral symmetry. According to Dr. Lucas this symmetry appears in the internal structure of elementary particles, atoms, nuclei, molecules, crystals, plant leaves, plant flowers, animal skeletons, and the structure of solar systems, galaxies, and the universe as a whole. The complete set of possible structures of these systems will depend on combinatorial geometry in conformance with the presumption of the ancient Greek philosophers.

According to Dr. Lucas the electrodynamic longitudinal vibration of organic molecules is the origin of life. In **The Universal Force Volume 4 Molecules and the Origin of Life**, he claims that the regulation of the cell is electrodynamic in nature. This approach also explains the observation that life can only come from life.

Furthermore Dr. Lucas claims that the symmetry of the universal force law will also affect the structure of the brain. Not just any type of language, thought or process can be entertained in the brain. This will cause a limited range of structures in language or linguistics as has already been observed by the Russian linguists. It will cause only certain types of mathematics to be conceivable as has already been observed by the French Bourbaki. Only certain types of ideas can be entertained by human brains in terms of economics, architecture, psychology, sociology, art, music, law, etc.

Finally Dr. Lucas claims that the symmetry of the universal electrodynamic force law is the source of all beauty in the universe. **Thus the nature of the proposed revolution in science is more far reaching than any previous revolution! And it has Platonic beauty!**

Gregory D. Volk - President, National Philosophy Alliance (NPA)

http://www.worldnpa.org
Editor, World Science Database
http://www.worldsci.org/php/

Preface

The grand aim of all science is to cover the greatest number of empirical facts by logical deduction from the smallest number of hypotheses or axioms.

Albert Einstein [1]

In this book series the way is shown to develop the universal force law and all of science from a more perfect union of the ancient Greek axiomatic method of "proof" used in Euclidean geometry and Newton's empirical scientific method to measure and mathematically define the minimal set of empirical force laws to explain nature. One might ask, "What is wrong with the current theories of physics and science in general?" The answer is that they are (1) based on idealizations and approximations instead of reality, (2) ignore truth developed in other areas of science and mathematics, and (3) fail to use the axiomatic method properly, as defined by Euclid and amended by Isaac Newton, to direct natural philosophy towards universal truth through the proper use of logic and experiment.

This problem with physics seems to have originated in the research universities with specialties in various disciplines such as chemistry, biology, astronomy, physics, sociology, psychology, economics, mathematics, geology, etc. Over time each of these disciplines became silos of information that developed somewhat independently of one another. The "truth" developed in one silo was sometimes ignored in another. The unity of natural philosophy and the concept of universal truth obtained through the diligent application of logic were lost. In this book series many examples will be given along the way where experiments and truth were overlooked for political expediency in order to maintain the current "politically correct" theories and world view.

Lee Smolin in his book **The Trouble with Physics** [2] claims that physics, the current foundation for many of the sciences, has lost its way. The pillars of modern physics, i.e. electrodynamics, quantum mechanics and relativity theory, have led scientists toward a string theory of everything that attempts to explain all the particles and forces of nature as well as how the universe began and evolved over time. But as Smolin reveals, there are fundamental flaws in string theory and its approach. No part of it has been tested. The string theories of everything appear to come in an infinite number of versions. This means that no experiment will ever be able to prove it false. Thus it fails to meet the fundamental testability criteria for being a scientific theory.

In addition the pillars of modern physics have an embarrassing problem of infinities. General relativity theory predicts the existence of black holes. Inside a black hole the density of matter and the strength of the gravitational field quickly become infinite. Quantum mechanics has a similar problem whenever it is used to describe fields, such as the electromagnetic field. The electric and magnetic fields have values at every point in space. This means that there are an infinite number of quantum variables for the field, because in any finite volume of space there are an infinite number of points. In quantum theory there are uncontrollable fluctuations in the values of every quantum variable. An infinite number of quantum variables, fluctuating uncontrollably, can lead to equations that get out of hand and predict infinite numbers when questions are asked about the probability of some event happening or the strength of some force.

Thus quantum mechanics and relativity theory appear to be an incomplete description of reality. According to Smolin, something is missing in these theories that will probably be discovered as progress is made in the effort to further unify physics.

This book is the first part of a series in which the ultimate goal of physics and science in general, will be accomplished via the derivation of the universal force law for real particles of finite-size and internal structure. The Copenhagen version of quantum mechanics and Einstein's special and general theories of relativity will be replaced with an improved version of the classical electrodynamic force law for finite-size particles with internal structure. **This improved version of the electrodynamic force law is being proposed as the "universal" force law.**

Smolin's **Theory of Principle** is identified in this book as the improved version of electrodynamics. The theory of elementary particles, atoms, nuclei, molecules, crystals, living cells, leaf structures, flowers, solar systems, galaxies, and the universe as a whole are described by the equations and models of **Constructive Theories** based and defined in terms of electrodynamics.

Currently in physics the quantum electrodynamic **Standard Model of Elementary Particles** attempts to describe the particles in nature in terms of 6 leptons, 6 quarks and four forces of nature mediated by bosons, i.e. gravity, electrodynamics, and the strong and weak interaction forces. However, for a fundamental theory, it has a big problem in that it must employ at least 19 adjustable constants in order to work. The fundamental theory cannot predict the values of these constants. Any value is consistent with the theory. These constants specify the properties of the particles, the masses of the quarks and the leptons, and the strength of the forces. The necessity of employing adjustable constants is a tremendous embarrassment for any fundamental theory. Each of these fundamental constants represents some basic information or physics that is missing from the theory.

The purely electrodynamic theory of finite-size elementary particles with internal structure, which is presented in the second book of this series on the universal electrodynamic force, is based on three-dimensional closed loop finite-size charge current structures being the building blocks of elementary particles. In this approach all elementary particle properties, decays, reactions, rest masses, and excited states are predicted from electrodynamics and combinatorial geometry for the internal structures of elementary particles alone **with no adjustable constants**. In particular the masses and excited states of an elementary particle are determined by the vibrations and rotations of the toroidal ring charge elements of the internal structure of the elementary particle.

According to Smolin the current **Standard Model of Cosmology** based on the pillars of modern science is unable to explain the velocity of the stars in the spiral arms of galaxies without inventing a new type of matter to rescue general relativity theory, called "dark matter". This special type of matter comprises over 90% of the matter in spiral galaxies and cannot be observed directly.

In the purely electrodynamic theory of gravity and inertia presented in this book, the dynamics of the spiral arms of galaxies is explained in terms of Mach's Principle and the electrodynamic concept of mass. Furthermore the electrodynamic theory of gravity explains the cosmic microwave background radiation at around 2.7^{O} K in terms of the decay of gravity. This logically superior and simpler description of the force of gravity does not require the invention of idealized "dark matter" or "dark energy" in order to successfully describe the motion of the outer spiral arms of galaxies. It also does not require the invention of the idealized **Big Bang** of the **Standard Model of Cosmology** to explain the cosmic background radiation.

According to Smolin, observations via the Hubble redshifts of the universe at larger scales than that of galaxies, corresponding to billions of light years, shows that the equations of general relativity are not satisfied even with the addition of "dark matter". According to the **Standard Model of Cosmology**, the expansion of the universe, set in motion by the **Big Bang** about 13.7 billion years ago, appears to be accelerating. However, given the amount of observed matter plus the calculated amount of "dark matter", it should be doing the opposite, i.e. decelerating. The chosen solution to rescue general relativity theory again is the invention of still another form of idealized "dark matter" or "dark energy" (using $E = mc^2$) that only becomes relevant on these very large scales. This new form of matter affects only the expansion of the universe.

In the purely electrodynamic theory of gravity presented in this book, the Hubble redshift data used to support the notion of the expansion of the universe is explained in terms of a decaying gravitational redshift. This logically superior and simpler description of the redshift data does not require the invention of idealized "dark matter" and "dark energy".

As Smolin points out in his book, the current pillars of modern science, i.e. quantum mechanics and relativity theory, have significant problems and need to be redone or replaced. Work in string theory has revealed the possibility and desirability of accomplishing that goal using a universal force law.

This book series is attempting to lead a reformation in the foundations of science. The reformation will extend beyond the universal force law to models for elementary particles, nuclei, atoms, molecules, crystals, solar systems, galaxies, and the structure of the universe as a whole. It will include the operation of living cells and the nature of life. Also the symmetry of the universal force will be shown to affect the molecular structure of the brain and the kinds of actions and notions that can be entertained by the brain. Then this unique structure of the brain will be used as the foundation for the fields of linguistics, sociology, psychology, economics, architecture and mathematics. Thus this book series is attempting to lead a most major reformation in science.

This first book in the series is attempting to reach the high standards of Euclid's **Elements of Geometry** and Newton's **Mathematical Principles of Natural Philosophy**. Euclid set the standard of natural philosophy, which includes the sciences, for developing theories using logic in the form of a proof. That proof is the heart of this book. Modern science has deviated from the rigorous proofs of Euclid. Under the Postmodern Philosophy of Science one guesses hypotheses or theories, determines consequences of the hypothesis, and performs experiments to confirm or falsify the hypothesis. Falsification in describing experimental data is the primary method for disproof of hypotheses. **Hypotheses are no longer accepted or rejected based on rigorous logical foundations called postulates, assumption or axioms!**

In the early chapters of this book the history of deviations of science from the logical model of Euclid are traced through various philosophies of science. Then a rigorous proof or derivation of a proper electrodynamic force law is presented. In order for most undergraduate math majors, physics majors, and electrical engineers to be able to follow the proof/derivation through every step, many smaller steps are given in the proof than most science journals will allow. The theoretical results of this derivation are model independent, although the radiation reaction term does require that the basic building block of all elementary particles be a closed charge current loop. Note that the derivation does not specify that only one closed current loop makes up the entire structure of the electron or any other elementary particle. They can consist of many charged loops in accordance with combinatorial geometry and the symmetry properties of the electrodynamic force as constraints.

Physical models of elementary particles, atoms, nuclei, and molecules are always based on theories. Theories are more fundamental than models. Theories are based on postulates or axioms. Following Newton all postulates in physics should be empirical laws. This work uses the complete set of the empirical laws of electrodynamics as the postulates for electrodynamics. No other approach, such as the relativistic version of Maxwellian electrodynamics, uses a complete set. Because the theory in this book is developed properly, it is not necessary to employ any approximations or idealizations in order to obtain a finished or complete theory. There is no longer a need to supplement an incomplete or imperfect version of electrodynamics with quantum mechanics and relativity theory.

Henri Poincaré, the mathematician and co-inventor of Special Relativity Theory with Einstein, argued from metatheory (the theory of theories) that no two fundamental theories may employ the same fundamental constant such as c for thye velocity of light. (See **Chapter 7** and **Appendix E**) Since electrodynamics, relativity theory and quantum mechanics all employ the fundamental constant c only one of them could be a fundamental theory. Poincaré predicted that all of them would eventually be discovered to be of electrodynamic origin as this book shows. [3] Poincaré also showed from meta-theory that no two fundamental theories could have the same mathematical form. Since Coulomb's electrostatic force is a $1/r^2$ force law and Newton's universal law of gravitation is a $1/r^2$ force law, Poincaré argued that gravity would also be discovered to be of electrodynamic origin as this book shows.

In the last chapters of the book the force of gravity and the force of inertia are derived from the electrodynamic force law. The purpose of these chapters is to go beyond the normal Euclidean proof and to show that the derived electrodynamic theories of gravity and inertia explain many more experimental phenomena than previous theories. The electrodynamic approach to gravity and inertia also explains that these forces are not action-at-a-distance forces, but local contact forces which natural philosophers have known for thousands of years. This work confirms the argument of Poincaré from metatheory that no two fundamental theories may employ the same mathematical form, i.e. $1/R^2$ of electrodynamics and gravity.[3]

In the chapters on gravity and inertia mass is defined electrodynamically. The Standard Model of Elementary Particles requires at least 19 adjustable fundamental parameters to explain the observed masses and properties of elementary particles. This electrodynamic approach requires no adjustable parameters (see **The Universal Force Volume 2 - An Electrodynamic Model of Elementary Particles**). In the electrodynamic approach mass is no longer a fundamental property of matter. It is a secondary electrodynamic property due to periodic motions in the charge structures which make up all matter.

Once mass is defined electrodynamically, Newton's Universal Gravitation constant G can be expressed entirely in terms of electrodynamics. No other theory in modern science can derive the value of G and mass from theory alone.

Modern physics, as currently taught in the scientific community, is based upon four fundamental forces, i.e. the electrodynamic force holding atoms, molecules and crystals together, the force of gravity holding the solar system and galaxies together, the strong force holding protons and neutrons together in the nucleus plus quarks together in heavy hadronic elementary particles, and the weak force governing nuclear and elementary particle decays. Using the highly accurate NIST atomic and nuclear data, it is shown that no role is played by the strong interactions in holding protons and neutrons together in the nucleus and the weak interaction in controlling the decay of nuclei.

Finally, based on the superior electrodynamic force plus the superior electrodynamic forces of inertia and gravity, the conjecture is made that this properly derived electrodynamic force is the universal force. The quest for the universal force has been the highest goal of science for the last 2500 years. This series of books is an attempt to complete that task.

Whenever new unifications or reformations are proposed in science, there are well known criteria to use to select the best one. Usually the best proposal for unification is the one that explains the most phenomena, can predict new phenomena never explained before and is the simplest. This series of books attempts to do just that with the additional claim to truth based on a proper logical union of the axiomatic and empirical scientific methods. Also this work attempts to show that the unique combination of spherical and chiral symmetry of the electrodynamic force is found in all structures on all size scales in the universe.

References

1. Albert Einstein http://www.einstein-quotes.com/Science.html
2. Lee Smolin, **The Trouble with Physics – the Rise of String Theory, the Fall of a Science, and What Comes Next** (Houghton Mifflin Company, Boston, 2006).
3. Poincaré, Henri, **Oeuvres de Henri Poincaré** edited by G. Darboux (Gauthier-Villars, Paris) **Vol. 9**, p. 497 (1954).

Acknowledgments

If I have seen further it is by standing on the shoulders of giants.
Isaac Newton [1]

The work on this book began in 1978 when **Dr. Thomas G. Barnes** of the Physics Department at the University of Texas at El Paso published a rudimentary paper entitled "A New Theory of the Electron"[2], that is summarized in his book **Physics of the Future: A Classical Unification of Physics**[3]. This paper showed the possibility of describing relativistic effects as being due to electrodynamic feedback effects on finite-size electrical particles of arbitrary shape. My attempt to solve all the fundamental laws of electrodynamics simultaneously using the Galilean transformation in order to obtain a more universal electrodynamic force law was based on Barnes suggestions. The work of Barnes was all the more remarkable when one considers that his Sc. D. from Hardin-Simmons University in 1950 was only honorary. Since I had a B.S. in physics with honors, an M.S. in physics with honors, a Ph.D. in theoretical physics, and post-doctoral work in elementary particle physics, I realized that I had the right background to continue and perfect his work. I had a wonderful overnight visit with Tom in his home the year before he died, and we happily reviewed all the research I had done in electrodynamics to continue his work.

In 1978 **Dr. Joseph Gonzalez Barredo** went to the Smithsonian Science Information Exchange in Washington, DC looking for help in publishing some scientific papers and a book on triondynamics entitled **The Subquantum, Ultramathematical Definition of Distance, Space, and Time [4]**. By chance I was selected to help this professor from the University of Madrid who was invited by Enrico Fermi to work with him in his laboratory in Chicago just before his death. Barredo introduced me to the possibilities of triune symmetry in explaining the internal structures of elementary particles and in explaining the process of formation and decay of many types of physical phenomena on various size scales in the universe.

In 1992 I met **David Bergman**, an Air Force electrical engineer, at a scientific conference at Northwestern College in Roseville, Minnesota where Dave presented a paper entitled "New Spinning Charged Ring Model of the Electron" [5]. I was very eager to apply my improved electrodynamic force law to his physical model of the electron. Dave and I founded the non-profit scientific educational foundation called **Common Sense Science** (www.commonsensescience.org) to promote work in electrodynamics. Many important papers were published in the Common Sense Science newsletter/journal entitled **Foundations of Science**.

In the fall of 1992 my son **Joseph Charles Lucas** entered the Oxon Hill Science and Technology High School in Oxon Hill, MD. One of the requirements of the high school was that each student had to design and develop a science project each year. Joseph was very ambitious and chose a theoretical and experimental physics project to develop a new theory of the atom using Dave Bergman's charged spinning ring model for the electron and combinatorial geometry. Dave was living in the Washington, DC area at the time and helped mentor the project. In 1995 the project won a first place in physics in the Oxon Hill Science and Technology High School Science Fair, a first place in physics and first place overall in the regional Prince

Georges County Science Fair, and a grand prize at the International Science Fair in Hamilton, Ontario, Canada.

At that time I thought that my work was merely eliminating special relativity from electrodynamics and probably all of science. Then at a scientific conference in the Canary Islands in 2002 I became acquainted with **Dr. Andre Assis** and his world-class work on the electrodynamics of Weber, especially his work on Weber's electrodynamics and gravity. [6, 7] I found that using a similar approach to that of Assis with my improved version of electrodynamics allowed me to derive a more satisfactory version of gravity than was possible from Weber's electrodynamics. This so encouraged me that I immediately set about to derive the force of inertia in a similar manner with good success.

In 2011 my work was criticized by **Dennis P. Allen, Jr.** for being inferior to the empirical science of Isaac Newton. I studied the work of Newton very closely. As a result I realized that I was perfecting the union of the axiomatic method of the ancient Greeks with the empirical force laws stemming from Newton's empirical method for doing science by using a complete set of these laws in electrodynamics. This seemingly minor idea was a major breakthrough in modern science, because the method or process for combining the axiomatic and empirical scientific methods had not been properly addressed since the time of Newton.

I would like to acknowledge the time and effort that **Jesse Musson** spent in improving the quality of the book cover using Photoshop.

I would like to acknowledge the time and effort that **George (Phil) Drake** and **Russel Moe** took to review the manuscript and correct numerous typos and some errors in the text. Any remaining errors or typos are my own fault.

References

1. Isaac Newton in a letter to Robert Hooke February 5, 1675.
2. Barnes, Thomas G., "A New Theory of the Electron", **Creation Research Society Quarterly 14**: 210-220 (1978)
3. Barnes, Thomas G., "Alternative to Einstein's Special Theory of Relativity", **Physics of the Future** (Institute for Creation Research, 1983) pp. 81-94.
4. Barredo, Joseph Gonzales, **The subquantum, Ultramathematical Definition of Distance, Space, and Time** (Maryland Institute of Advanced Study, Washington, DC, 1983).
5. Bergman, David L., "Spinning Charged Ring Model of Electron Yielding Anomalous Magnetic Moment", **Galilean Electrodynamics, 1 (5)** 63-67 (1990).
6. Assis, A. K. T., "Deriving Gravitation from Electromagnetism", **Canadian Journal of Physics 70**: 330-340 (1992).
7. Assis, A. K. T., "Gravitation as a Fourth Order Electromagnetic Effect" in **Advanced Electromagnetism - Foundations, Theory and Applications** (World Scientific, Singapore, 1995) pp. 314-331.

Chapter 1 The Development of Classical Science

Let no one ignorant of geometry enter here.

Slogan over the door of Plato's Academy [1]

The development of science from ancient times was based on some intuitively obvious assumptions about the universe. These are as follows:

1. **Determinism** – There are natural causes for everything that happens in the universe.
2. **Objective Truth** – Observations of the universe can be made independent of the observer.
3. **Consistency** – The same causes produce the same effects everywhere in the universe.

These assumptions have been challenged by the theories of modern science. For instance the Copenhagen version of quantum mechanics claims that the universe is governed 100% by random statistical processes and that there is no Law of Cause and Effect thereby denying **Determinism** and **Consistency**. Also this version of quantum mechanics, according to Heisenberg, claims that reality is in the "observation process" and so is not independent of the observer and there is no **Objective Truth**.

The approach of this book will be to uncover the fallacies of the main pillars of modern science, i.e. Maxwellian electrodynamics, Einstein's relativity theories, and the Copenhagen version of quantum mechanics, and to replace them with an improved version of electrodynamics consistent with the fundamental assumptions of science above.

In order to do this it will be useful to review some key methods of natural philosophy that were developed over time to ensure progress in science toward ultimate truth and understanding of the universe. In particular it is necessary to understand the role of deductive logic as defined by the axiomatic scientific method as well as experimentation and inductive logic of the empirical scientific method in order to achieve the goals of science and natural philosophy.

1.1 The Axiomatic Method

The axiomatic method was invented by the ancient Greeks as the proper way to organize and demonstrate deductive reasoning in the pursuit of natural philosophy. The axiomatic method is a logical procedure by which an entire system of natural philosophy (e.g. a branch of science or mathematics) is generated in accordance with specified rules of logical deduction from certain basic propositions (axioms or postulates), which in turn are constructed from a few terms (charge, mass, length, velocity, acceleration) taken as primitives. These terms and axioms are to be defined and constructed according to some method by which some warrant for their truth is felt to exist. One of the oldest examples of an axiomatic system is the ancient Greek Euclidean geometry.

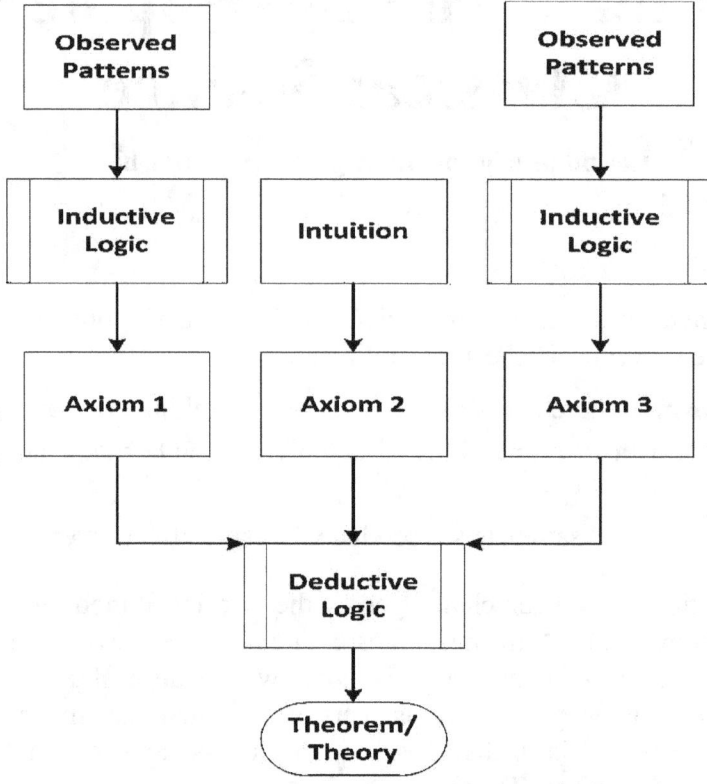

Figure 1-1 Axiomatic Scientific Method

Euclid, in the process of developing geometry, defined the axiomatic method of proof to be used in logically establishing theorems in geometry. To the extent that the axioms or postulates he chose were valid, his logically developed theorems would be valid.

The ancient Greeks were so impressed by the work of Euclid that they put the slogan "Let No One Ignorant of Geometry Enter Here" over the door of their academies of natural philosophy. The modern world has also been impressed by Euclid to the extent that his book **Elements (of Geometry) [2]** has been published in more languages and editions than any other natural philosophy or scientific book in the history of the world. [3]

1.2   Limitations of the Axiomatic Method

Euclid's approach worked well in geometry where the propositions could be imagined or justified by simple geometrical constructions using a straight edge and compass, but in physics and other areas of Natural Philosophy, the ancient Greek natural philosophers were not able to discover the appropriate axioms or postulates so easily. This was due to the fact that the axiomatic method is a method of logical organization of abstract proofs of theorems or theories, but not a method of postulate or axiom discovery of objective reality.

1.3   Newton's Empirical Method of Axiom Discovery

When Isaac Newton published his **Principia**, he stated that he intended to illustrate a new way of doing natural philosophy that overcomes some of the limitations of the axiomatic method. This method he called the empirical scientific method. The goal of Newton's method was to find empirically the forces of nature.

> **And therefore our present work sets forth mathematical principles of natural philosophy. For the whole difficulty of philosophy seems to be to find the forces of nature from the phenomena of motions and then to demonstrate the other phenomena from these forces... For many things lead me to have a suspicion that all phenomena may depend on certain forces by which the particles of bodies, by causes yet unknown, either are impelled toward one another and cohere in regular figures, or are repelled from one another and recede. Since these forces are unknown, philosophers have hitherto made trial of nature in vain. But I hope that the principles set down here will shed some light on either this mode of philosophizing or some truer one. [4]**

Newton claims that in the past natural philosophers tried to understand nature in vain, because they did not use an empirical approach to find the axioms or the fundamental forces of nature based on experimentation. The empirical approach is more effective and efficient in discovering the causes and effects of nature. As a result he argues that the empirical approach combined with the axiomatic method is a more secure path toward truth in natural philosophy. The problem faced by the ancient Greek philosophers was that they could not guess or discover the relevant propositions and appropriate primitive terms for natural philosophy upon which to apply logic to derive the theorems or theories of natural philosophy outside of geometry and mathematics. These need to be discovered by experiment.

Before Newton, Kepler discovered three empirical laws for the motions of the planets about the sun.

1. **The planets orbit the Sun in ellipses with the Sun at one focus.**
2. **The line joining the Sun and a planet sweeps through equal areas in an equal amount of time.**
3. **The square of the period of a planet's orbit (P) is directly proportional to the cube of the semi major axis (A) of its elliptical path, i.e. $P^2 = kA^3$**

Although these empirical laws were practical and useful, the fundamental cause of the motions of the planets was not revealed by them. Newton's emphasis on empirical forces turned out to be much more useful than Kepler's Laws and to give a better and simpler understanding of the mechanics of the solar system that could be applied even to processes on the Earth. From his empirical force laws of equations (1-1) and (1-2) below Newton was able to deduce Kepler's Laws. However, in 1766, Titius Bode revealed his empirical law showing the quantum periodicity of the orbits of the planets. This indicated that Newton's empirical force laws were incomplete. Perhaps there were additional smaller terms in the force laws that accounted for this periodicity.

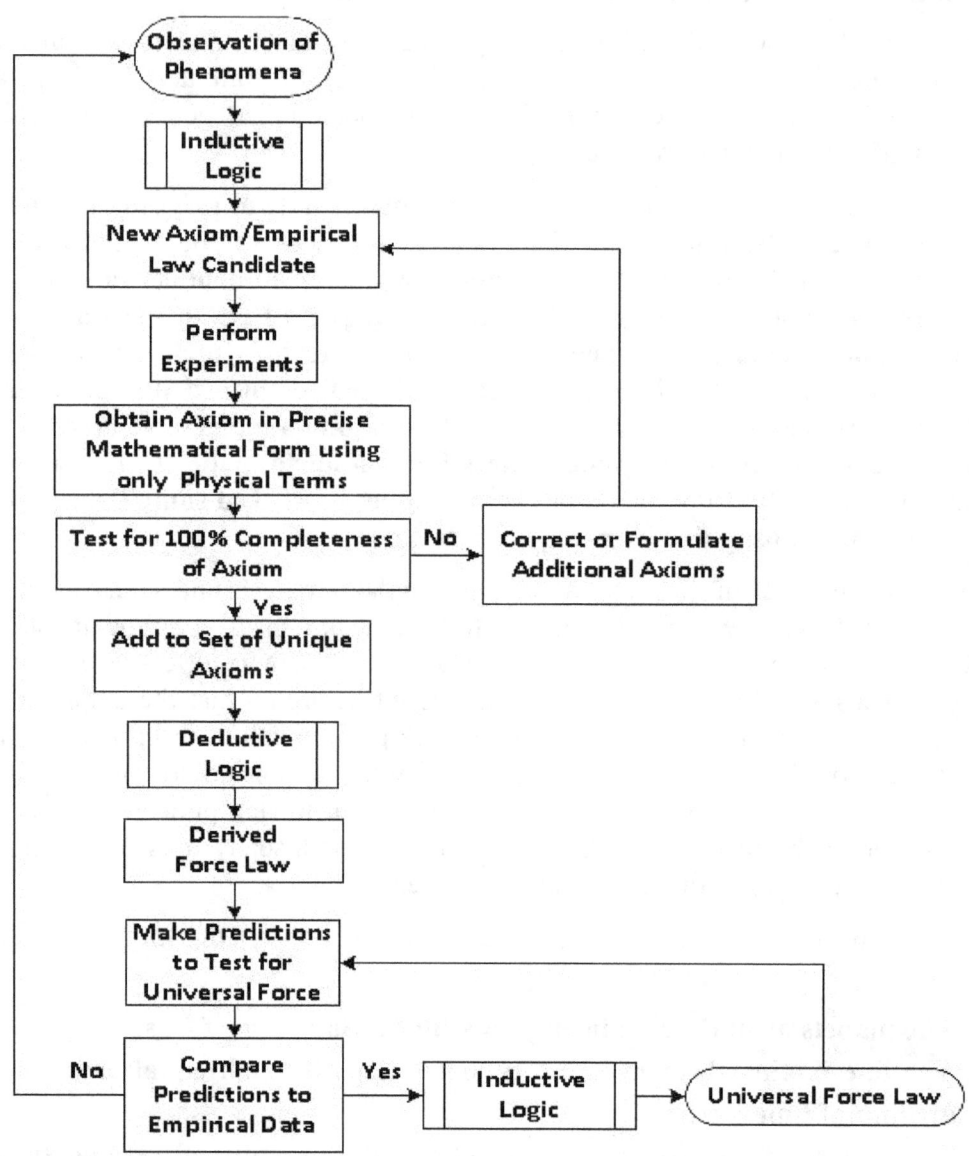

Figure 1-2 Newton's Empirical Scientific Method

Newton's empirical approach emphasizing forces does not lead to all truth at once, as Newton himself recognized with regard to his study of inertia and gravity. He never claimed to understand the causes and nature of inertia and gravity, even though he could define the empirical Force of Inertia and the empirical Force of Gravity as shown below.

$$\textbf{\textit{Force of Inertia }} \vec{F}_I = m_I \vec{A} \qquad (1-1)$$

$$\textbf{\textit{Force of Gravity }} \vec{F}_G = G \frac{m_{G1} m_{G2}}{R_{12}{}^2} \hat{R}_{12} \qquad (1-2)$$

When Newton was asked what inertial mass m_I was, he replied that inertial mass was a measure of some characteristic of matter that caused the force of inertia and that increased as the amount of matter increased. When Newton was asked what gravitational mass m_G was, he replied that gravitational mass was a measure of some characteristic of matter that caused the force of gravity between bodies of matter and increased as the amount of matter increased. When the ratio of the experimental inertial and gravitational masses were found to be equal in magnitude for the same two bodies, Newton realized that instead of the force of inertia and the force of gravity being different fundamental forces, they might have a common cause. As the quote from the **Principia** below shows, Newton did not know the cause of the force of gravity.

> **I have not as yet been able to deduce from phenomena the reason for these properties of gravity, and I do not feign hypotheses. For whatever is not deduced from the phenomena must be called a hypothesis; and hypotheses, whether metaphysical or physical, or based on occult qualities, or mechanical, have no place in experimental philosophy. In this experimental philosophy, propositions are deduced from the phenomena and are made general by induction.** [5]

Later Newton softened his renunciation of hypotheses by adding "unless as conjectures or questions proposed to be examined by experiments". [6] He had come to realize that generalized force laws obtained by induction could be used to predict additional phenomena which could be examined by experiments and serve as a check on the veracity of the generalized force law such as the Universal Force of Gravitation.

Newton's approach was evaluated by his critic Christiaan Huygens, the foremost figure in science at the time, in his **Discourse on the Cause of Gravity.**

> **One finds in this subject a kind of demonstration which does not carry with it so high a degree of certainty as that employed in geometry; and which differs distinctly from the method employed by geometers in that they prove their propositions by well-established and incontrovertible principles, while here principles are tested by the inferences which are derivable from them. The nature of the subject permits no other treatment. It is possible, however, in this way to establish a probability which is little short of certainty. This is the case when the consequences of the assumed principles are in perfect accord with the observed phenomena, and especially when these verifications are numerous; but above all when one employs the hypothesis to predict new phenomena and finds his expectations realized.** [7]

Here the test of generalized force laws is to use them to predict new phenomena which can be tested empirically. This is quite different from the role of hypotheses in modern science today. In modern science the hypotheses do not have foundations based on empirical laws and are not derived by deductive logic from the empirical foundations.

Newton took a very practical approach to forces. He assumed the total force on a body was due to the sum or linear superposition of the individual forces of the particles making up that body.

For it is reasonable that forces directed toward bodies depend on the nature and the quantity of matter of such bodies, as happens in the case of magnetic bodies. And whenever cases of this sort occur, the attractions of the bodies must be reckoned by assigning proper forces to their individual particles and then taking the sums of these forces. [8, Scholium at end of Book I, Section II]

Newton also realized that mathematics is a good tool to enable an analysis of forces, to help identify the causes of forces and to argue more securely.

Mathematics requires an investigation of those quantities of forces and their proportions that follow from any conditions that may be supposed. Then, coming down to physics, these proportions must be compared with the phenomena, so that it may be found out which conditions of forces apply to each kind of attracting bodies. And then, finally, it will be possible to argue more securely concerning the physical species, physical causes, and physical properties of these forces. [9]

In the mechanical philosophy of Newton's time all forces had to be contact forces due to causality. According to Descartes [10] the mechanical philosophy could only allow contact forces between physical bodies, if there were some sort of medium or aether to convey the force between the bodies. Newton realized, however, that no hypothetical aether contact mechanism seems even imaginable to effect "attractive" forces among particles of matter generally. In the face of criticism from Huygens and others, Newton claimed that he is employing mathematically formulated theory in physics in a new way in which forces are treated abstractly, independently of physical cause or contact mechanism. In other words the two functions could be performed separately with progress being made on the one when no progress could be made on the other.

The first type of proposition in Newton's Principia is a mathematical proposition that links parameters in rules characterizing forces to parameters of motion. As one can easily see, measurement is very important to the methodology of the **Principia**. The second type of proposition in the **Principia** consists of combinations that contrast different conditions of force in terms of different conditions of motion.

By contrast, an examination of the mathematical theories of Galileo and Huygens shows that the propositions that they were pursuing were ones that made a distinctive empirical prediction that provided an answer to some practical question, or explained some known phenomena. Newton in the **Principia** was not so interested in conjecturing hypotheses and then testing the implications of those hypotheses, but rather to use mathematics to provide a basis for specifying experiments and observations by which the empirical world could provide answers to more general questions.

In Definition 8 for force at the beginning of the **Principia** Newton says "this concept is purely mathematical, for I am not considering the physical causes and sites of forces". Thus we could say that Newton differed from his predecessors in that he treated forces from a mathematical point of view instead of the physical. From the mathematical point of view any unbalanced force acting on a body is a quantity with magnitude and direction.

In Book 1 Newton considers centripetal forces with a direction specified toward a center and the magnitude taken to vary as a function of distance from that center. In Book 2 Newton considers resistive forces with the direction specified opposite to the direction of motion and the magnitude varying as a function of velocity. In Book 3 Newton considers gravitational forces and resistive forces arising from the inertia of the fluid from the physical point of view. Newton requires five conditions to be met for a component of a mathematically characterized force to be considered a physical force as follows:

1. **The direction of the force must be determined by some material body other than the one it is acting on**
2. **All aspects of the force's magnitude must be given by a general law such that the action and reaction forces are always the same magnitude but in opposite direction**
3. **Some of the physical quantities in a force law must pertain to the other body in a way that determines the direction of the force**
4. **The force law must hold for some forces that are indisputably real**
5. **If the force acts on a macroscopic body, then it must be composed of forces acting on the microphysical parts of that body**

Notably absent from this list is anything about the mechanism or process effecting the force. Adherents of the "mechanical philosophy" such as Descartes and Huygens would have required not only a mechanism causing the force, but also a contact mechanism for delivering the force. Newton believed that progress could be made in determining the properties of the force mathematically even though not all aspects of the force were known, such as its cause and the mechanism by which it was delivered.

Newton was not the first to conclude that the forces between real bodies in the universe are very complex. He believed that an investigation of the microstructural forces within bodies was key to understanding the macro forces between bodies. This program is described in detail in the unpublished portion of the **Preface** to the **Principia** as given below

> **I therefore propose the inquiry whether or not there be many forces of this kind, never yet perceived, by which the particles of bodies agitate one another and coalesce into various structures. For if Nature be simple and pretty conformable to herself, causes will operate in the same kind of way in all phenomena, so that the motions of smaller bodies depend upon certain smaller forces just as the motions of larger bodies are ruled by the greater force of gravity. It remains therefore that we enquire by means of fitting experiments whether there are forces of this kind in nature, then what are their properties, quantities, and effects. For if all natural motions of great or small bodies can be explained through such forces, nothing more will remain than to inquire the causes of gravity, magnetic attraction, and the other forces. [12]**

Although Newton was somewhat vague in his writings about how to make the transition from mathematically characterized forces to physically characterized forces, he did realize the potential of the microscopic forces for this purpose. Of course, he did specify the use of predictions of new or additional phenomena as a way of checking force laws. Here the process is to address the complexity of real forces in a sequence of successive approximations. Each force approximation is based upon certain idealizations with systematic deviations from it being used to improve the next version of the force law. Before Newton the small residual discrepancies between idealized theory and the real world were dismissed as being of no practical importance. After Newton every systematic deviation from current theory automatically has the status of a pressing unsolved problem.

Newton views these successive approximations for forces as exact. His fourth Rule for Natural Philosophy says:

In experimental philosophy, propositions gathered from phenomena by induction should be considered either exact or very nearly true notwithstanding any contrary hypotheses, until yet other phenomena make such propositions either more exact or liable to exceptions. This rule should be followed so that arguments based on induction may not be nullified by hypotheses. [13]

Attempting to proceed in successive approximations in this way involves restrictions on how second-order phenomena are to be marshaled as evidence. In the case of orbital motions in the solar system, any systematic discrepancy from the idealized theoretical motions had to be identified with a specific physical force – if not a gravitational force, then one governed by some other generic force law. However, not just any kind of force was permissible. Newton's first Rule for Natural Philosophy that "no more causes … should be admitted than are both true and sufficient to explain their phenomena", has the effect of confining the number and type of forces to no more than the experimental data clearly demands. Requiring the force laws to be deduced from phenomena is a way of meeting this Rule. This approach is an attempt to limit risk in developing force theories as much as possible to just "inductive generalization". For example this restriction would preclude inventing unobservable forces due to dark matter and dark energy to "rescue" Einstein's Theory of General Relativity.

Newton acknowledged the risk of inductive generalization in his famous methodological passage in the **Opticks**, in the discussion of the methods of "analysis and synthesis" in the next to last paragraph of the final Query, which was added in 1706:

The Analysis consists in making Experiments and Observations, and in drawing general Conclusions from them by Induction, and admitting of no Objections against the Conclusions, but such as are taken from Experiments, or other certain Truths, for Hypotheses are not to be regarded in experimental Philosophy. And although the arguing from Experiments and Observations by Induction be no Demonstration of general Conclusions; yet it is the best way of arguing which the Nature of Things admits of, and may be looked upon as so much the stronger, by how much the Induction is more general. And if no Exceptions occur from Phenomena, the Conclusion may be pronounced generally. But if at any time afterwards any Exception shall occur from Experiments, it may then begin to be pronounced with such Exceptions as occur. By this way of Analysis we may proceed from Compounds to Ingredients and from Motions to the Forces producing them; and in general, from Effects to their Causes, and from particular Causes to the more general ones, till the Argument end in the most general. [14]

Newton's arguments for a universal force of gravity and a universal force of inertia illustrated his new empirical approach to natural philosophy and physics in general. This new approach was based on a generic mathematical theory, the contrast between mathematical and physical points of view, the roles of deduced theory and idealizations in ongoing research, and the insistence on pushing theory far beyond its original experimental domain.

1.4   Limitations of the Empirical Method in Mechanics

Newton's empirical approach to science was quite successful in his time in answering many questions in natural philosophy. However the failure of Newton's empirical method to address certain philosophical problems related to the forces of gravity and inertia would eventually be discovered. Some of these are summarized below.

The first philosophical problem that Newton had was that he could not explain via the empirical method what mass was. Newton could show that mass was a property of matter that was cut in half if the lump of matter was cut in half, but its real nature (electric, magnetic, etc.) was unknown. Newton's empirical method assumed that the fundamental forces of inertia and gravity were independent of one another and the gravitational and inertial masses for the same body should be different. Then ratios of the gravitational and inertial masses for the same two bodies were found to be the same empirically. This indicated that the force of inertia and gravity were probably just two different aspects of some more general force law which the empirical method had missed up to that time.

Newton's second philosophical problem resulting from his empirical method was that he could not explain the nature or cause of the forces of inertia and gravity. Without knowing the nature or cause of these forces, he could not explain how the forces were conveyed through empty space in our solar system. Later geologists would discover that the earth, the moon, the planets, and their moons were all expanding over time. Newtonian mechanics did not provide any explanation for how the earth could expand in size so much and so quickly. Presumably the planets and moons came from a star and that is why the interior is still so very hot. Over time the planets and moons cool off. Since most materials shrink when they cool, the planets and moons should be shrinking unless something like the force of gravity is decaying. Without knowing the nature or cause of gravity, the empirical method cannot explain why gravity is decaying over time.

Even Einstein's General Theory of Relativity would have this same problem. Astronomers have discovered that the velocities of the outer spiral arms of spiral galaxies do not obey Newtonian mechanics, i.e. the forces of gravity and inertia. The velocities are constant and too high, but they should be the same as Newton's forces predict in the non-relativistic limit.[15] Since mass was not understood in either Newtonian mechanics nor General Relativity Theory, scientists resorted to inventing an additional type of matter, called "dark matter", to explain the phenomena. This was not just a small effect. Dark matter had to comprise at least 95% of the matter in the universe and be undetectable in laboratory experiments. Also astronomers observe that the universe as a whole is expanding instead of contracting under the force of gravity. In General Relativity Theory the only way to explain this was to invent "dark energy" which according to $E = mc^2$ is a second form of dark matter. All of these phenomena that required the invention of dark matter can be simply explained by the decay of gravity, once the true nature and cause of gravity is known. These things will be explained in later chapters as well as the role that Mach's Principle plays in the process.

1.5   The Existential Method in Electrodynamics

Despite Newton's great success, a number of natural philosophers became dissatisfied with the lack of logical rigor in Newton's approach toward the end of the nineteenth century. They felt that it did not describe the "real" world. For instance Newton postulated the existence of gravitational and inertial mass, but nobody knew what mass was or could explain it. Newton, himself, admitted that he did not know what mass was. Why were the ratios of gravitational and inertial masses equal for the same two bodies? Presumably they should be fundamentally different, since they were associated with fundamentally different forces. Newton's force of gravity was an action-at-a-distance type force with no known mechanism to produce spatial contact from the cause to the effect. Newton had tried to use the aether as a medium to transmit the force from the cause to the effect, but the aether was not fully satisfactory in explaining attractive and repulsive forces and the force of inertia.

The description of the universe in terms of apparently fictitious quantities such as mass, aether, and action-at-a-distance forces eventually led to the creation of the philosophy of existentialism. This philosophy amended the axiomatic and empirical approach to natural philosophy to purposely allow science to be developed in terms of idealizations and fictitious terms. The existentialist philosophers accepted the purposelessness and absurdity of the world that Newton so successfully described with fictitious forces and masses. They abandoned the strict role of deductive logic in the scientific method and substituted the much weaker criterion of falsifiability of hypotheses.

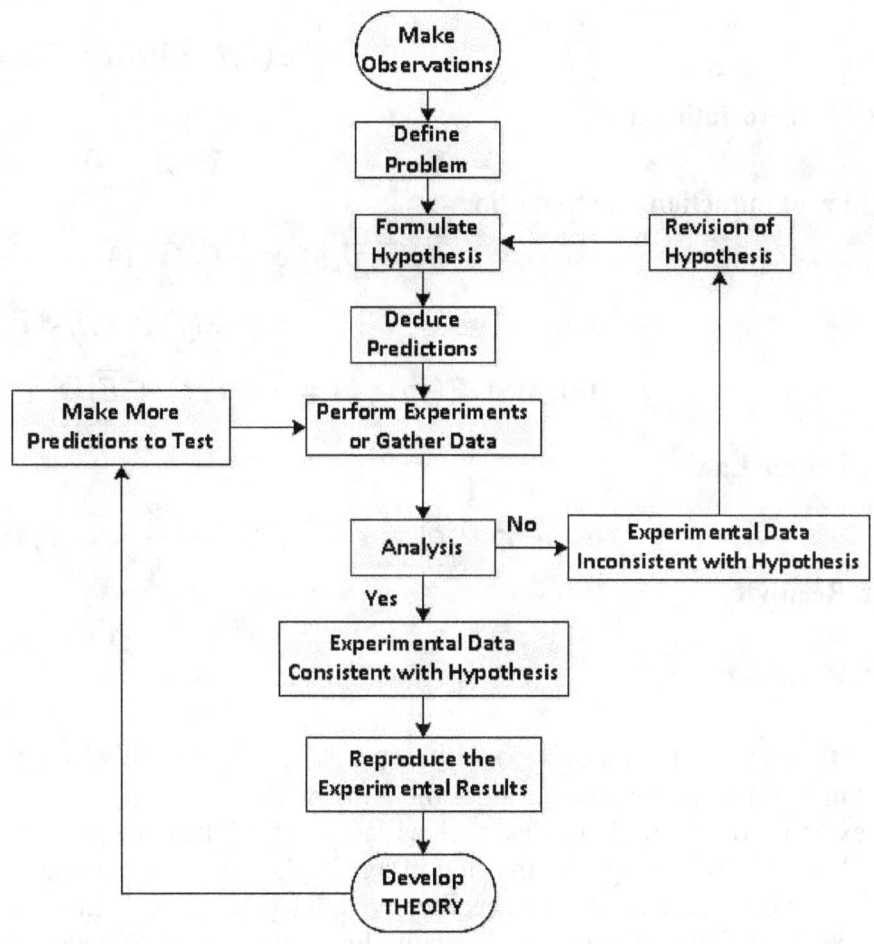

Figure 1-3 Existential & Post-Modern Scientific Method

In the area of electric and magnetic phenomena, later investigators used Newton's approach to make significant progress in defining empirical mathematical laws. Ampere defined his circuital law relating magnetic fields to electric currents that produce them in 1826. Faraday defined his law of electromagnetic induction in 1831. Lenz defined his law for induction of non-linear electromagnetic fields due to motion in 1834. Gauss defined his laws for static electric and magnetic fields in 1835.

After much effort and experimentation, the number of empirical laws related to electrodynamics was reduced to 8 unique laws involving 8 parameters. They are listed below.

1. **Generalized Ampere's Law for force between circuits**

$$\vec{B}_i(\vec{r}, t) = \frac{\vec{v}}{c} \times \vec{E}_0(\vec{r}', t') \quad (1-3)$$

2. **Faraday's Law of electromagnetic induction**

$$\int \vec{E}(\vec{r}', t') dl' = -\frac{1}{c}\frac{d}{dt}\int \vec{B}(\vec{r}, t) \cdot \hat{n} da \quad (1-4)$$

3. **Gauss's Electrostatic Law**

27

$$\int \vec{E}(\vec{r}, t) \cdot \hat{n} da = 4\pi q \quad (1-5)$$

4. **Gauss's Magnetostatic Law**

$$\nabla \cdot \vec{B} = 0 \qquad (1-6)$$

5. **Lenz's Law of induction due to motion**

$$\vec{E}_i(\vec{r}, \vec{v}, t) \propto -\vec{E}_0(\vec{r}, t)$$

$$= -\lambda(\vec{r}, \vec{v}, t)\vec{E}_0(\vec{r}, t) \quad (1-7)$$

$$where \ \vec{E}(\vec{r}, \vec{v}, t) = \vec{E}_0(\vec{r}, t) + \vec{E}_i(\vec{r}, \vec{v}, t)$$

6. **Lorentz's Force Law**

$$\vec{F}(\vec{r}, \vec{v}, t) = q\vec{E}(\vec{r}, \vec{v}, t) - \frac{\vec{v}}{c} \times \vec{B}(\vec{r}, \vec{v}, t) \quad (1-8)$$

7. **Galilean Relativity**

$$\vec{r}' = \vec{r} - \vec{v}t \qquad (1-9)$$

8. **Galilean Relativity**

$$t' = t \qquad (1-10)$$

The 8 unknown parameters are F, E_0, Ei, B_0, Bi, $\lambda(r, v, t)$, r', t'. Note that the proportionality sign $\propto$ in Equation (1-7) mathematically implies the $= \lambda(r, v, t)$.

Many textbooks on electrodynamics, such as Jackson [16], take the point of view that the primary role of Lenz's Law is to specify the sign in Faraday's Law. One of many things Lenz's Law states is that the induced current (and accompanying magnetic flux) is in such a direction as to oppose the change of flux through the circuit. From this point of view Lenz's Law is subsumed by Faraday's law and is not an independent empirical law.

However, both Einstein and Feynman commented on the necessity to make the physical distinction between whether the magnet or the conductor are in motion. Einstein said

It is known that Maxwell's electrodynamics, as usually understood at the present time when applied to moving bodies, leads to asymmetries which do not appear to be inherent in the phenomena. Take, for example, the reciprocal electrodynamic action of a magnet and a conductor. The observable phenomena here depend only on the relative motion of the conductor and the magnet, whereas the customary view draws a sharp distinction between the two cases in which either the one or the other of these bodies is in motion. For if the magnet is in motion and the conductor is at rest, there arises in the neighborhood of the magnet an electric field with a certain definite energy, producing a current at the places where the parts of the conductor are situated. But if the magnet is stationary and the conductor is in motion, no electric field arises in the neighborhood of the magnet. In the conductor, however, we find an electromotive force, to which in itself there is no corresponding energy, but which gives rise – assuming equality of relative motion in the two cases discussed – to electric currents of the same path and intensity as those produced by the electric forces in the former case. [17]

Feynman said

> **We know of no other place in physics where such a simple and accurate general principle requires for its real understanding an analysis in terms of** *two different phenomena.* **[18]**

In this work the point of view is taken that Faraday's Law and Lenz's Law describe different distinct phenomena and are therefore both independent empirical laws. From Chapter 4 equations (4-39 and 4-43) we derive the more complete form of Lenz's Law as

$$\vec{E}_i(\vec{r},\vec{v}) = -\lambda(\vec{v})\vec{E}_0(\vec{r}) = \frac{1-\beta^2}{(1-\beta^2 sin^2\theta)^{\frac{3}{2}}}\,\vec{E}_0(\vec{r}) - \vec{E}_0(\vec{r}) \quad (1-7)$$

This interpretation will lead in Chapter 4 to the notion that all relativistic velocity effects are due to electrical feedback effects on finite-size charge structures and not on Einstein's Special Relativity Theory.

In 1861 Maxwell developed a theory of electrodynamics in the axiomatic fashion using 4 of the 8 empirical equations of electrodynamics. By this means he succeeded in deriving a wave equation for light. This version of electrodynamics is defined in terms of Maxwell's set of 4 differential equations. Maxwell also defined in 1861 a form of the Lorentz force law, which was later published by Lorentz in 1892. The Lorentz electrodynamic force law replaced the separate electric and magnetic force laws when combined with Maxwell's equations.

In order to eliminate 3 of the 8 unknowns in the empirical equations of electrodynamics, Maxwell had to use a number of idealizations or approximations as follows:
1. **Point particle approximation**
2. **Displacement currents exist in capacitors**
3. **Only linear field effects exist**
4. **Induced and static fields are identical, i.e. $E_0(r, t)=E_i(r, v, t)$ and $B_0(r, t)=B_i(r, v, t)$**
5. **Action-at-a-distance forces are valid**

According to Newton, Maxwell should have used the complete set of the empirical laws of electrodynamics with no approximations or idealizations that violated empirical results. **Of course some of those empirical results came much later in time!**

Because Maxwell was unable to use the complete set of empirical laws of electrodynamics, the version of electrodynamics he developed was incomplete. The theories of quantum mechanics and Special Relativity were developed later to supplement Maxwell's electrodynamics.

After Maxwell's death in 1879 some self-styled "Maxwellians" consisting of G. F. FitzGerald (1851-1901), Oliver Heaviside (1850-1925) and Oliver Lodge (1851-1940) reinterpreted Maxwell's **Treatise on Electricity and Magnetism** of 1873 for their own ends. They located energy in the field around an electrical conductor and emphasized energy above the mechanical structures giving rise to it. See Equation (1-11) for the covariant version of Maxwell's electrodynamics that replaced Maxwell's version based on quaternions.

$$\nabla^2 \Phi - \frac{1}{c^2}\frac{\partial^2 \Phi}{\partial t^2} = -\frac{\rho}{\epsilon_0} \qquad (1-11)$$

$$\nabla^2 A - \frac{1}{c^2}\frac{\partial^2 A}{\partial t^2} = -\mu_0 J$$

William Thomson (Lord Kelvin) and others treated these newer electromagnetic views with contempt for neglecting the mechanical structures giving rise to the energy as reflected in the quote below.

> **It is mere nihilism, having no part or lot in Natural Philosophy, to be contented with two formulas for energy, electrostatic and electromagnetic, and to be happy with a vector (potential A) and delighted with a page of symmetrical formulas.** Lord Kelvin in a letter to FitzGerald April 9, 1896 [19]

1.6 Limitations of the Existential Method in Electrodynamics

The reinterpreted Maxwell's equations, the Lorentz force law, and Galilean relativity defined electrodynamics for the next 50 years. However, the reinterpreted Maxwellian approach to electrodynamics was based upon the well-known approximation or idealization known as the point particle approximation. (See equation (1-12) where the second term on the right is not zero for finite-size particles but it was dropped) Also this Maxwellian approach did not use Lenz's empirical law which gave electrodynamics support for Newton's 3rd Law and conservation of energy for dynamic magnetic fields.

$$\nabla \times \vec{B}(\vec{r},t) = \frac{4\pi}{c}\vec{j}(\vec{r},t) + \frac{1}{c}\nabla \oint \frac{\nabla' \cdot \vec{j}(\vec{r}',t)}{|\vec{r}-\vec{r}'|}d^3r' \approx \frac{4\pi}{c}\vec{j}(\vec{r},t) \quad (1-12)$$

Point Particle Idealization

One of the reinterpreted Maxwellian equations (1-6) attributed to Gauss for magnetism states that the divergence of the magnetic field is zero or more commonly understood as there are no magnetic monopoles. In 2009 scientists discovered for the first time magnetic monopoles. [20] This appears to invalidate the revised Maxwellian wave equation and the general method of obtaining the relativistic covariant electrodynamics using the vector potential approach. The vector potential approach is also used by quantum mechanics and relativity theory to extend electrodynamics.

Perhaps the most significant limitation of Maxwell's electrodynamics is that the force law it generates is not sufficient to explain the future state of a charged particle, because it only applies to constant velocity phenomena. In general charged particles, such as the electron, will undergo acceleration, will emit radiation, and will experience radiation reaction or recoil effects requiring terms involving the acceleration "a" and "da/dt" in the electrodynamic force law. These terms are missing in the classical Maxwell equations, even after they are improved by uniting with Special Relativity Theory. Chapters 4, 5, and 6 will derive these terms in the more complete electrodynamic force law.

1.7 References

1. http://plato-dialogues.org/faq/faq009.htm
2. Euclid, **Elements** (about 300 BC).
3. Lucas Nicolaas Hendrik Bunt, Phillip S. Jones, Jack D. Bedient, **The Historical Roots of Elementary Mathematics** (Dover Publications Inc., New York , 1988) page 142.
4. Isaac Newton, **The Principia, Mathematical Principles of Natural Philosophy: A New Translation**, translators I Bernard Cohen and Anne Whitman (University of California Press, Berkeley, 1999) p. 382.
5. **Ibid,** p. 943.
6. Isaac Newton, "An Account of the Book Entitled Commercium Epistolicum", reprinted in A. Rupert Hall, **Philosophers at War: The Quarrel between Newton and Leibniz** (Cambridge University Press, Cambridge, 1980), p. 312.
7. Christiaan Huygens, "Traite de la Lumiere", in **Oeuvres completes de Christiaan Huygens, vol. 19** (The Hague: Martinus Nijhoff, 1937), p. 454.
8. Newton, **Principia**, p. 588.
9. **Ibid,** p. 588f.
10. Rene Descartes, **Principles of Philosophy**, trans. Valentine Rodger Miller and Reese P. Miller (D. Reidel, Dordrecht, 1983) , gravity in Propositions 20 – 27 and magnetism in Propositions 133-183 in Part 4.
11. Newton, **Principia**, p. 588.
12. A. Rupert Hall and Marie Boas Hall (editors) **Unpublished Scientific Papers of Isaac Newton** (Cambridge University Press, Cambridge, 1962), p. 307.
13. Curtis Wilson, "Predictive Astronomy in the Century after Kepler" in Rene Taton and Curtis Wilson editors **Planetary Astronomy from the Renaissance to the Rise of Astrophysics, Part A: Tycho Brahe to Newton** (Cambridge University Press, Cambridge, 1989), pp. 172-185.
14. Isaac Newton, **Opticks: or A Treatise of the Reflections, Refractions, Inflections and Colours of Light** (Dover, New York, 1952) p. 404.
15. Milgrom, Mordehai, "A Modification of the Newtonian Dynamics as a Possible Alternative to the Hidden Mass Hypothesis", **Astrophysical Journal, 270**, 365-370 (1983)
16. J. D. Jackson, **Classical Electrodynamics-Third Edition** (John Wiley and Sons, Inc., New York, 1999), page 209.
17. A. Einstein, "Zur Elektrodynamik bewegter Körper" (On the Electrodynamics of Moving Bodies), **Annalen der Physik 322** (10): 891–921 (June 30, 1905). www.fourmilab.ch/etexts/einstein/specrel/specrel.pdf
18. Richard P. Feynman, Robert B. Leighton, and Matthew Sands, **The Feynman Lectures on Physics, Volume 2 Third Printing** (Addison-Wesley Publishing Company, Inc., Reading, MA) Chapter 17, page 2 (1965).

19. Crosbie Smith, **The Science of Energy – A Cultural History of Energy Physics in Victorian Britain** (University of Chicago Press, Chicago, 1998) p. 289.
20. Fennell, T. et al., "Magnetic Coulomb Phase in the Spin Ice $Ho_2Ti_2O_7$", **Science 326**, 415 (2009).

Chapter 2 The Development of Modern Science

There is nothing new to be discovered in physics now. All that remains is a few small dark clouds on the horizon to be resolved. Lord Kelvin [1]

Kelvin's small dark clouds on the horizon to be resolved were a set of key experiments that formed the basis for the development of two future pillars of modern science, i.e. quantum mechanics and relativity theory. These experiments included the blackbody radiation experiments, the Michelson-Morley 1886 and 1887 experiments, and the photoelectric experiments.

All of these experiments were of the electromagnetic type. Maxwell's equations for electrodynamics were incomplete and could not explain these phenomena. This was due to the imperfect way in which the fundamental empirical equations of electrodynamics were turned into axioms and used to derive an electrodynamic wave equation. Maxwell had not used the complete set of empirical electrodynamic force equations as specified by Newton. He had failed to use Lenz's Law in order to satisfy conservation of energy and momentum. Furthermore he had used a number of idealizations including the point-particle idealization, an idealized displacement current in capacitors, the idealization of only linear superposition of fields, the idealized notion that induced and static fields are identical, the idealized notion that quantum effects do not exist, and the idealized luminiferous aether as the medium for propagating light waves and to convey the electrodynamic force between two bodies.

Instead of perfecting the union of the empirical force laws of electrodynamics with the axiomatic method and removing the unreal idealizations to derive an improved version of electrodynamics, scientists of that time chose to invent additional fundamental theories to supplement Maxwell's electrodynamics. They also chose to continue the practice of incorporating unreal idealizations in their theories. In this chapter we will focus on Einstein's Special Theory of Relativity and Heisenberg's Copenhagen version of Quantum Mechanics. These two theories are currently known as the "pillars of modern science". They prop up Maxwell's electrodynamics. However, the intent of this book is to show that electrodynamics alone is the proper foundation of modern science.

2.1   Special Relativity Theory

Time and Space... It is not nature which imposes them upon us. It is we who impose them upon nature, because we find them convenient. Henri Poincaré [2]

In 1905 Einstein published a number of significant scientific papers. [3-6] The first of these papers [3] gave an explanation of the photoelectric effect in which light was identified as a stream of particles, later called photons, moving at velocity c. The idea was that monochromatic light consists of a stream of photons of specific energy. Electrons are bound to atoms by various binding energies. If the energy of a photon was absorbed by an electron, the electron could escape the atom if the energy was equal to or larger than the electron atomic binding energy.

In his second 1905 scientific paper [4] Einstein showed that the identification of light as a stream of particles appeared to solve some "problems" in electrodynamics. The Michelson Morley experiment of 1887 attempted to detect the motion of the earth through the luminferous aether. No effect of the earth moving through the luminferous aether was observed.

In his third 1905 scientific paper [5] Einstein introduced his theory of special relativity and explained the results of the Michelson-Morley experiment of 1886. In this experiment the velocity of light was measured when moving through a moving medium consisting of water. It was measured moving in the direction of motion of the medium and against the direction of motion. Classical Galilean relativity predicted that the velocity of light was c + v and c − v for these two cases, but this prediction did not agree with what was observed. Einstein's relativity theory appeared to predict what was observed. The velocity of light was a constant value c in all frames of reference.

According to the Theory of Special Relativity, the discovery that light has a constant velocity c in all inertial reference frames makes it necessary to correct Maxwell's equations. This correction is not based on the discovery of a new force, as Newton required, but on the hypothesis that light is a particle (photon) and that the speed of light is finite and independent of the motion of the source. This correction was made to Maxwell's equations resulting in a relativistic covariant formulation of electrodynamics. [7] (See equations 2-1, 2-2, 2-3, and 2-4 where β = v/c) **Note that equation (2-4) suggests that mass might be an electrodynamic quantity, since mass is associated with c^2.**

$$\vec{E}(\vec{r},\vec{v}) = \frac{q(1-\beta^2)\hat{r}}{r^2(1-\beta^2 sin^2\theta)^{3/2}} \quad (2-1)$$

$$\vec{B}(\vec{r},\vec{v}) = \frac{\vec{v}}{c} \times \vec{E}(\vec{r},\vec{v}) = \frac{q(1-\beta^2)\vec{\beta} \times \hat{r}}{r^2(1-\beta^2 sin^2\theta)^{3/2}} \quad (2-2)$$

$$\vec{F}(\vec{r},\vec{v}) = q'\left[\vec{E}(\vec{r},\vec{v}) + \frac{\vec{v}}{c} \times \vec{B}(\vec{r},\vec{v})\right] \quad (2-3)$$

$$E = mc^2 \quad (2-4)$$

The theory of relativity transformed theoretical physics and astronomy during the 20th century when it superseded a 200-year-old theory of mechanics and astronomy introduced by Isaac Newton. Light was identified with the transmission of information from one reference frame to another at velocity c. These notions proposed significant changes in perceptions about motion.

The theory of relativity denied the concept of motion as described by Galilean invariance or Galilean relativity from Newton's day, by positing that all motion is relative to moving frames of reference instead of relative coordinates in one frame of reference. Time was no longer uniform and absolute. Now there was more than one reference frame. Each reference frame had its own space and time. Physics could no longer be understood as space by itself and time by itself. Time was now an added dimension to be associated with each reference frame. A comparison of time in different reference frames depended on the relative velocity of the reference frames. This could cause four-dimensional space-time to appear curved as the relative velocity of the reference frames approached the velocity of light c.

In the field of physics relativity theory made a significant impact on the science of elementary particles and their fundamental interactions as well as ushering in the nuclear age based on the energy mass formula $E = mc^2$. Some of the idealized fundamental assumptions or postulates or axioms of Einstein's Special Relativity Theory are

1. **Idealized inertial frames of reference move at constant velocity a=da/dt=0**
2. **Maxwell's idealized electrodynamics is valid**
3. **Idealized action-at-a-distance forces**
4. **Idealization that no information can be passed faster than the speed of light**
5. **Idealized point particles**
6. **Idealized space is homogeneous and isotropic**
7. **Idealization that the velocity of light is c everywhere in the universe**
8. **Idealized non-Euclidean geometry**
9. **Idealization that quantum effects do not exist**
10. **Idealization that Mach's Principle does not need to be satisfied**
11. **Idealized notion of mass**
12. **Idealization that the velocity of light is independent of the velocity of the source**

Einstein's Special Theory of Relativity gave rise to the mathematical equations (2-1,2,3,4) for relativistic electric and magnetic fields and forces which have been confirmed in accelerator experiments at velocities close to the speed of light. The success of these equations in describing the experimental observations has caused physicists to ignore problems with the idealized foundational axioms. Below problems with the idealizations of each of the foundational axioms is reviewed.

1. **Scientists have never been able to make any measurements in an inertial reference frame moving at constant velocity. No inertial reference frames exist in reality. The Earth is always rotating on its axis plus accelerating around the sun which is always accelerating around the center of the Milky Way galaxy which is accelerating around the center of the universe.**
2. **Maxwell's equations do not fully support conservation of energy and momentum due to not including Lenz's Law. Also Maxwell's equations are based upon a number of idealizations which do not correspond to the real world.**
3. **Philosophers for thousands of years have known that all forces in nature must be local contact forces in order to affect a distant body.**
4. **In 1935 Einstein, himself, introduced the Einstein-Podolsky-Rosen Paradox [8] in which scattering particles are passing information significantly faster than the**

speed of light using an invented particle that can move faster than the speed of light called the tachyon. Since that time many different types of experiments appear to have observed information being passed faster than the speed of light.

5. Accelerator particle scattering experiments have shown that massive particles, such as the proton and neutron, have both a finite-size and an internal charge structure consisting of at least three primary structures. See Hofstadter's electron scattering data in Figure 2-1 below. [9]

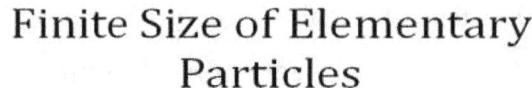

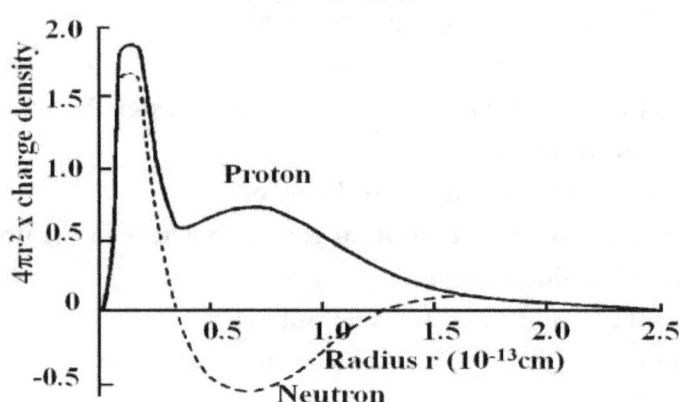

Figure 2-1 Scattering Data Showing Proton and Neutron Internal Structure and Finite-Size [9]

6. With regard to space being homogeneous and isotropic, Bode [10] was the first to show that space in our solar system is quantized with the sun at the center. The modern version of Bode's Law was first published by Stanley Dermott [11, 12] in the 1960s as shown in Figure 2-2. Dermott's work was extended by Tifft [13, 14] and Humphreys [15] who studied the quantization of red shifts of light from the stars and showed that the universe has a center with quantized bands of galaxies following the modern version of Bode's Law as shown in Figure 2-3.

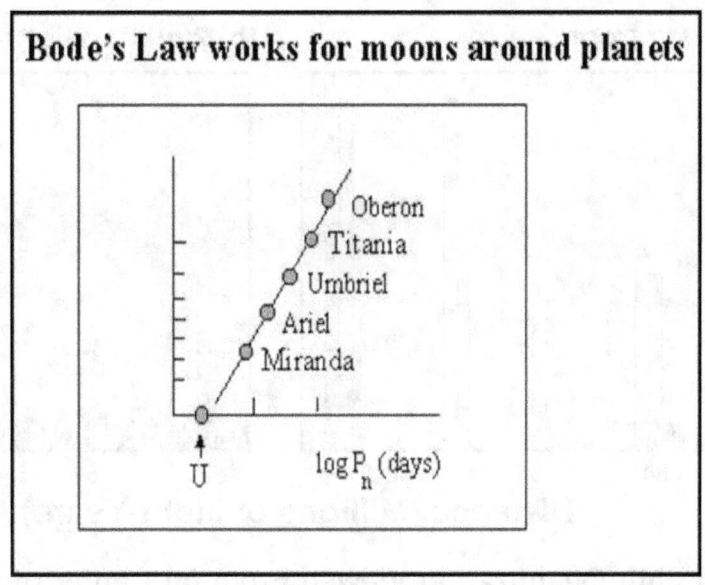

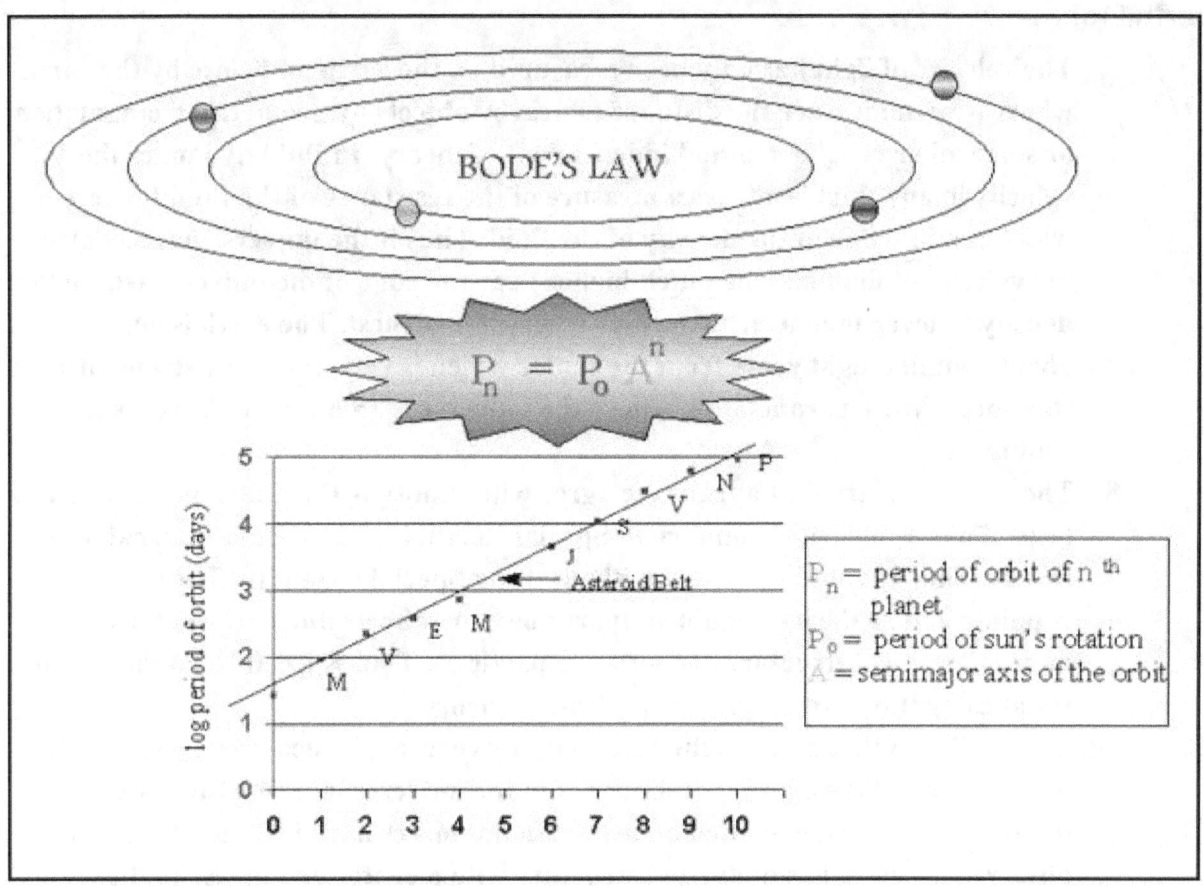

Figure 2-2 Planetary and Moon Data Supporting the Modern Version of Bode's Law [16]

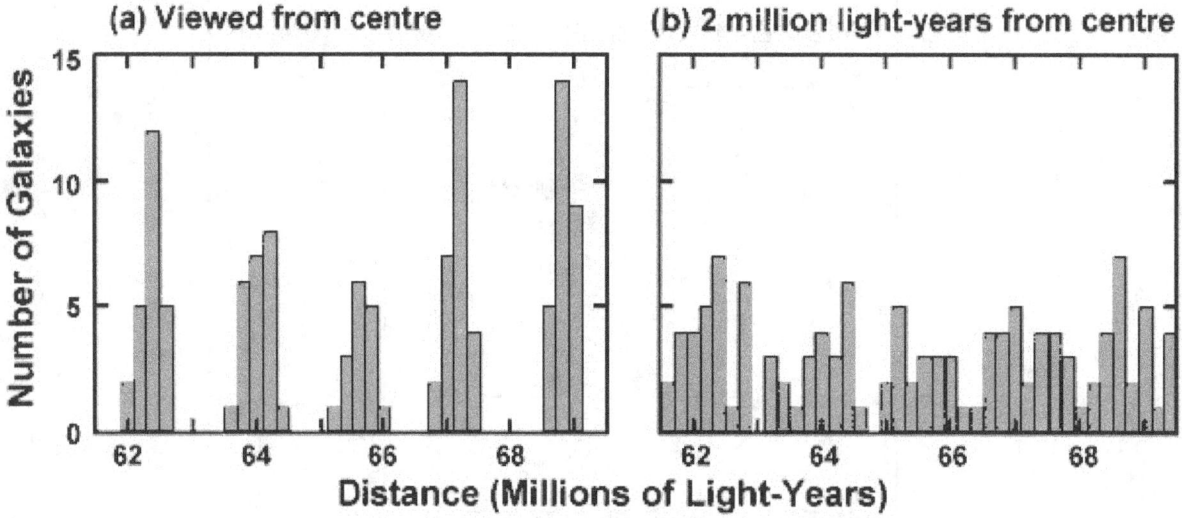

Figure 2-3 Red Shift Quantization Viewed From (a) Center of the Universe (b) the Earth [15]

7. The velocity of light has only been measured on the earth and close by the earth where we can measure the distance to nearby objects by geometric triangulation or some other reliable method independent of theory. In fluid dynamics the velocity in any fluid is always a measure of the resistance of the fluid to motion which is a function of the density of the fluid. Thus if the universe has a center, the velocity of light may be much higher near the edge of the universe where the density is lower than near the center where it is highest. The earth is currently about 2 million light years from the center which is very close by astronomical standards. Most astronomers believe the universe is 15 billion light years in radius.

8. The only geometry that appears to agree with reality is Euclidean geometry. The tests of non-Euclidean geometry in Special Relativity Theory are internal to the theory. They cannot be verified without using Special Relativity Theory.

9. A point particle theory cannot support quantum effects due to the details of the internal physical structure of finite-size particles. Thus Special Relativity Theory is not compatible with a physical quantum theory.

10. Mach's Principle expresses the idea that any correct physical theory must take into account all the charges and masses in the universe in a consistent way, because the electromagnetic and gravitational forces have infinite range. The observed universe has quantized structure with a center giving rise to the observed force of inertia on all matter. Special Relativity Theory assumes all of space is homogeneous and isotropic with no center. Therefore it cannot properly support Mach's Principle.

11. Special Relativity Theory does not explain what mass is. It assumes that gravitational and inertial mass are the same without justification. Henri Poincaré , the co-inventor of Special Relativity Theory, showed from meta-theory (the theory of theories) that independent fundamental theories cannot use the fundamental constants of other fundamental theories. Since Special and General Relativity Theories, as well as quantum mechanics and electrodynamics, use the fundamental constant c for the velocity of light, they cannot all be fundamental theories. Poincaré argued that all these theories must be electrodynamic in origin.

12. The application of the Extinction Effect and the Doppler shift changed the results of the Michelson-Morley Experiment of 1886 such that the data now agrees with Galilean Relativity Theory and not Special Relativity Theory.

Besides having problems with its fundamental idealized axioms, Special Relativity Theory now has problems with the fundamental experiments upon which the theory was initially developed such as the Michelson-Morley experiments of 1886 and 1887 plus the photoelectric experiment. These experiments are reviewed in the following sections.

2.1.1 Michelson-Morley Experiment of 1886

In 1851 Hippolyte Fizeau performed an experiment to measure the relative speeds of light in moving water. [18] A modified version of the Fizeau experiment was performed by Michelson and Morley in 1886. [19] The original analysis of these experiments appeared to indicate that Galilean relativity was inadequate to explain the results. When Einstein introduced his Theory of Special Relativity in 1905 [5], he was able to explain the Michelson-Morley experiment of 1886 and the explanation of the experiment seemed to be satisfactory.

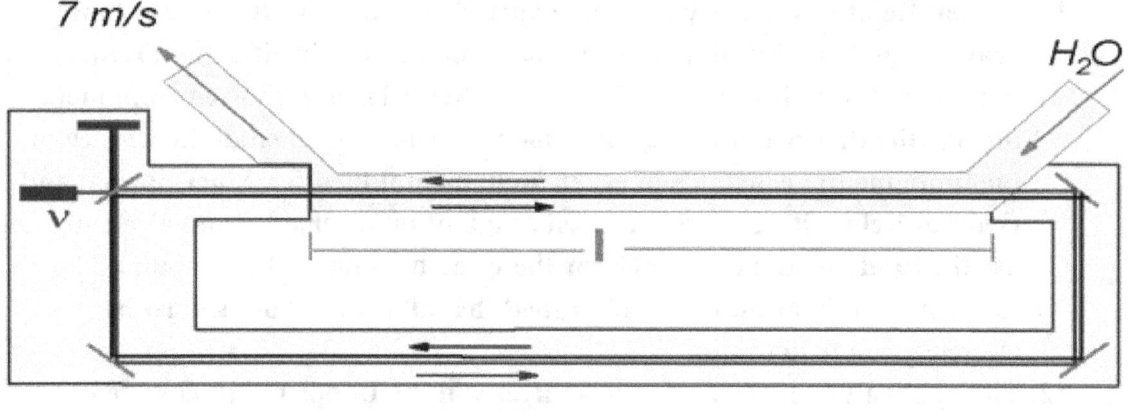

$$\lambda'_1 \sim \lambda(c - v)/c = \text{Wavelength of clockwise light at water entry}$$

$$\lambda'_2 \sim \lambda(c + v)/c = \text{Wavelength of counter-clockwise light at water entry}$$

Figure 2-4 Modified Fizeau Experiment Measuring the Velocity of Light through Moving Water

Then Ewald in 1912 and Oseen [20] in 1916 discovered the extinction effect whereby light is absorbed by the atoms in all media including the near vacuum of outer space and then re-emitted. The extinction theorem states that the speed of light will approach the speed c/n relative to the medium where n is its index of refraction. The Extinction Theorem also defines the minimum path length for its applicability. This distance depends strongly on the index of refraction of the medium and the wavelength of light. For visible light and optical glass it is less than a micron. For visible light in air it is about a millimeter. For visible light in the intergalactic medium of space it is a few parsecs.

No one applied the extinction theorem to the analysis of the Michelson-Morley experiment until Fox [21, 22] did it in the 1960s. By the 1960s Special Relativity had received the status of a politically correct theory, and no one paid attention to Fox's work. So Renshaw [23] published a more detailed analysis of the Michelson-Morley experiment in 1996. In that paper he showed that taking into account the extinction effect and the resulting Doppler shift in the analysis of the data makes Special Relativity Theory's assumption that c' = c invalid and Galilean relativity c' = c ± v correct after all. The results can be seen in Figure 2-5.

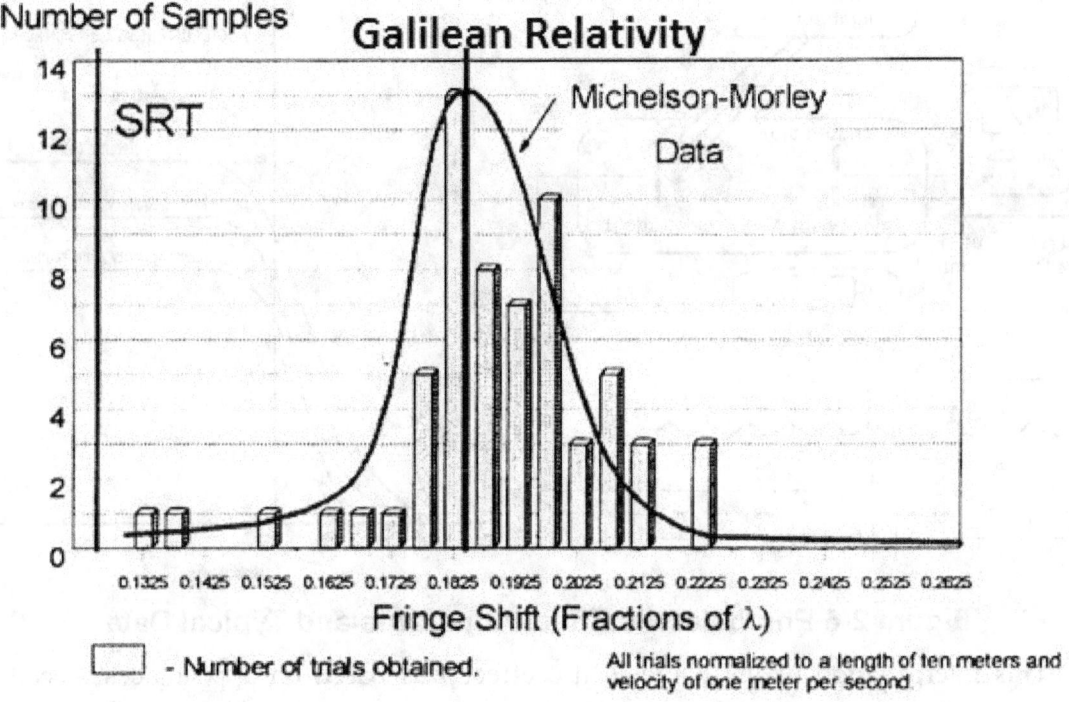

Figure 2-5 Extinction Effect Changes Results of Michelson-Morley Experiment [23]

2.1.2 Photoelectric Effect

All the fifty years of conscious brooding have brought me no closer to answer the question, 'What are light Quanta?' Of course today every Tom, Dick and Harry thinks he knows the answer, but he is mistaken. Albert Einstein [24]

In 1887 Hertz and Hallwachs [25-27] discovered simultaneously that ultraviolet light incident upon crystalline metallic sodium (Na) surfaces caused ejections of negatively charged particles later identified as electrons. This was the discovery of the photoelectric effect.

In 1905 Einstein's wife published the Nobel Prize winning paper "**On a Heuristic Viewpoint Concerning the Production and Transformation of Light**"[3], based on her laboratory research with Einstein's name on it in order to get it published. (Women were not allowed to publish in most science journals of that day.) In the paper she suggested the existence of discrete quanta of light now called photons. The original experiments on the distribution in energy of photoelectrons showed that the apparent maximum kinetic energy of emission was independent of the intensity but was a function of the frequency of the radiation. These results led the Einsteins to the hypothesis that photoelectric emission was a quantum effect in which the energy hυ of a quantum of radiant energy was "absorbed" by an electron in the metal which thereby increased its kinetic energy by this amount. The observed distribution in energy of the photoelectrons was assumed to result from energy losses suffered by the electrons in escaping from the metal.

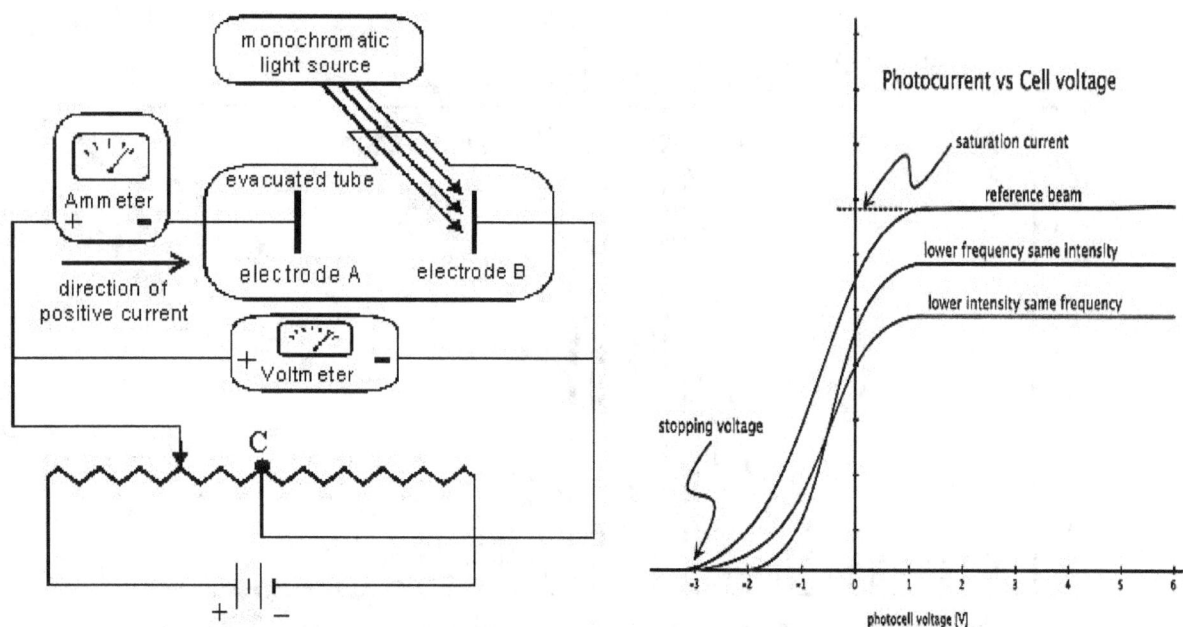

Figure 2-6 Photoelectric Effect Apparatus and Typical Data

This interpretation of the photoelectric effect was based on appearances, because the photoelectric experiments were incomplete at that time. The photoelectric effect had been measured on crystalline metals such as sodium, but not yet on the amorphous (non-crystalline) form of the same metals. According to Einstein's explanation of the photoelectric effect, it should make little difference whether the metal was in the crystalline or the amorphous state. The binding energy of the electron would be approximately the same.

Robert A. Millikan performed microscopic charged oil drop experiments in 1916 that confirmed the energy states of the electron as hυ. This experiment also provided an independent value of Planck's constant h. However, Millikan wrote "Einstein's photoelectric equation... cannot in my judgment be looked upon at present as resting upon any sort of a satisfactory theoretical foundation for the existence of photons," even though "it actually represents very accurately the behavior" of the photoelectric effect. [28] Millikan realized that the energy states of the electron might be due to its own internal structure. Even though Millikan had his misgivings about Einstein's interpretation of the photoelectric effect in terms of photons, his work was instrumental in Einstein receiving the Nobel Prize for the photoelectric effect in 1922.

The development of the toroidal ring model of the electron began with André-Marie Ampère, who in 1823 proposed tiny magnetic "loops of charge" to explain the attractive force between current elements. [29] Since the time of Ampere many scientists have conceived of the electron as a toroidal ring type structure. Of those working on some sort of toroidal model of the electron Parson [30] stands out. Then in 1990 Bergman and Wesley [31] perfected the early toroidal models of the electron by taking into account its spin and magnetic moment. An essential weakness of previous toroidal models of the electron had been that forces of unknown origin had to be postulated ad hoc in order to hold the electron together against electrostatic repulsion. Bergman and Wesley's model was the first to be completely stable under the action of classical electromagnetic forces alone.

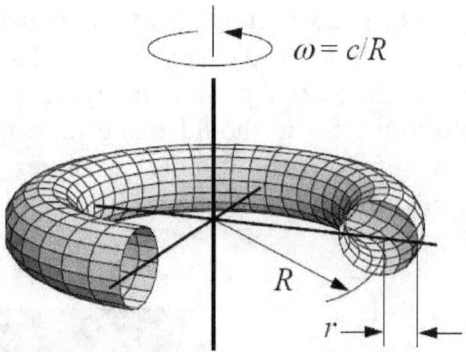

$$\omega = c/R$$

$$R$$

$$r$$

Figure 2-7 Bergman & Wesley Spinning Ring Model of Electron

Three features of the spinning charged ring model of electrons and protons are especially important to the structure of the atom. The dominating characteristics provided by the ring model are first, the physical size of each particle; second, the magnetic dipole exhibited by each particle; and third, the property that a charged spinning ring, which is surrounded by static electric and magnetic fields, does not radiate continuously.

Planck's constant h, the fundamental constant of quantum mechanics, was derived by Bergman and Wesley [31] for their physical electrodynamic toroidal model to be

$$h = \frac{e^2}{2\pi\varepsilon_0 c} ln\left(\frac{8R}{r}\right) \quad (2-5)$$

where the value of h is determined from the ring structure by the balance of the electric and magnetic forces. Since Bergman's model is a physical model, it allows one to predict from first principles Planck's constant h, spin, magnetic moment, mass, and other physical properties of elementary particles. The spinning charge in the ring can have various standing wave configurations corresponding to various excited states ε = hv.

Thus a satisfactory physical structure for the electron of finite-size is capable of explaining the quantization of the photoelectric effect and blackbody radiation. [32] Now scientists are faced with a choice. There is a physical model of the electron where the structure gives rise to quantum effects and there is a purely mathematical model based on many idealizations that explains many of the same quantum effects. **It seems obvious that physics should go with the physical model!**

Later experiments found that the photoelectric effect using ultraviolet light was significantly reduced or unobservable on the same metals in the amorphous state instead of the crystalline state. These experiments seemed to suggest that light does not exist as discrete quanta, but as waves. The crystalline lattice serves as an antenna array to receive sufficient energy from the waves to eject an electron from an atom. If the antenna is too small, as in the case of amorphous metals, the photoelectric effect occurs much more weakly, if at all, and at a different wavelength.

Then in 2011 work to develop optical circuits for computer chips [33] showed that it was necessary to build antenna arrays in these microscopic circuits (nanometer size) for the photoelectric effect to occur. Note that in Figure 2-8 the light in an optical chip is only absorbed when the wavelength of the light is closely correlated with the microscopic size of the nanometer crystalline antenna arrays. Antennas that are too small or too large do not absorb a significant amount of energy. According to Einstein's photon interpretation of light, the absorption is occurring on single atomic electrons. So it should make little difference whether the atomic electrons are in crystalline structures or amorphous structures of the same size. **Thus the photoelectric effect has been misinterpreted for over a hundred years to support light as a particle called the photon!**

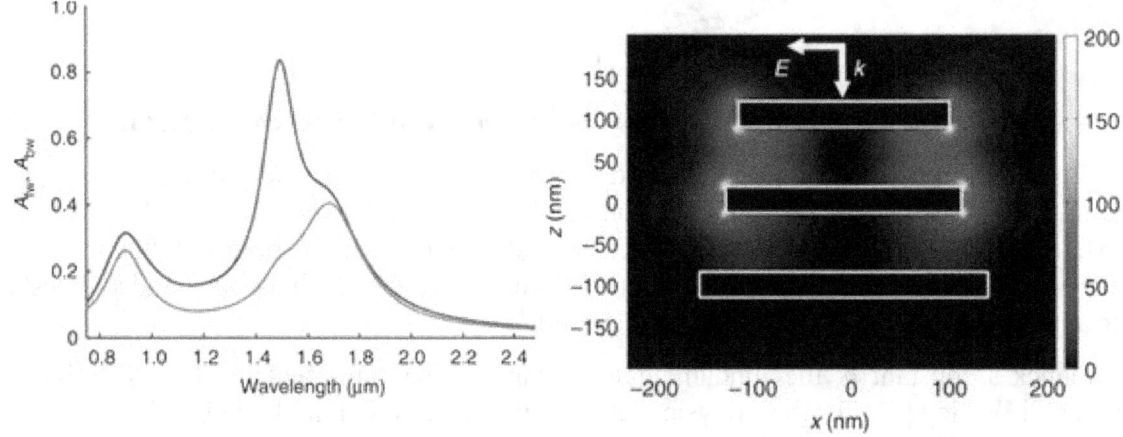

Figure 2-8 Absorption of Light on Optical Antenna Arrays [33]

2.1.3 Fields of Moving Charges

According to Cullwick [34] and the experiments of Hooper [35] the fields of a moving charge remain "attached" to the moving charge. (See Figure 2-9 with electric fields of static and moving charges and Figure 2-10 with iron filings showing the fields attached to a magnet) They are modified by the feedback effect according to Lenz's Law and Mach's Principle. If the magnet of Figure 2-10 is moved, the field patterns of the iron filings move with the magnet.

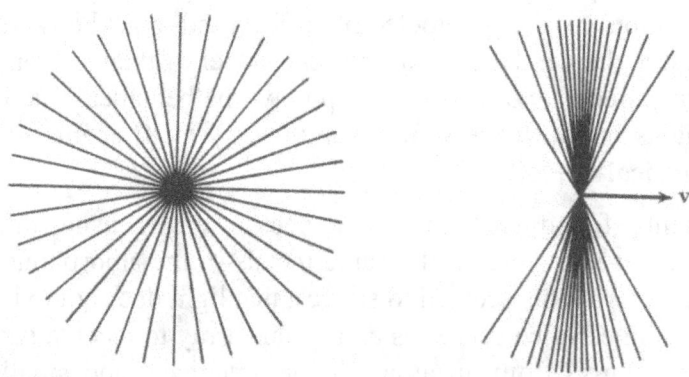

Figure 2-9 Static and Moving Electric Fields of a Charge [35]

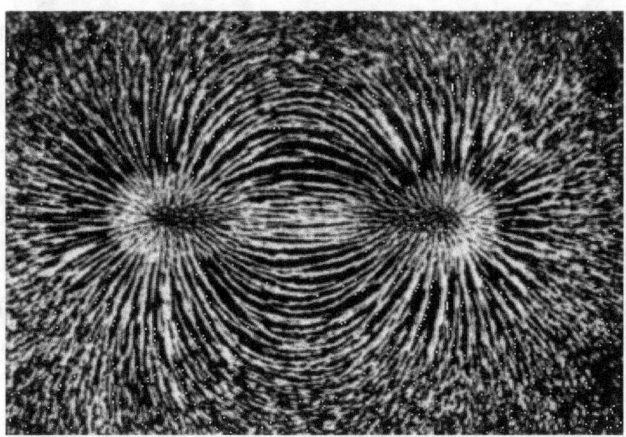

Figure 2-10 Iron Filings Showing Pattern of Magnetic Fields Attached to a Magnet

Note that this is in stark contrast to the relativistic Maxwellian electrodynamics which pretends that the fields are not "attached" to the charged particle, but travel through some medium independent of the originating charged particle's motion and the resistance of the rest of the universe as expressed by Lenz's Law and Mach's Principle. The relativistic notion of retarded fields is not applicable for contact forces but only fictitious **"action-at-a-distance"** type forces. For a charged particle moving at constant velocity there are no radiation or retardation effects. Thus the retarded field methodology of relativistic electrodynamics is inconsistent with experimental data and causality. Only contact forces agree with the experiments of Hooper and causality.

2.1.4 Michelson-Morley Experiment of 1887

In 1887 Albert Michelson and Edward Morley performed a light experiment to verify the existence of Maxwell's luminiferous aether for the transmission of electromagnetic waves. [36-38] Since waves in water need the water in which to move just like sound waves need air or some other medium in which to move, it was believed that light waves also needed something in which to move. The apparatus is shown in Figure 2-11 below.

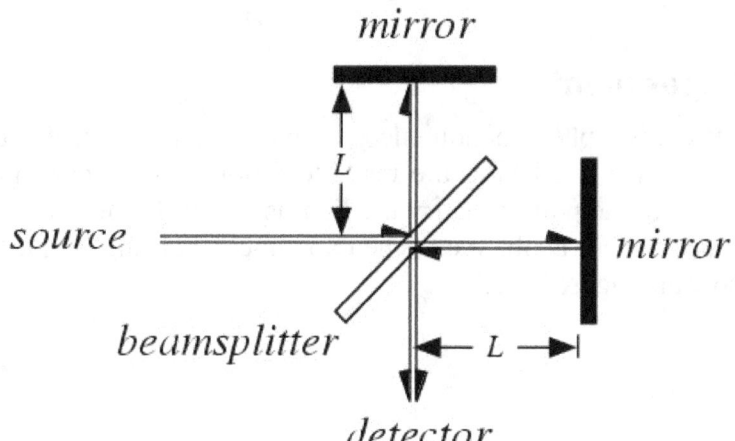

Figure 2-11 Michelson-Morley Experiment of 1887

The apparatus used in the experiment is known as a Michelson interferometer. It consists of a monochromatic light source, a half-silvered glass plate that acts as a beam splitter, two regular mirrors, and a telescope for the detector. The mirrors are placed a right angles to each other and at equal distance from the glass plate beam splitter, which is obliquely oriented at an angle of 45^0 relative to the two mirrors. The original apparatus was mounted on a rigid base that rotated freely in a basin filled with liquid mercury in order to reduce friction and allow smooth rotation of the apparatus in the horizontal plane.

The prevailing scientific theories at the time held that the aether formed an absolute reference frame with respect to which the center of the universe was stationary. It should therefore appear to be moving from the perspective of an observer on the rotating Earth orbiting the sun. Thus, for particular orientations of the apparatus, light would sometimes travel in the same direction as the aether, and at other times in the opposite direction. The goal of the experiment was to measure the speed of light in different directions in order to measure the speed of the aether relative to the Earth. A reasonable value would establish the existence of the aether.

Michelson and Morley were able to measure the minute differences in the speed of the monochromatic light through the two perpendicular arms of their apparatus by bringing the beams from each arm back together in the telescope detector and looking for interference fringes between the beams to see if they were out of phase due to traveling faster in one direction than the other.

Although Michelson and Morley were expecting to find a significant number of fringe shifts as they rotated the orientation of the apparatus with respect to the surface of the earth and its annual orbit around the sun, they did not find any such fringe shifts. This null result appeared to deny the classical concept of aether.

However, in light of the work of Cullwick [34] and the experiments of Hooper [35] the fields of a moving charge remain "attached" to the moving charge and light is a ripple in the fields of the charge. The "aether" is the actual fields of the charged particles that are emitting light. Thus the "aether" of light sources on earth is effectively attached to the earth. The rotation of the earth and the orbit of the earth around the sun should not cause any fringe shifts in the Michelson interferometer.

2.1.5 Sagnac Experiment

The Sagnac effect is a phenomenon encountered in interferometry caused by rotation. A beam of light is split and the two beams are made to follow a trajectory in opposite directions around a ring. On return to the point of entry the light is allowed to exit the apparatus in such a way that an interference pattern is obtained. The experimental arrangement, known as a Sagnac interferometer, is shown in Figure 2-12.

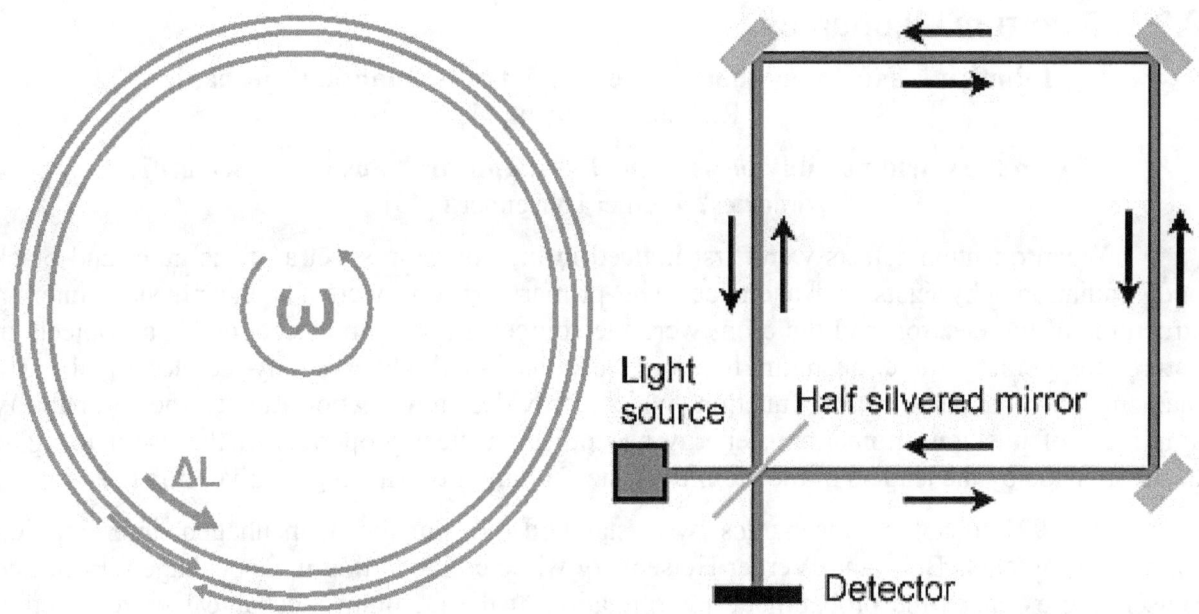

Figure 2-12 Sagnac Interferometer and Path of Light

The first suggestion to undertake such an interferometry experiment was given by Oliver Lodge in 1897, and then by Albert Abraham Michelson in 1904. The purpose of the experiment was to measure the rotation of the Earth by optical means in order to decide between the notion of a stationary aether and one that is completely dragged by the Earth. In 1911 Max von Laue [39, 40] predicted from both Special Relativity and the stationary aether that the angular velocity of the earth would be measured by such an experiment, because in both theories the speed of light is independent of the velocity of the source. He also predicted that only complete-aether-drag models would give a null result.

The first interferometry experiment aimed at observing the correlation of angular velocity and phase-shift was performed by the Frenchman Georges Sagnac in 1913. [41, 42] The effect that he observed is called the Sagnac Effect in his honor. He was able to measure a positive angular velocity for the earth.

In 1926 Albert Michelson and Henry Gale [43] set up a very large ring interferometer with a perimeter of 1.9 kilometers to measure the angular velocity of the earth. Their experiment was able to measure the angular velocity of the Earth to within the measured accuracy from astronomical experiments.

The theoretical results of Cullwick [34] and the experimental results of Hooper [35] that the fields of the charges remain attached to the light source charges also explain this experiment. Here the Sagnac effect in a circular loop can be understood as follows. When the loop is rotating, the point of light entry/exit moves during the transit time of the light. The backwards-propagating beam covers less distance than the forwards-propagating beam and arrives earlier as shown in the left side of Figure 2-12. This creates a shift in the interference pattern. The shift of the interference fringes is proportional to the platform's angular velocity. **It should be noted that the approach of Cullwick and Hooper is the only one that can explain all the experiments above 2.1.1-2.1.5 that originally supported Special Relativity Theory!**

2.2   Quantum Mechanics

I think it is safe to say that no one understands Quantum Mechanics.
Richard Feynman [44]

Can the world possibly be as absurd as it appears to us in our scientific theories? Werner Heisenberg [45]

When quantum effects were first noticed in the emission spectra of the atom and black body radiation, physicists had a choice. The primary options were (1) the physical internal structures of the electron and the atom were the source of quantum effects or (2) all objects in nature are point particle quantum harmonic oscillators which can only be described by a mathematical quantum wave function which provides information about the probability amplitude of position, momentum, energy and other physical properties of the oscillator. The decision made by the leaders of the atomic physics community was to go with option (2).

In 1927 quantum mechanics was standardized on the Copenhagen interpretation formulated by Niels Bohr and Werner Heisenberg while collaborating in Copenhagen. Bohr and Heisenberg extended the probabilistic interpretation of the quantum mechanical wave function originally proposed by Max Born. In this interpretation questions like "Where was the particle before I measured its position?" are meaningless. The measurement process randomly picks out exactly one of the many possible states allowed by the quantum state's wave function in a manner consistent with the defined probabilities that are assigned to each possible state. According to this interpretation, the interaction of an observer or apparatus that is external to the quantum system is the cause of wave function collapse to a specific state. Thus, according to Heisenburg, "reality is in the observations, not in the structure (of the electron or the atom)". [46]

The Copenhagen interpretation of quantum mechanics departs from classical physics primarily at the atomic and subatomic scales where the dynamics of the systems can be described in terms of Planck's constant h. It provides a mathematical description of the dual particle and wave-like behavior and interactions of matter and energy. The name "quantum mechanics" is derived from the observation that some physical quantities can only change by discrete amounts, or quanta in Latin. The angular momentum of an electron bound to an atom is quantized. The energy of an atomic electron is quantized resulting in discrete atomic emission line spectra.

This version of quantum mechanics is significantly different from classical science in that there is no basis for the law of cause and effect. All of nature is based 100% on random statistical probabilities. Since no internal physical structure of quantum oscillators is considered in this approach, all elementary particles are treated as point oscillators with no internal structure. Within the Copenhagen version of quantum mechanics the wave-particle duality of energy and matter and the uncertainty principle attempt to provide a unified view of the behavior of photons, electrons and other atomic-scale objects.

The history of quantum mechanics [47] began with a number of different scientific experiments. In 1838 Michael Faraday discovered the existence of cathode rays later identified as beams of electrons by J. J. Thompson in 1897. Next in the winter of 1859-1860 Gustav Kirchoff made statements about black body radiation. In 1877 Ludwig Boltzmann suggested that the energy states of a physical system could be discrete instead of continuous. Then in 1887

Heinrich Hertz discovered the photoelectric effect. Finally in 1900 Max Planck made the quantum hypothesis that any radiating atomic system could be divided into a number of discrete energy elements such that the energy ε of each of these energy elements is proportional to the frequency ν with which each of them individually radiate energy, i.e. ε = hν where h is called Planck's constant.

Then in 1905 Albert Einstein explained the photoelectric effect previously reported by Heinrich Heinz in 1887 by postulating that light itself is made of individual quantum particles in order to be consistent with Max Planck's quantum hypothesis. In 1926 these quantum particles were called photons by Gilbert N. Lewis. The phrase "quantum mechanics" was first used in Max Born's 1926 paper "Zur Quantenmechanik der Stoßvorgänge". [48]

In the field of physics quantum mechanics has made a significant impact on the science of the elementary particles, the atom, the nucleus and molecules. Some of the fundamental idealizations in the assumptions or postulates or axioms of the Copenhagen version of quantum Mechanics are

1. **The state of any system is determined by an idealized universal wave function $\Psi(x,t)$**
2. **$\Psi^* \Psi$ gives rise to an idealized statistical interpretation of the universe**
3. **There is no Law of Cause and Effect in the universe**
4. **The Heisenberg Uncertainty Principle $\sigma_x \sigma_p \geq h/4\pi$**
5. **All particles are point oscillators**
6. **There is no deformation or elasticity of particles**
7. **All processes obey the linear superposition principle**

These are all idealizations. Problems with these idealizations are reviewed below.

1. **Experiments have not been able to detect any sort of physical aether to support a universal wave. Thus the universal wave function has been declared non-physical and just mathematical, i.e. probabilistic in nature.**
2. **A non-physical mathematical universal wave function cannot give rise to anything but a statistical interpretation of the universe. The statistical interpretation of the universe is in disagreement with the well-established law of cause and effect.**
3. **The primary purpose of science is to explain effects in terms of causes.**
4. **The quantum uncertainty arises in quantum mechanics due to the idealized matter wave nature of all quantum objects. The uncertainty relationship between any pair of non-commuting self-adjoint operators such as position and momentum are subject to uncertainty limits. These are entirely mathematical in origin, since the universal wave function is not a physical wave. It allows violations of conservation of energy and momentum for short periods of time.**
5. **Accelerator particle scattering experiments have shown that massive particles such as the proton and neutron have both a finite-size and an internal charge**

structure consisting of at least three primary structures. See Hofstadter's electron scattering data in Figure 2-1 above. [9]

6. **All experimentally observed massive physical particles have finite-size and are elastic and can deform. Only idealistic unphysical point particles are not deformable.**

7. **Many physical phenomena, such as that giving rise to the laser (Light Amplification by Stimulated Emission of Radiation), are based on non-linear electrodynamic processes. However idealistic point particles cannot participate in non-linear processes.**

Besides having problems with its fundamental axioms, Quantum Mechanics now has problems with the fundamental experiments upon which the theory was initially developed such as the photoelectric experiment, the Blackbody Radiation experiment and the atomic emission spectra. For the Photoelectric experiments see section 2.1.2 above. The Blackbody Radiation experiment and atomic emission spectra are reviewed in the next two sections.

2.2.1 Blackbody Radiation

In 1858 Balfour Stewart performed experiments on the thermal radiative emissive and absorptive powers of polished plates of various substances compared with the radiative emissive and absorptive powers of lamp-black surfaces at the same temperature. [49] This was the first measurement of black body radiation. In 1859, not knowing of Stewart's work, Gustav Robert Kirchoff reported the coincidence of the wavelength of spectrally resolved lines of absorption and of emission of visible light at the same temperature. [50-52]

In 1900 Max Planck explained black body radiation by suggesting that electromagnetic energy could only be emitted in quantized form [53, 54], i.e. the energy could only be a multiple of an elementary unit $E = h\nu$, where h is Planck's constant and ν is the frequency of the radiation. Planck's law of black-body radiation states that

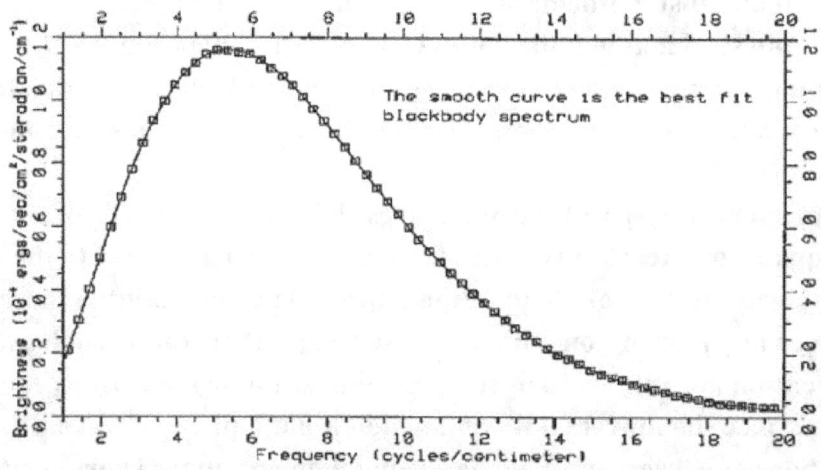

Figure 2-13 Blackbody Radiation Intensity I vs. Frequency [55]

$$I(v, T) = \frac{2hv^3}{c^2} \frac{1}{e^{hv/KT} - 1} \quad (2-6)$$

where

I (v, T) is the energy per unit time radiated per unit area of emitting surface in the normal direction per unit solid angle per unit frequency by a black body at temperature T

h is the Planck constant

c is the speed of light in a vacuum

K is the Boltzmann constant

v is the frequency of the electromagnetic radiation emitted

T is the temperature of the body in Kelvins

Planck's successful formula for black body radiation did not specify what oscillator was quantized. The structure of a finite-size electron oscillator could be quantized or a quantum point oscillator or the light itself could be quantized into a light particle called the photon. Planck's constant h must be determined empirically.

In 1990 Bergman and Wesley [31] perfected the early toroidal models of the electron by taking into account its spin and magnetic moment. An essential weakness of previous toroidal models of the electron had been that forces of unknown origin had to be postulated ad hoc in order to hold the electron together against electrostatic repulsion. Bergman and Wesley's model was the first to be completely stable under the action of classical electromagnetic forces alone.

Four important characteristics provided by the ring model are (1) the physical size of each particle, (2) the magnetic dipole exhibited by each particle, (3) the property that a charged spinning ring, which is surrounded by static electric and magnetic fields, does not radiate continuously, and (4) Planck's constant h, the fundamental constant of quantum mechanics, is derived by Bergman and Wesley [31] for their physical electrodynamic toroidal model to be equation (2-5) where the value of h is determined from the ring structure by the balance of the electric and magnetic forces. Since Bergman's model is a physical model, it allows one to predict from first principles Planck's constant h, spin, magnetic moment, mass, and other physical properties of elementary particles. The spinning charge in the ring can have various standing wave configurations corresponding to various excited states ε = hv.

In 2003 Lucas [56] used the spinning charge ring toroidal model for the electron to calculate the same formula Equation (2-7) for black body radiation. (See **Appendix C** for the derivation.) Now there is a non-physical mathematical theory and a physical theory that explains the same phenomena using the same formula. The physical theory also predicts the value of Planck's constant h. Which theory should be used? **In physical science the physical model should always be given priority over the purely mathematical model, if it can explain the same data, give rise to the same empirically confirmed equations describing the observed experimental data, and predict the fundamental constant h!**

2.2.2 Atomic Emission Spectra

By the late 1900s scientists such as Balmer and Rydberg began to notice that when a sample of gaseous atoms of an element at low pressure is subjected to an input of energy, such as an electric discharge, the atoms themselves begin to emit electromagnetic radiation. If the light is passed through a very thin slit and then through a prism, the light emitted by the excited atoms is separated into its component frequencies. The spectral lines in the atomic emission spectrum for hydrogen are shown in Figure 2-14 below.

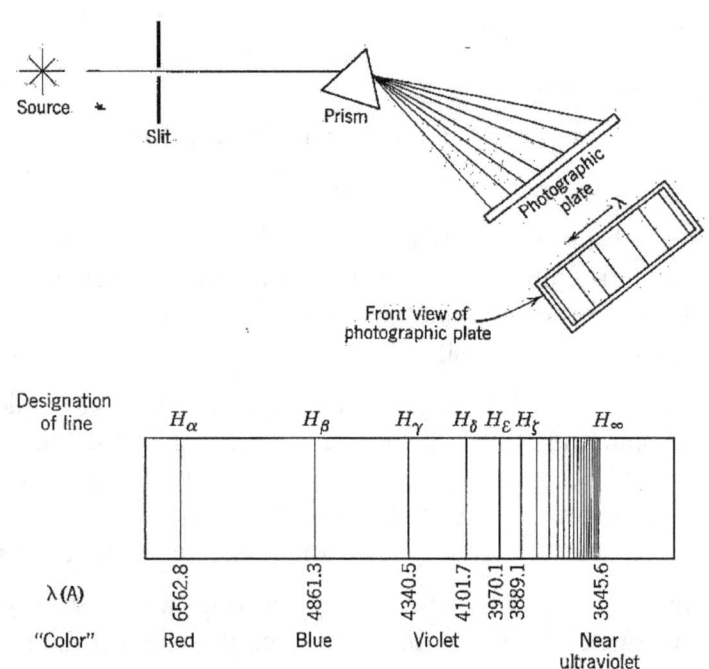

Figure 2-14 Atomic Emission Spectra for Hydrogen Atoms

The spectral lines occur in series in the different regions (infra-red, visible and ultra-violet) of the spectrum of electromagnetic radiation. The spectral lines in each series get closer together with increasing frequency. In Figure 2-14 frequency increases from left to right. Each element has its own unique atomic emission spectrum. Balmer developed an empirical formula that described one of these series. Rydberg developed an empirical formula that described all the series for the hydrogen atom in 1888.

In 1901, Sutherland [57] deduced that the origin of the lines in atomic emission spectra arise from kinematical effects within the atom. He explained the spectral lines in terms of the nodal subdivisions of a circle. This work was an extension of Ampere's work [58] in 1825 in which he suggested that electric currents consist of moving magnetic molecules. These magnetic molecules (electrons) interact like tiny loops of charge.

In 1911, Rutherford [59] determined by analyzing alpha particle scattering off of thin metal foils that most of the mass of the atom and all of its positive charge was concentrated in a very small region in the atom called the nucleus.

In 1913 Neils Bohr proposed a quantum model of the atom to explain the Rydberg formula in which a point-like electron moved in a circular orbit around the nucleus of the atom obeying the laws of Classical Mechanics. Bohr chose to attribute Sutherland's circle to the circular orbit of a point-like electron instead of a finite-size electron in the shape of a toroidal ring.

This choice caused a number of problems in logic for the Bohr model. In order to deal with these problems, Bohr postulated the following idealizations: [60]

1. **A point-like electron in an atom moves in a circular orbit about the nucleus under the influence of the Coulomb attraction between the electron and the nucleus, and obeying the laws of Classical Mechanics.**

2. **However, instead of an infinite number of orbits, which would be possible in Classical Mechanics, it is only possible for an electron to move in an orbit for which its angular momentum is an integral multiple of Planck's constant h divided by 2π.**

3. **Despite the fact that the electron is constantly accelerating and must continually emit radiation according to electrodynamics, a point electron moving in such an allowed orbit does not radiate electromagnetic energy. Its total energy E remains constant.**

4. **Electromagnetic radiation is emitted if the point electron, initially moving in an orbit of total energy E_i, instantly and discontinuously changes its motion so that it moves in an orbit of total energy E_f. The frequency of the emitted radiation ν is equal to the quantity (E_i - E_f) divided by Planck's constant h.**

In the Bohr model of the atom the electron in the hydrogen atom exists only in certain definite energy levels. These levels are called Principle Quantum Levels, denoted by the Principal Quantum Number n. Principal Quantum Number n = 1 is closest to the nucleus of the atom and of lowest energy. When the electron occupies the energy level of lowest energy, the atom is said to be in its "ground state". An atom can have only one ground state. If the electron occupies one of the higher energy levels, then the atom is in an "excited state". An atom has many excited states E_n as given by the Equation (2-7)

$$E_n = \frac{Z^2 R_E}{n^2} \quad (2-7)$$

where R_E is the Rydberg constant and n = 1, 2, 3 … is the Principal Quantum Number. The energy E of a photon emitted by a hydrogen atom is given by the difference of two hydrogen energy levels

$$E = E_i - E_f = R_E \left[\frac{1}{n_f{}^2} - \frac{1}{n_i{}^2} \right] \quad (2-8)$$

where n_f is the final energy level quantum number and n_i is the initial energy level quantum number.

When a gaseous hydrogen atom in its ground state is excited by an input of energy, its electron is "promoted" from the lowest energy level to one of higher energy. The atom does not remain in its excited state, but re-emits energy as electromagnetic radiation. This radiation is the result of an electron "falling" from a higher energy level to one of lower energy. This electron transition results in the release of a photon from the atom of an amount of energy ($E = h\upsilon$) equal to the difference in energy of the electronic energy levels involved in the transition. In a sample of gaseous hydrogen where there are many trillions of atoms, all of the possible electron transitions from higher to lower energy levels will take place many times. A glass prism can be used to separate the emitted electromagnetic radiation into its component frequencies (wavelengths or energies). These are then represented as spectral lines along an increasing frequency scale or wavelength scale to form an atomic emission spectrum.

Bohr's classical quantum model of the atom was able to explain many of the atomic emission spectral lines of hydrogen and other one-electron atoms. However, it was based upon many idealizations such as the solar system model in which the electron orbited the nucleus of the atom in fixed orbits. According to electrodynamics any charged particle that is being accelerated is required to radiate energy in the form of light continuously. This would cause the electron to radiate energy until it fell into the nucleus and was annihilated. Also this model had problems in that it was only valid for the hydrogen atom which is a one electron atom. Furthermore, this model had problems with the binding of molecules. The electron had to orbit all the nuclei in each molecule in order to keep the atoms in the molecule bound together. This is not very feasible. Finally, the Copenhagen version of quantum mechanics incorporated the Heisenberg Uncertainty Principle which said that it is impossible to determine the position and momentum of a particle simultaneously with certainty. In the Bohr model it is assumed that there are fixed angular momentums for each quantized orbital in violation of the Uncertainty Principle.

In 1915 Alfred Lauck Parson [61] reintroduced Ampere's notion that the electron was not a small point-like sphere as Bohr assumed, but a very thin ring about 1.5×10^{-9} cm in radius on which the negative charge revolves at a velocity of approximately the velocity of light c. In Parson's model the electrons did not orbit the nucleus. The electrons are in stable equilibrium due to the balance of electric and magnetic forces at some finite distance from the nucleus with the electrons arranged in spherically symmetric stable configurations as shown in Figure 2-15 below. Note the magnetic flux lines through the ring centers and the great circles for spherical symmetry.

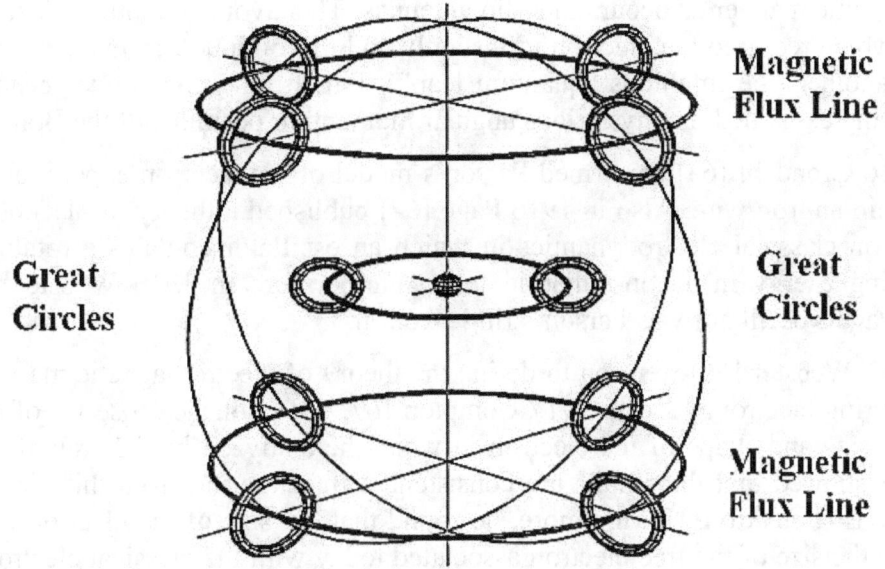

Magnetic
Flux Line

Great
Circles

Great
Circles

Magnetic
Flux Line

Figure 2-15 Parson's Model of Atom for Neon

Parson used this stable configuration to explain various types of chemical bonds and other chemical phenomena. Gilbert Lewis used Parson's model to define the Lewis dot diagrams for chemical bonding to form molecules. [62] This approach explained the angles of molecular bonds in a simple physical geometrical way.

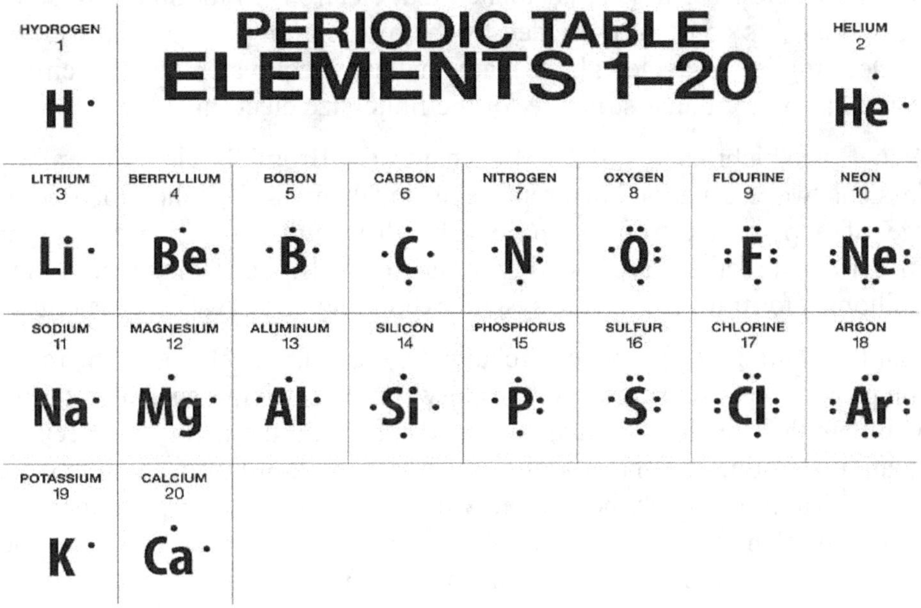

PERIODIC TABLE ELEMENTS 1–20

HYDROGEN 1							HELIUM 2
H·							He·
LITHIUM 3	BERRYLLIUM 4	BORON 5	CARBON 6	NITROGEN 7	OXYGEN 8	FLOURINE 9	NEON 10
Li·	Be·	·B·	·C·	·N·	·O·	·F·	·Ne·
SODIUM 11	MAGNESIUM 12	ALUMINUM 13	SILICON 14	PHOSPHORUS 15	SULFUR 16	CHLORINE 17	ARGON 18
Na·	Mg·	·Al·	·Si·	·P·	·S·	·Cl·	·Ar·
POTASSIUM 19	CALCIUM 20						
K·	Ca·						

Figure 2-16 Lewis Dot Diagrams

Parson's continuous ring model of the electron has electromagnetic energy stored in the form of non-radiating standing waves thereby avoiding the idealization that Bohr had to overcome with an illogical postulate. Parson also noted that the alternating force of a light wave can induce charge oscillations on the ring capable of absorption and emission of electromagnetic energy in the same manner as occurs in radio antennas. This avoided another idealization of the Bohr model which required the electron to instantly and discontinuously make a transition from one orbit to another (the infamous "quantum leap"). Finally, the condition for stable standing waves in the ring explained the mysterious angular momentum postulate of the Bohr model.

In 1916 Grondahl [63] confirmed Parson's model of the electron experimentally for free electrons within an iron wire. Also in 1916 Page [64] published a theory of blackbody radiation based purely on classical electrodynamics in which an oscillator contains a rotating degree of freedom storing energy in a non-radiating form. Furthermore in 1916 Webster [65] at MIT showed that Page's oscillator was Parson's ring electron.

In 1917 Webster [66] went on to define the theory of electromagnetic mass for Parson's magnetron or ring electron. Also in 1917 Compton [67, 68] published a series of experimental papers on the size and shape of the electron in which he analyzed hard X-ray and gamma ray scattering. He showed that the results are consistent with scattering from thin flexible rings of electricity, i.e. ring electrons. Furthermore, he found that the size of the electron in the atom is different from the size of the free electron associated today with the classical electron radius and the Compton wavelength. Compton [69] also derived Owen's experimental law for fluorescent absorption of X-rays based on the electron ring model.

In 1919, Allen [70] presented the case for a ring electron to the Physical Society of London. After the adoption of the Copenhagen interpretation of quantum mechanics there were no more papers published on the Ring Model for 40 years.

In 1923 Louis De Broglie [71] postulated that electrons in motion have associated wave lengths given by Planck's constant h divided by the momentum p = mv of the electron, i.e. λ = h/mv = h/p. De Broglie's wavelength is mathematically equivalent to the circumference of Sutherland's and Parson's circular structure for the finite-size electron.

In 1926 Erwin Schrödinger [72-75] used the De Broglie's electron wave postulate to develop a classical "wave equation" that represents "mathematically" the probability distribution of the charge of a point electron in space. The distribution is spherically symmetric and prominent in certain directions representing valance bonds. Schrödinger also introduced the classical Hamiltonian form for the atomic system expressing conservation of energy.

In 1928 Paul Dirac [76] introduced relativistic quantum mechanics. The relativistic Dirac wave equation provided a description of elementary spin ½ particles, such as electrons, that were consistent with both the principles of quantum mechanics and the theory of special relativity. In the Dirac equation spin does not have a physical interpretation, but in the improved models of the finite-size physical model of Sutherland, it will have a physical interpretation. The electron is a continuous loop of charge on the surface of a toroidal ring in which there is a rotation of the charge about the cross section of the toroidal ring (spin) in addition to the rotation around the circumference of the ring.

The Dirac equation made relativistic corrections to classical quantum mechanics. It improved upon Schrödinger's wave equation by accounting for the fine structure of the hydrogen atom emission spectrum. The Dirac wave equation implied the existence of antimatter particles, such as the positron, which were later discovered.

Starting in 1956 Winston Bostick, one of Compton's last graduate students, began developing plasma-focus devices and plasma-jet devices in the plasma fusion effort. With these devices he was able to demonstrate the existence of plasmoids, i.e. force-free, minimum-free-energy structures (like spherical droplets) carrying their electric currents in slender, force-free, tensile-strength-possessing vortex strings.

In 1966, Bostick [77, 78] proposed a revision of Sutherland's model such that the electron is a string-like submicroscopic force-free plasmoid constructed by the self-energy forces of electric E and magnetic H vectors. He found that a string of charge that makes up the electron naturally assumes the configuration of a helical spring that is connected end-to-end to form a deformable ring or torus (see Figures 2-17 and 2-18).

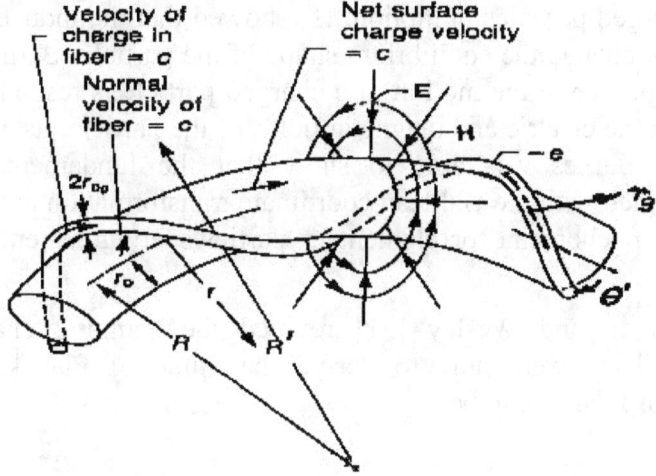

Figure 2-17 Bostick Plasmoid Helical Spring Fiber on Toroidal Surface [78]

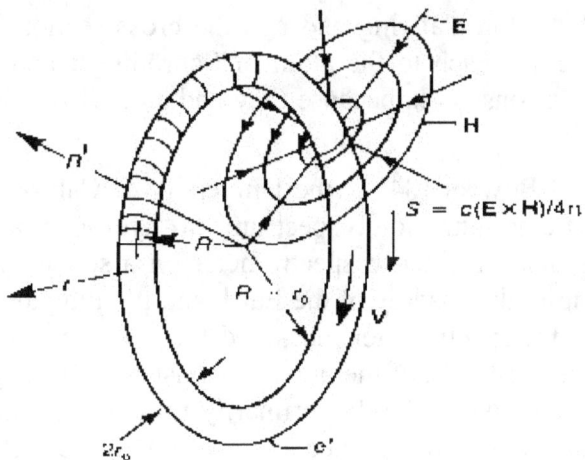

Figure 2-18 Bostick E and H Fields of Torus [78]

The cornerstone of this completely electrodynamic model of the electron was electric charge in the form of an extremely slender, electrically charged, electromagnetic closed-loop fiber that is in stable equilibrium by its own self forces-and whose electromagnetic energy is 2.5

x 10^{18} GeV. [78] Bostick proposed this physical structure as the origin of elementary particle structure in terms of charge strings and gave arguments that

1. **All mass, momentum, and energy are electromagnetic in character**
2. **The strong force is due to the electromagnetic forces between two finite-size toroidal particles**
3. **The transverse deformation waves on the filament are equivalent to the De Broglie waves of Quantum Mechanics**

In 1978, Barnes [79] showed that if one takes into account the finite-size and elasticity of charged elementary particles, then Faraday's Law, Ampere's Law, and Gauss's Laws predict a feedback effect on charged particles in motion. He showed that the induced magnetic fields due to the particle's motion change the equilibrium shape of the particle. Barnes derivations showed that this change in shape due to the motion of a charged particle is responsible for the so-called "relativistic" change in the electric and magnetic fields of the particle, its change in mass, and its decay half-life. Thus Barnes was able to show that the fundamental empirical laws of electrodynamics combined with the Galilean coordinate transformation are able to predict all the observed "relativistic" phenomena for elementary particles in agreement with Poincaré's [80] argument from logic.

In 1990 Bergman and Wesley [31] revived the spinning charge Ring Model of elementary particles. They were able to derive the value of Planck's constant from the electrodynamics of a toroidal ring to be

$$h = \frac{e^2}{2\pi\varepsilon_0 c} ln\left(\frac{8R}{r}\right) \quad (2-9)$$

where R is the radius of the toroidal ring and r is the cross sectional radius of the ring. The quantum mechanical wave approach to the atom of Schrödinger and Dirac was never able to derive the value of Planck's constant h, because they had an idealized mathematical model not a physical model.

In 1991, Labov and Bowyer [84] at the University of California at Berkeley devised a way to measure the extreme ultraviolet spectrum of hydrogen and helium from 80-650 Angströms. They put a grazing incidence spectrometer on a sounding rocket to get above the Earth's atmosphere. Flying in the shadow of the Earth and pointing away from the sun toward a dark area of the universe, the spectrometer measured the spectrum from 80 to 650 Angströms. Astronomers believe that at least 95% of the universe consists of hydrogen and helium gas. Thus the spectrum observed, if any, would be due primarily to hydrogen and helium atoms. A large number of spectral lines or peaks were obtained as shown in Figure 2-19 below.

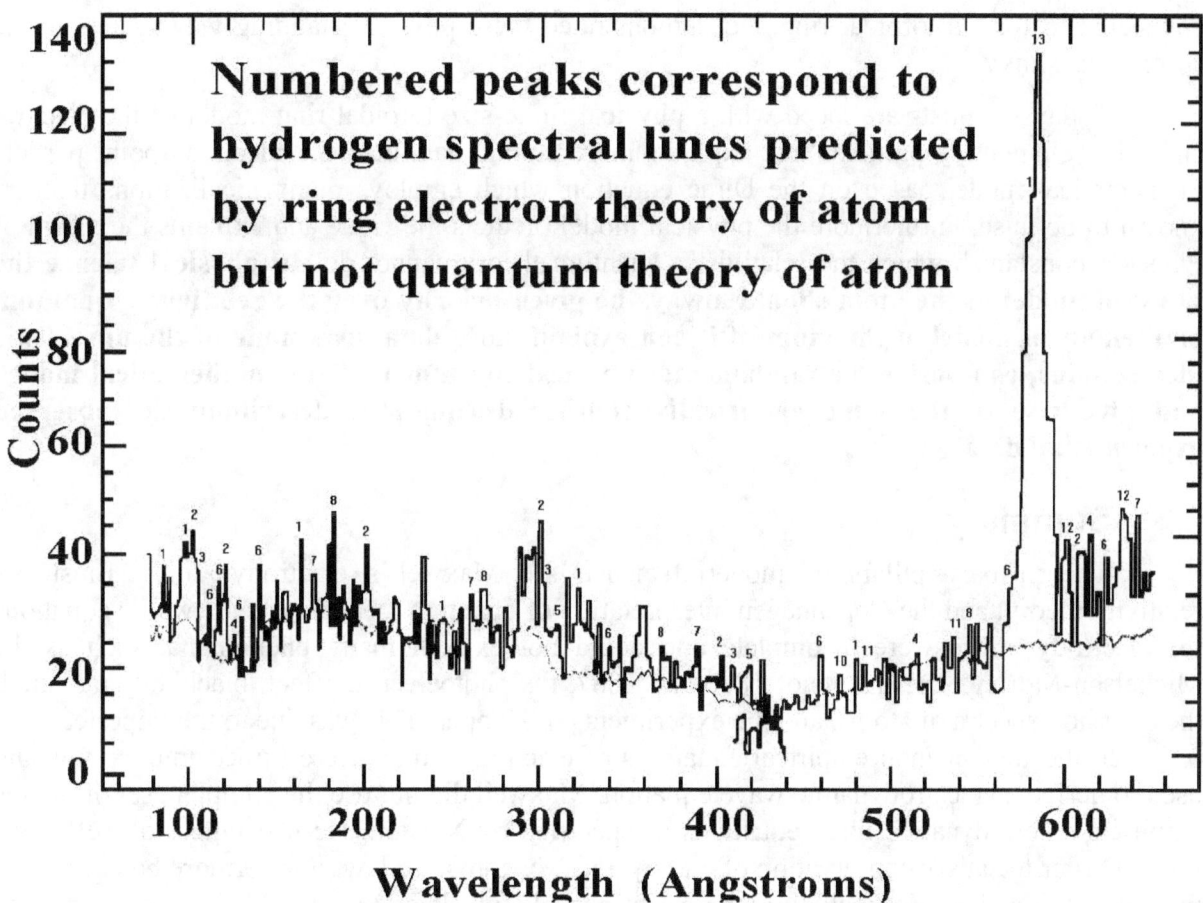

Figure 2-19 Extreme Ultraviolet Spectrum for Helium and Hydrogen [84]

The relativistic Dirac quantum theory of the atom does not predict that there are any spectral lines from hydrogen or helium to be observed in this spectral range, but the observed data is not random noise. There are up to 7 orders of 12 different spectral lines of hydrogen in the extreme ultraviolet as can be seen in the numbered peaks.

In 1996-1997, Charles and Joseph Lucas [85] explained the fundamental phenomena that established Quantum Theory, i.e. blackbody radiation, the photoelectric effect, and the emission spectra of atoms, in terms of the toroidal Ring Model of the electron. The explanation of these phenomena turned out to be logically superior to that of the Copenhagen interpretation of Quantum Physics. It did not incorporate any of the idealizations associated with quantum mechanics. It did not require any postulates to overrule classical electrodynamics. Furthermore, the new physical model of the atom predicted emission spectral lines for hydrogen and helium in the extreme ultraviolet. The toroidal Ring Model of the Atom predicted up to 7 orders of 12 different spectral lines for hydrogen in the spectrum from 80 to 650 Angströms. All were observed as predicted within experimental error. These emission spectral lines correspond to Principle Quantum Numbers n = 1/2, 1/3, 1/4, etc. These states are not possible in the point-particle wave mechanics of the Dirac quantum theory of the atom. However, in the continuous toroidal ring model with tensile forces maintaining its structure, it is possible to have these standing wave states. Joseph performed mechanical experiments on a continuous slinky connected to form a toroidal ring and demonstrated these physical standing wave states in the continuous slinky.

Today scientists are faced with a physical finite-size toroidal ring model of the electron and other elementary particles that explains more atomic data than the idealized point particle mathematical model based on the Dirac equation which employs many idealizations that are known to be false. Furthermore the physical model predicts the value and explains the origin of Planck's constant h which the relativistic quantum theory cannot do. **In physical science the physical model of the atom should always be given priority over the relativistic quantum mathematical model of the atom, if it can explain more data, uses none of the unrealistic idealizations, can derive the fundamental physical constant h of the mathematical model, and give rise to the same empirically confirmed equations describing the observed experimental data!**

2.3   Summary

The primary pillars of modern science are Maxwell's electrodynamics, Einstein's relativity theory, and the Copenhagen interpretation of quantum mechanics. Maxwell's equations for electrodynamics were incomplete and could not explain many phenomena, such as the Michelson-Morley experiments of 1886 and 1887, the photoelectric effect, blackbody radiation, the emission spectra of atoms, and the experiments of Hooper. This was due to the imperfect way in which the fundamental empirical equations of electrodynamics were turned into axioms and used to derive an electrodynamic wave equation. Maxwell did not use the complete set of unique empirical electrodynamic force equations as specified by Newton. He had failed to use Lenz's Law in order to satisfy conservation of energy and Newton's 3rd law. Furthermore he had used a number of idealizations including the point-particle idealization, an idealized displacement current in capacitors, the idealized linear superposition of fields, the idealized notion that induced and static fields are the same, the idealized notion that quantum effects do not exist, and the idealized notion of an action-at-a-distance force with no means to convey the force between two bodies.

In 1978 and 1992 Lucas [81, 82] pointed out two major problems in logic with the union of special relativity with Maxwell's equations. The first logical problem is that Faraday's empirical law shown in Equation (2-10) includes field transformation information between the moving prime and stationary lab observer's unprimed reference frames. In forming Maxwell's equations this field transformation information is discarded. [73]

$$\oint \vec{E}'(\vec{r}') \cdot d\vec{l}' = -\frac{1}{c}\frac{d}{dt}\oint \vec{B}(\vec{r}) \cdot \hat{n}da \quad (2-10)$$

The second logical problem is that the point-particle idealization is used to simplify Ampere's Law in Equation (2-11). [82, pp. 137-139] This is necessary in order to obtain the covariant form of Maxwell's equations.

$$\nabla \times \vec{B}(\vec{r}, t) = \frac{4\pi}{c}\vec{J}(\vec{r}, t) + \frac{1}{c}\nabla \int \frac{\nabla' \cdot \vec{J}(\vec{r}', t')}{|\vec{r} - \vec{r}'|} dr \approx \frac{4\pi}{c}\vec{J}(\vec{r}, t) \quad (2-11)$$

Point Particle Idealization

Note that the second term on the right of Equation (2-11) is not zero for finite-size particles in general. Thus the four-vector version of Maxwell's Equations is not proper or rigorous. Furthermore, the application of the Theory of Special Relativity to electrodynamics is an unwarranted duplication, since electrodynamics already has in Faraday's Law and Ampere's Law the transformation from the observer's frame of reference to the charged particle's rest frame.

Instead of perfecting the union of the empirical force laws of electrodynamics with the axiomatic method and removing the unreal idealizations to derive an improved version of electrodynamics, scientists of that time chose to invent additional fundamental theories to supplement Maxwell's electrodynamics, i.e. quantum mechanics and special relativity theory. They also chose to continue the practice of incorporating unreal idealizations in their theories.

In the last two centuries, starting with Ampere in 1825 [58], there has been a physical model for magnetic molecules (electrons) in the form of a continuous current loop. This model was developed in parallel with quantum mechanics and relativity theory, but without incorporating idealizations which are not true to nature. It explained the key experiments supporting quantum mechanics and relativity theory better than they did. Furthermore it went on to explain additional data such as the extreme ultraviolet emission spectrum of the hydrogen atom, the photoelectric effect for amorphous metals and nanometer optical antenna arrays, the Michelson-Morley experiments of 1886 and 1887, the Sagnac Effect, and the fields of charged particles remaining attached to all distances which they could not explain.

In physical science the physical model should always be given priority over the mathematical model, if it can explain the same data, and give rise to the same empirical equations describing the observed experimental data! The leaders of the scientific community in the past missed this important distinction and emphasized primarily mathematical theories based on idealizations. This was compatible with the Existential Philosophy which was dominant in science at the time. The next chapter will review the history of science from the perspective of various philosophies and show the role of structure as the source of reality and meaning in science.

2.4   References

1. Lord Kelvin in an address to the British Association for the Advancement of Science in 1900.
2. Henri Poincaré, **The Value of Science, Introduction**, p. 13 (1905).
3. Einstein, Albert, "Über einen die Erzeugung und Verwandlung des Lichtes betreffenden heuristischen Gesichtspunkt" (On a Heuristic Point of View about the Creation and Conversion of Light), **Annalen der Physik, Vol. 17, No. 6**, pp. 132-148 (1905)

4. Einstein, Albert, "Über die von der molekularkinetischen Theorie der Wärme geforderte Bewegung von in ruhenden Flüssigkeiten suspendierten Teilchen" (Investigations on the Theory of Brownian Movement), **Annalen der Physik, Vol. 17, No. 8,** pp. 549-560 (1905)

5. Einstein, Albert, "Zur Elektrodynamik bewegter Körper" (On the Electrodynamics of Moving Bodies), **Annalen der Physik, Vol. 17, No. 10,** pp. 891-921 (1905)

6. Einstein, Albert, "Ist die Trägheit eines Körpers von seinem Energieinhalt abhängig?" (Does the Inertia of a Body Depend upon Its Energy Content?), **Annalen der Physik Vol. 18, No. 13**, pp. 639-641 (1905).

7. Jackson, John David, **Classical electrodynamics – Third Edition**, (John Wiley & Sons, Inc., New York, 1999), p. 560.

8. A. Einstein, B. Podolsky, and N. Rosen, "Can the Quantum-Mechanical Description of Physical Reality Be Considered Complete?", **Phys. Rev. Vol. 47**, pp. 777-780 (1935).

9. Littauer, Schopper, and Wilson, "Structure of the Proton and Neutron", **Phys. Rev. Letters, Vol. 7,** pp. 144-147 (1961).

10. http://en.wikipedia.org/wiki/Titius%E2%80%93Bode_law#References

11. S.F. Dermott, "On the origin of commensurabilities in the solar system - II: the orbital period relation", **Mon. Not. RAS vol. 141,** pp. 363–376 (1968).

12. S. F. Dermott, "On the origin of commensurabilities in the solar system - III: the resonant structure of the solar system", **Mon. Not. RAS vol. 142**, pp. 143–149 (1969).

13. Tifft, W. G., "Discrete States of Redshift and Galaxy Dynamics. I. Internal Motions in Single Galaxies", **Astrophysical Journal. Vol. 206**, pp. 38-56 (1976).

14. Tifft, W. G. and Cocke, W. J., "Global Redshift Quantization", **Astrophysical Journal, Vol. 287**, pp. 492-502 (1984).

15. Humphreys, D. Russell, "Our Galaxy is the Centre of the Universe - Quantized Red Shifts Show", **The Journal, Vol. 16, No. 2**, pp. 95-104 (2002).

16. http://www.astro.cornell.edu/academics/courses/astro201/bodes_law.htm

17. http://en.wikipedia.org/wiki/Mach's_principle

18. H. Fizeau, "The Hypotheses Relating to the Luminous Aether, and an Experiment which Appears to Demonstrate that the Motion of Bodies alters the Velocity with which Light Propagates itself in their Interior", **Philosophical Magazine Vol. 2**, pp. 568-573 (1851).

19. Michelson, A. A. and Morley, E. W., "Influence of Motion on the Medium on the Velocity of Light", **Am. J. Science Vol. 31**, pp. 377-386 (1886).

20. C. W. Oseen, "Uber die Wechselwirkung Zwischen Zwei Elektischen Dipolen der Polarisationsebene in Kristallen und Flussigkeiten" (On the Interactions between Two Electric Dipole Planes of Polarization in Crystals and Liquids), **Ann. Phys. Vol. 48**, pp. 1-56 (1915).

21. J. G. Fox, "Experimental Evidence for the Second Postulate of Special Relativity" **Am. J. Phys. Vol. 30**, pp. 297-300 (1962).

22. J. G. Fox, "Experimental Evidence for the Second Postulate of Special Relativity" **Am. J. Phys. Vol. 30**, pp. 297-300 (1962).

23. Curt Renshaw, "Fresnel, Fizeau, Hoek, Michelson-Morley, Michelson-Gale and Sagnac in Aetherless Galilean Space", **Galilean Electrodynamics Vol. 7**, pp. 103-108 (1996).

24. Einstein in a letter to Michael Bessi in 1954 http://www.spaceandmotion.com/quantum-theory-albert-einstein-quotes.htm

25. W. Hallwachs, **Wiedmann's sche Annalen Vol. 29**, pp. 1-12 (1886).

26. W. Hallwachs, **Wiedmann's sche Annalen Vol. 32**, pp. 64-74 (1887).

27. H. Hertz, **Berliner Berichte**, pp. 487-490 (1887).

28. R. A. Millikan, "A Direct Photoelectric Determination of Planck's h", **Phys. Rev., Vol. 7**, p. 362 (1916).

29. André-Marie Ampère, **Mém. de l'Académie des Sciences, V6**, p. 175 (Paris: 1823).

30. Alfred L. Parson, "A Magneton Theory of the Structure of the Atom", **Smithsonian Miscellaneous Collection, Pub 2371**, 80pp (Nov 1915) {Reprinted **Pub 2419, V65, N11** (1916)}.

31. David L. Bergman and J. Paul Wesley, "Spinning Charged Ring Model of Electron Yielding Anomalous Magnetic Moment", **Galilean Electrodynamics, Vol. 1**, pp. 63-67 (1990).

32. David L. Bergman and Charles W. Lucas, Jr., "Physical Models of Matter" **Foundations of Science Vol. 1, No. 2**, pp. 1-23 (1996).

33. Daniel Dregely, et al., "3D Optical Yagi-Uda Nanoantenna Array" **Nature Communications Vol. 2, Article 267**, April 5, (2011).

34. E. G. Cullwick, **Electromagnetism and Relativity** (Longmans Green and Co., London, 1957) p. 245.

35. W. J. Hooper, **New Horizons in Electric, Magnetic, and Gravitational Field Theory** (Electrodynamic Gravity, Inc., 543 Broad Blvd., Cuyahoga Falls, OH 44221, 1974), preface.
http://www.rexresearch.com/hooper/horizon.htm

36. Michelson, A. A. "The Relative Motion of the Earth and the Luminiferous Aether." *Amer. J. Sci.* **22**, pp. 120-129, (1881).

37. Michelson, A. A. and Morley, E. W. "On the Relative Motion of the Earth and the Luminiferous Ether." *Amer. J. Sci.* **34**, pp. 333-345, (1887).

38. Michelson, A. A. and Morley, E. W. "On the Relative Motion of the Earth and the Luminiferous Aether**." Philos. Mag. 24**, pp. 449-463, (1887).

39. Laue, Max von, "On an Experiment on the Optics of Moving Bodies". **Münchener Sitzungsberichte**: pp. 405–412 (1911).

40. Guido Rizzi, Matteo Luca Ruggiero, "The relativistic Sagnac Effect: two derivations". In G. Rizzi and M.L. Ruggiero. **Relativity in Rotating Frames.** (Kluwer Academic Publishers, Dordrecht, 1981).

41. Sagnac, Georges, "The demonstration of the luminiferous aether by an interferometer in uniform rotation". **Comptes Rendus 157**: pp. 708–710 (1913).

42. Sagnac, Georges, "On the proof of the reality of the luminiferous aether by the experiment with a rotating interferometer". **Comptes Rendus 157**, pp. 1410–1413 (1913).

43. Albert Abraham Michelson, Henry G. Gale "The Effect of the Earth's Rotation on the Velocity of Light*"*, **The Astrophysical Journal 61**, pp. 140–145 (1925).

44. Richard P. Feynman, **The Character of Physical Law** (Messenger Lectures, Cornell University, 1964) published (2001).

45. Heisenberg, Werner, **Philosophy and Physics, Vol. XIX of World Perspectives Series**, 1958. Quoted from the Harpers Torchbook edition, p. 42 (1962).

46. http://www.naturalthinker.net/trl/texts/Heisenberg,Werner/Heisenberg,%20 Werner%20-%20Physics%20and%20philosophy.pdf

47. Mehra, J.; Rechenberg, H., **The Historical Development of Quantum Theory**. (Springer-Verlag, New York, 1982).

48. Max Born, "Zur Quantenmechanik der Stoßvorgänge", **Zeitschrift für Physik A Hadrons and Nuclei, Vol. 37, No. 12**, pp. 863-867 (1926).

49. Siegel, D. M., "Balfour Stewart and Gustav Robert Kirchoff: two independent approaches to Kirchoff's radiation law", **Isis Vol. 67**, pp. 565-600 (1976).

50. Kirchoff, G., "Uber die Fraunhofer'schen Linien", **Monatsberichte der Koeniglich Preussichen Akademie der Wissenschafften su Berlin**, pp. 662-665 (1860).

51. Kirchoff, G., "Uber den Zusammenhang zwischen Emission und Absorption von Licht und Waerme", **Monatsberichte der Koeniglich Preussichen Akademie der Wissenschafften su Berlin**, pp. 783-787 (1860).

52. Kirchoff, G., "On the relation between the radiating and absorbing powers of different bodies for light and heat", **Philosophical Magazine Series 4, Vol. 20**, pp. 1-21 (1860).

53. Planck, M., "Über eine Verbesserung der Wienschen Spektralgleichung" (On an Improvement of Wien's Equation for the Spectrum), **Verhandlungen der Deutschen Physikalischen Gesselschaft Vol. 2**, pp. 202–204 (1900).

54. Planck, M., "Zur Theorie des Gesetzes der Energieverteilung im Normalspektrum" (On the Theory of the Energy Distribution Law of the Normal Spectrum), **Verhandlungen der Deutschen Physikalischen Gesselschaft Vol. 2**, p. 237 (1900).

55. http://ned.ipac.caltech.edu/level5/Tyson/Figures/figure4.gif

56. Charles W. Lucas, Jr, "A Physical Model for Atoms and Nuclei — Part 4; Blackbody Radiation and the Photoelectric Effect," **Foundations of Science, Vol. 6, No. 3**, (August 2003).

57. Sutherland, **Philosophical Magazine, Vol. 11**, p. 245 (1901).

58. Ampère, André-Marie, **Mém. De l'Acad,** p. 175 (1825).

59. E. Rutherford, "The Scattering of α and β Particles by Matter and the Structure of the Atom", **Philosophical Magazine Series 6, Vol. 21**, pp. 669-699 (1911).

60. Eisberg, Robert Martin, **Fundamentals of Modern Physics,** (John Wiley and Sons, Inc., New York), pp. 113-115 (1961).

61. Alfred Lauck Parson, **Smithsonian Miscellaneous Collections, Vol. 65, No. 11** (1915).

62. Gilbert N. Lewis, "The Atom and the Molecule", **Journal of the American Chemical Society, Vol. 38**, pp. 762-785 (1916).

63. Grondahl, American Physical Society address in December 1916 (printed in **Physical Review Series II**, p. 9, 1917).

64. Leigh Page, "The Distribution of Energy in the Normal Radiation Spectrum". **Physical Review Vol. 7, No. (2)**, pp. 229–240 (1916).

65. Webster, David L., "Notes on Page's Theory of Heat Radiation," **Physical Review Series II, Vol. 8 No. 1**, pp. 66-69 (1916).

66. Webster, David L., "The Theory of Electromagnetic Mass of the Parson Magneton and other Non-Spherical systems," **Physical Review Series II, Vol. 9, No. 6**, pp. 484-499 (1917).

67. Compton, Arthur H., American Physical Society address December 1917, **Physical Review Series II**, p. 330 (1918).

68. Compton, Arthur H., **Physical Review Series II, Vol. 14 No. 1**, pp. 20-43 (1919).

69. Compton, Arthur H., **Physical Review Series II, Vol. 14 No. 3**, pp. 247-259 (1919).

70. Allen, H. S., "The Case for a Ring Electron," **Proceedings of the Physical Society of London, Vol. 31**, pp. 49-68 (1919).

71. Louis de Broglie, "Ondes et Quanta / Waves and Quanta", **Comptes Rendus, Vol. 177**, pp. 507-510 (1923).

72. E. Schrödinger, "Quantisierung als Eigenwertproblem (Erste Mitteilung)", **Annalen der Physik, Series 4, Vol. 79**, pp. 361-376 (1926).

73. E. Schrödinger, "Quantisierung als Eigenwertproblem (Zweite Mitteilung)", **Annalen der Physik, Series 4, Vol. 79**, pp. 489-527 (1926).

74. E. Schrödinger, "Quantisierung als Eigenwertproblem (Dritte Mitteilung)", **Annalen der Physik, Series 4, Vol. 80**, pp. 437-490 (1926).

75. E. Schrödinger, "Quantisierung als Eigenwertproblem (Vierte Mitteilung)", **Annalen der Physik, Series 4, Vol. 81**, pp. 109-139 (1926).

76. Dirac, P. A. M., "The Quantum Theory of the Electron". **Proceedings of the Royal Society A: Mathematical, Physical and Engineering Sciences, Vol. 117**, p. 610 (1928).

77. Bostick, Winston H., W. Prior, L. Grunberger, and G. Emmert, "Pair Production of Plasma Vortices", **Physics of Fluids, Vol. 9, Issue 10**, pp. 2078-2080 (1966).

78. Bostick, Winston H., "Mass, Charge, and Current: The Essence and Morphology," **Physics Essays, Vol. 4, No. 1**, pp. 45-59 (1991).

79. Barnes, Thomas, "Alternative to Einstein's Special Theory of Relativity," **Physics of the Future**, ICR, pp. 81-94 (1983).

80. Poincaré, Henri, **Oeuvres, Vol. 9**, p. 497 (1954).

81. C. W. Lucas, Jr., "Is Relativity Necessary for Electrodynamics?", **Bulletin of the American Physical Society II Vol. 23** p 544 (1978).

82. Lucas, Jr., Charles W. and Joseph Lucas, "Electrodynamics of Real Particles vs. Maxwell's Equations, Relativity Theory, and Quantum Mechanics," **Proceedings of the Twin Cities Conference, Minneapolis, MN**, pp. 243-248 (1992).

83. Jackson, John David, **Classical Electrodynamics**, (John Wiley & Sons, Inc., London, pp. 170-173 (1962).

84. Labov, Simon E. and Stuart Bowyer, "Spectral Observations of the Extreme Ultraviolet of Background," **The Astrophysical Journal, Vol. 371**, p. 810 (1990).

85. Lucas, Jr., Charles W. and Joseph Lucas, "A New Foundation for Modern Science," **Proceedings of the International Conference on Creationism**, held at Geneva College in Beaver Falls, PA, USA (August 3-8, 1998).

Chapter 3 Structuralism – Key to Reality and Meaning in Science

Structure is a concept that is both amenable to mathematical manipulation and definition, and derives its power from mathematics. ... Structuralism deals with relationships between parts and the whole. Totality takes logical priority over individual parts, and the relationships are more important than the entities they connect. The hidden structure is thus much more important than what is obvious or apparent in any given situation. It is the symbolism that matters, rather than the entities symbolized. [1]

The purpose of this chapter is to briefly review the various philosophies under which science has developed in order to see why the current Post-Modern philosophy is preventing significant progress in science at the present time. Also this review will help us discover the best way to continue in order to perfect science to a higher level.

3.1 Foundations of Natural Philosophy

Atomism, one of the earliest foundations of natural philosophy, was developed by Leucippus and his student Democritus in the fifth century BC. These atomists theorized that the natural world consists of atoms and void where *void* is a mere nothing. Atoms are intrinsically unchangeable and move about the void combining into different clusters or substances. Atoms are reality's very small, indestructible building blocks. [2] The word atomism is derived from the ancient Greek adjective *atomos*, which literally means 'uncuttable' (*a - tomos* (*not cuttable*) - *tomos* a conjugate of the Greek verb *temnein* (*to cut*)).

The law of cause and effect, a second foundation of natural philosophy, was a dominant part of natural philosophy or science up until the twentieth century. Causality denotes a necessary relationship between event A (called cause) and another event B (called effect) which is the direct consequence or result of event A. There are three assumptions normally associated with causality:

1. **There are physical laws by which the occurrence of an event B depends on the occurrence of an event A.**
2. **The cause A must be prior to, or at least simultaneous with, the effect B.**
3. **The cause A and effect B must be in spatial contact.**

The physical laws above are normally assumed to be force laws. The force must be applied before the effect is seen. The forces in nature must be local or contact forces in order to reach out and cause an effect.

3.2 Axiomatic Philosophy

Aristotle and other ancient Greeks developed Syllogism or the logic of inference. Syllogism is a kind of logical argument in which one proposition or conclusion is inferred from two or more other propositions known as premises. Syllogism became the core of deductive reasoning, where facts are determined by combining existing statements using logic. By contrast inductive reasoning is where the facts are determined by repeated observations.

The axiomatic method was invented by the ancient Greeks as the proper way to organize and demonstrate deductive reasoning in the pursuit of natural philosophy. The axiomatic method is a logical procedure by which an entire system of natural philosophy (e.g. a branch of science or mathematics) is generated in accordance with specified rules of logical deduction from certain basic propositions (axioms or postulates), which in turn are constructed from a few terms (charge, mass, length) taken as primitives. These terms and axioms are to be defined and constructed according to some method by which some warrant for their truth is felt to exist. One of the oldest examples of an axiomatic system is the ancient Greek Euclidean geometry.

Euclid, in the process of developing geometry, defined the axiomatic method of proofs to be used in logically establishing theorems in geometry. To the extent that the axioms or postulates he chose were valid, his logically developed theorems would be valid.

3.3 Empirical Philosophy

In 1687 when Isaac Newton published his famous book **Mathematical Principles of Natural Philosophy**. [3, 4], he stated that he intended to illustrate a new way of doing natural philosophy that overcomes some of the limitations of the axiomatic method. This method is now called the empirical scientific method. The goal of Newton's method was to find empirically the forces of nature by induction. Thus Newton was expanding the axiomatic method to include both inductive and deductive logic.

Newton's book is considered by many as the most important contribution to science in the history of the world, because it was the first to show how to describe the physical world in terms of the precise language and equations of mathematics which would become the laws of science. Newton's work laid the groundwork for classical mechanics, which dominated the scientific view of the physical universe for the next three centuries

The axiomatic method was logically rigorous, but it was not broad enough. It lacked a reliable method to discover the axioms of science and the most appropriate terms for the axioms.

3.4 Existential Philosophy

Despite Newton's great success, a number of natural philosophers became dissatisfied with his slighting of the logical rigor and inductive methodology of Continental philosophers. They felt that Newton's empirical method did not describe the "real" world. For instance Newton postulated the existence of gravitational and inertial mass, but nobody knew what mass was or could explain it. Newton, himself, admitted that he did not know what mass was. Why were the ratios of gravitational and inertial masses equal for the same two bodies? Presumably they should be fundamentally different, since they were associated with fundamentally different forces. Newton's force of gravity was an action-at-a-distance type force with no known mechanism to produce spatial contact from the cause to the effect. Newton had modified the ancient aether of Greek cosmology and made it the medium to transmit the force from the cause to the effect, but the aether was not fully satisfactory in explaining attractive and repulsive forces and the force of inertia.

The description of the universe in terms of apparently fictitious quantities such as mass, aether, and action-at-a-distance forces eventually led to the creation of the philosophy of existentialism. This philosophy amended the axiomatic and empirical approach to natural philosophy to purposely allow science to be developed in terms of idealizations and fictitious terms. The existentialist philosophers accepted the purposelessness and absurdity of the world that Newton so successfully described with fictitious forces and masses that did not exist. They abandoned the strict role of deductive logic in the scientific method and substituted for it the much weaker criterion of falsifiability of predictions of hypotheses.

During the time that existentialism was a dominant force in philosophy, many major theories of modern science were developed. These included quantum mechanics, especially the Copenhagen interpretation of quantum mechanics, in which the particles of nature are all point-like. The unphysical mathematical universal wave function of quantum mechanics describes point particles as governed 100% by random statistical processes instead of the law of cause and effect. Thus in the quantum realm it is not possible to determine that action A caused result B. Also in the words of Heisenberg, "reality is in the observation process, not in the structure of the atom or electron". Previously natural philosophers had believed from experience with the real world that the universe is not totally random in nature, but has a certain degree of order, and the law of cause and effect is dominant. Also modern natural philosophers realized from scattering experiments that every particle in nature has a finite-size and an internal charge structure contrary to the assumptions of quantum mechanics. (See Figure 2-1 above) Existential philosophers believed that progress in science was to be achieved by following in the footsteps of what Newton did, rather than in what he said to do.

Einstein's theory of relativity was also introduced during this time. It too was based on the point-particle idealization combined with the second idealization that the spatial universe was homogeneous and isotropic. The lumpiness of stars and galaxies in space and the quantization of orbits of planets about the sun, stars about the Milky Way galaxy, and galaxies about the center of the universe seemed to deny the latter assumption. Also relativity theory introduced the notion of four-dimensional space, where time is the fourth dimension. No wonder the existentialist philosophers found the universe confusing without purpose and meaning. Their scientific theories described the universe using nonsensical notions and idealizations that defied the reality of the ancient natural philosophers.

During this time the theory of evolution of life and the physical universe was introduced. The evolutionary process was one that led from disorder to order to more order. This was in disagreement with the experimentally based laws of thermodynamics that all systems tend to disorder over time and energy is conserved. The nonsensical evolutionary process did not agree with common every day experience of the natural philosophers and the laws of thermodynamics.

3.5   Poincaré's Philosophy of Structural Realism in Science

Poincaré was considered to be one of the last of the great philosophers of science that covered all of natural philosophy. He built his philosophy of science in many ways upon Newton's empirical philosophy of science. He attempted to find the proper balance between the existential philosophy and Newton's empirical philosophy. For Poincaré a defensible scientific realism must be structured in the sense that it attributes reality to the relational structure of scientific theories as expressed in mathematical equations. He thought that it was a mistake to think that we can ever understand the nature of the basic things of which the universe is made. In all his writings Poincaré stresses that which is knowable by people with the particular faculties they have as human beings.

3.5.1 Goal of Science is Mathematical Relations

According to Poincaré in the preface to his book **Science and Hypothesis** of 1902,

> **The aim of science is not** (understanding) **things in themselves, but the** (mathematical) **relations between things; outside those** (mathematical) **relations there is no reality knowable. [5]**

Later in the book Poincare says,

> **Science is built up of facts, as a house is built of stones; but an accumulation of facts is no more a science than a heap of stones is a house. [5, p. 141]**

Employing a different analogy, Poincaré compares the function of mathematical physics to that of a library catalogue, where the experimental facts serve as the "books". [5, pp. 144-145] It is the role of mathematical physicists to take new data generated by experiment and to order them usefully. This cataloguing process directs scientists toward new sorts of experiments necessary for supplementing the library with interesting new books. Thus scientific facts must be given structure. This is what it is to do science and to construct scientific theories.

3.5.2 Scientific Theories Are Structures Describing Empirical Data

According to Poincaré scientific facts are placed into structures to create scientific theories. In other words, scientific theories "order the facts". Our knowledge of the "scientific facts" is knowledge of relations as expressed mathematically.

Poincare distinguishes between the "brute facts" and the facts that serve as the subject matter of knowledge or science. He explains in his book **The Value of Science** of 1905,

> **The scientific fact is only a crude fact translated into a convenient** (mathematical) **language...All the scientist creates in a fact is the** (mathematical) **language in which he enunciates it. [6]**

Scientific facts according to Poincaré involve an ineliminable human contribution. Thus statements concerning truth, reality and objectivity should not be given a straightforward realist interpretation as shown in the quote below.

> **If truth be the sole aim worth pursuing, may we hope to attain it? It may well be doubted. Readers of my little book Science and Hypothesis already know what I think about the question. The truth we are permitted to glimpse is not altogether what most men call by that name. [6, p 12]**

A few pages after the quote above, Poincaré says,

> **Does the harmony the human intelligence thinks it discovers in nature exist outside of this intelligence? No, beyond all doubt, a reality completely**

independent of the mind which conceives, sees, or feels it is an impossibility. A world as exterior as that, even if it existed, would be for us forever inaccessible. But what we call objective reality is, in the last analysis, what is common to many thinking beings, and could be common to all; this common part, we shall see, can only be the harmony expressed by mathematical laws. It is this harmony then which is the sole objective reality, the only truth we can attain. [6, p. 14]

According to Poincaré, only the mathematical relations occurring in a unified and empirically successful theory mirror the ontological order of things. As for the nature of the related things, it will forever remain hidden from us.

This is well illustrated by the theory of the atom. The atom consists of things called electrons, protons, and neutrons. In the modern theory of the atom these things are idealized as point particles and the mathematical relations between these things are given by the Dirac equation. The Dirac equation describes the emission spectra of atoms and other atomic observables. This same idealized approach is also used in the Standard Model of Elementary Particles where all elementary particles are point particles constructed of smaller things called quarks and leptons.

3.5.3 Arithmetic Based on a Priori Intuition

Poincaré grounded arithmetic in synthetic a priori intuition. He argues that "indefinite repetition of the same act", and thus reasoning "by recurrence", is essential to arithmetical reasoning, allowing us to pass from particular results to general theorems. This rule of reasoning by recurrence is obtained neither from experience nor from logic (it does not follow from the principle of non-contradiction), but is an a priori synthetic intuition. "The mind has a direct intuition of this power, and experiment can only be for it an opportunity of using it". [**5,** p. 13]

3.5.4 Geometry Based on a Priori Intuition

Poincaré also grounded geometry in a priori synthetic intuition. In Chapter III of **Last Essays** of 1913 Poincaré explicitly states that we have an intuition of the spatial continuum that has the same status as our arithmetical intuition.

I shall conclude that there is in all of us an intuitive notion of the continuum of any number of dimensions whatever, because we possess the capacity to construct a physical and mathematical continuum; and that this capacity exists in us before any experience because, without it, experience properly speaking would be impossible and would be reduced to brute sensations, unsuitable for any organization; and because this intuition is merely the awareness that we possess this faculty. And yet this faculty could be used in different ways; it could enable us to construct a space of four just as well as a space of three dimensions. It is the exterior world; it is experience which induces us to make use of it in one sense rather than in the other. [7]

Mathematics, like any science, must seek after truth. Truth means more than mere consistency. In mathematics Poincaré takes truth to mean that the axioms cohere with our intuitions, that is, with the form of experience.

Poincaré believes that the very possibility of our experiences of the world containing empirical objects depends on spatial intuition, when he says

This capacity exists in us before any experience because, without it, experience properly speaking would be impossible and would be reduced to brute sensations, unsuitable for any organization. [7, p. 44]

In other words, spatial intuition is that through which our sensations are constituted into our experiences of physical objects; objects which endure through space and time.

3.5.5 Role of Generalizations in Empirical Science

A science of empirical objects goes beyond mere experience of objects. We must form generalizations concerning these objects, and our ability to form these generalizations is itself grounded in a priori intuition. In his book **Science and Hypothesis** Poincaré asserts that the business of science concerns generalizations – to move from premises to conclusions that are "in a sense more general than the premises". [5 p. 4] His position may be summed up by his own slogan: "There is no science but the science of the general". [5, p. 4] Consider first the nature of this generalization, and then what grounds our ability to form such generalizations.

Poincaré distinguishes between mathematics and the physical sciences in the following way. He highlights the similarity between (i) reasoning by recurrence in arithmetic and (ii) inductions in physical science. He points out that induction in the physical sciences is uncertain, whereas reasoning by recurrence is not. [5, p. 13] The reason that he gives is that induction depends for its success on "an order which is external to us" whereas proof by recurrence depends for its success on "a property of the mind itself".

3.5.6 Three Ways Generalizations Are Based on a Priori Intuition

What grounds our ability to generalize in the physical sciences? In mathematics the ability to generalize is grounded in arithmetical intuition. A priori intuition, or the form of experience, is that via which we understand, by our sensory perceptions, an experience of a single object enduring through space and time, despite the inevitably incomplete character of experience. It is also that via which we understand certain rules as characterizing infinite, yet determinate, collections. A priori intuition acts as a "glossing over" faculty which glosses over the incomplete character of both empirical and mathematical experience. It provides a procedure whereby we ignore all the elements which could be generated by a rule, and we disregard or "smooth out" the disparate character of perception. [5, p. 86]

There is a second place where a priori intuition plays a role in the physical sciences. Disregarding certain features of sensory experience as irrelevant is necessary but not sufficient for us to form a generalization. The fallibility of induction lies in the fact that we can make mistakes when we decide which aspects of the particular physical objects to ignore as irrelevant, and which to take into account when forming the generalization.

Also Poincaré believed that the concept of indefinite iterability is foundational, not only for arithmetic, but for all systematic thinking. Its epistemological basis is synthetic a priori intuition. This concept underlies all systematic thinking, because it underlies our ability to generalize.

According to Poincaré generalization is involved in physical science at two different stages. The first is when the empirical data are organized. We must draw our line through the dots on a graph that records our experimental results. This choice goes beyond mere generalization.

> **However timid we may be, there must be interpolation. Experiment only gives us a certain number of isolated points. They must be connected by a continuous line, and this is a true generalization. But more is done. The curve thus traced will pass between and near the points observed; it will not pass through the points themselves. Thus we are not restricted to generalizing our experiment, we correct it… Detached facts cannot therefore satisfy us, and that is why our science must be ordered, or better still, generalized. [5, pp. 142-143]**

Thus the interpolation, the act of drawing a curve to fit the data, is a moment in which Poincaré claims we not only generalize the data, but also correct it. This additional feature of empirical generalization (beyond that found in mathematics) arises from the different nature of objects that serve a subject-matter for the generalizations, and leads to the fallibility of those generalizations.

The second type of generalization takes the results of this first stage (drawing curves through the data points) as input in order to generate empirical laws. Thus empirical laws are grounded from the start on generalizations. However, it is these laws that enable us to achieve the desired generality that the human mind seeks, and which enables us to progress in science.

> **Who gives us the right of attributing to the principle itself more generality and more precision than to the experiments which have served to demonstrate it? This is asking if it is legitimate to generalize, as we do every day with empirical data … One thing alone is certain. If this permission were refused to us, science could not exist; or at least would be reduced to a kind of inventory, to the ascertaining of isolated facts. It would no longer be to us of any value, since it could not satisfy our need of order and harmony, because it would be at the same time incapable of prediction. [5, pp. 129-130]**

In summary it seems that a priori intuition clearly grounds our ability to construct the generalizations that form the very substance of physical scientific theories. Furthermore scientific theories are just structures that we impose on the facts.

There is a third role for a priori intuition. It is not just our ability to construct generalizations that is grounded in a priori intuition, but our ability to apply the resulting generalizations is similarly grounded. In order to understand the abstract characterization of a rule or theory, we must understand an arbitrary instance of it. Applying a rule requires that the application possesses the same essential structural properties as the arbitrary instance given in the schematic characterization of the rule. The aspects which are structural are those which an arbitrary instance possesses. A priori intuition supplies the ability to understand what these are.

For Poincaré laws capture the relations between things. Thus to recognize that objects offer an instance of a given law is to recognize that they follow the relations that are described by the law. In order to do this, we must ignore the non-relational features of objects, if any such are presented to us in experience. A priori intuition is therefore not only that which enables us to generalize, but also that which enables us to apply the resulting generalizations.

For structural scientific realism, the central idea is that scientific theories do provide information unavailable to us in observation and experimentation, but that information is about the form or structure, rather than the nature or content, of what is unobservable. When one theory is replaced by another, it is information about the essential nature of what is unobservable that is replaced, rather than information about the structure of the unobservable. This is illustrated by the replacement of Newtonian mechanics by Einstein's relativistic mechanics. Only small changes are made to the mathematical equations of classical mechanics. These changes are due to the directly unobservable nature of the things of the universe.

3.5.7 Changes in the Size of Things with Technology

During the times of the ancient Greeks, the smallest physical thing in the universe was an atom, where the word atom means uncutable or indivisible. During Poincaré's life (1854-1912) the electron was discovered. After Poincaré's death Earnest Rutherford discovered the proton to be the hydrogen nucleus in 1920. In 1932 James Chadwick discovered the neutron as a subatomic particle. At this point the indivisible thing called the atom consisted of electrons, protons, and neutrons. In theories of the atom these particles were idealized as points, since their internal structure and size were unknown. After that the electron, proton and neutron were discovered to have a magnetic moment indicative of current flowing in some sort of structure.

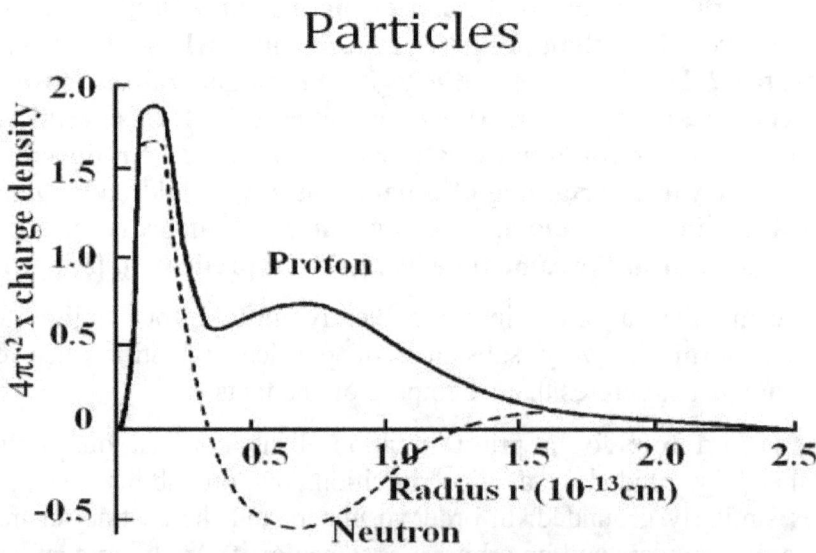

Figure 3-1 Scattering Data Showing Proton and Neutron Internal Structure and Finite-size [9]

Then in 1956 Robert Hofstadter [8, 9] began a series of high energy electron scattering experiments in which he measured the size and internal structure of the proton and neutron as shown in Figure 3-1. This data shows not only the finite size but also the internal structure of the proton and neutron. Both the proton and neutron consist of at least three sub-structures.

In Chapters 4, 5, and 6 of this book an improved version of the electrodynamic force is derived by perfecting the union of the Axiomatic and Empirical scientific methods. In Chapter 6 the radiation reaction force is derived. In order to obtain the observed radiation reaction force, a boundary condition on the internal charge structures of elementary particles is obtained. The boundary condition is that the charge in a finite-size elementary particle exists only in closed loops.

The significance of this detail is greatly enhanced in Chapters 9 and 10 where improved versions of the force of gravity and the force of inertia are derived that appear to be superior to all such theories in the past in that they describe many new phenomena. Also it is significant to note that the only other remaining so-called "fundamental forces" in nature, i.e. the weak and strong interaction forces, have a range that is on the order of the size of an elementary particle suggesting that they are really just mocking up the finite-size electrodynamic effects of elementary particles. All this lends support to the notion that the improved electrodynamic force is actually the universal force.

Thus it appears that the only force existing within an elementary particle is the electrodynamic force. This combined with the information that all elementary particles consist of closed charge loop structures enables us to determine the structural relations between the closed charge loops within elementary particles. This is the subject of the next book in the series **The Universal Force Volume 2 - An Electrodynamic Model of Elementary Particles**. Thus the unknowable "things" of Poincare are reduced in size from elementary particles to the closed charge loops within elementary particles. Now the question is "What is a closed charge loop?"

In 1990 Bergman and Wesley [10] examined the interior structure of the electron assuming that the structure was determined by electrodynamic forces alone. They were able to derive equation (2-9) giving the relation between the large radius R of the toroidal charge ring and the small cross sectional radius r in terms of Planck's constant h. The balance of electric and magnetic forces within the toroidal charge ring determines this ratio and the value of Planck's constant. Thus the question now is "What is the thing called charge?"

3.6   Structural Philosophy

In the 1930s a new philosophy called structuralism [11-13] was developed by the Bourbaki (a secret society of French mathematicians) extending Poincaré's structuralism using some new ideas from linguistics. According to this new philosophy

1. **Every system has a structure**
2. **There are laws responsible for the structure of systems**
3. **There are unique elements that make up systems**
4. **Meaning is derived from the structures of systems which act as signs**

The items making up any particular system exemplifying the structures are based on axioms which comprise the barest set of first principles. The theorems of mathematics or the theories of science are obtained by the rigorous application of deductive logic to these axioms in a manner similar to the way proofs of theorems are done in Euclidean geometry. There are underlying structures in science and mathematics and the relationship of these structures is the source of meaning and reality that was missing in existential mathematics and science.

Structuralism is closely related to semiotics which is also called semiotic studies or semiology. Semiotics is the study of sign processes (semiosis or signification and communication), signs and symbols, both individually and grouped into sign systems. It includes the study of how meaning is constructed and understood.

Structuralism appeared in academia in the second half of the 20th century. It was an approach to the human sciences that attempted to analyze a specific field (for instance, mythology) as a complex system of interrelated parts. It began in linguistics with the work of Ferdinand de Saussure. But many French intellectuals perceived it to have a wider application, and the model was soon modified and applied to other fields, such as anthropology, psychoanalysis, sociology, literary theory, mathematics, economics and architecture. This ushered in the dawn of structuralism as not just a method, but also an intellectual movement that came to take existentialism's pedestal in France in the 1960s. [13]

The secret French mathematical society known as Bourbaki, in a series of 10 volumes, revolutionized most of mathematics by establishing an axiomatic basis for all of mathematics and showing its common structures. The Bourbaki believed that every fact in mathematics must have an explanation. Using set theory they attempted to show the unity and universality of mathematics in terms of axioms, logic, and structures. Structure was seen as the mathematically describable portion of reality that has meaning. This meaning can be expressed in terms of mathematical symbols and equations. Structuralism was perceived as the method of intellectual inquiry that provides a framework for organizing and understanding areas of human study that enables the discovery of meaning. Structuralism replaced existentialism which regarded human existence and science as unexplainable and without meaning, i.e. not in agreement with logic and common sense.

Members of the Bourbaki worked with researchers in many fields and applied structuralism to their studies. In the areas of linguistics [14], literary theory [15], psychology [16], anthropology [17, 18], and economics [19] the Bourbaki were able to help researchers move their study from a descriptive phase to one based on mathematical symbols and mathematical equations with laws and theories derived by rigorous logic from a finite set of axioms. Many of these researchers won Nobel prizes for their work. One fundamental assumption of structuralism is that all of human behavior arises from the innate structures in the human brain. The ultimate goal in the social sciences is to discover and understand the cause and nature of the innate structures of the brain.

3.7 Postmodern Philosophy

The success of structuralism in the social sciences and mathematics was not matched in the hard sciences of physics, astronomy, geology, chemistry and biology. The scientists in these fields were content with the previously developed existentialist type theories in their fields. As a result they refused to participate in the structural revolution. Thus the grand goal of structuralism was slowly abandoned and replaced by postmodernism. [20, 21, 22]

The postmodernists did not like the conclusions of metatheory as developed by Poincaré, because metatheory proved that the Copenhagen version of quantum mechanics and Einstein's theories of special and general relativity theory were not fundamental theories. They were ultimately just add-on theories to compliment the incomplete theory of electrodynamics.

Since Postmodernists liked the existential theories of electrodynamics, relativity, quantum mechanics and evolution which agreed with their worldview, they did not want to participate in the structural reformation of science. So they developed a new philosophy of science to accomplish their goals. Postmodernists decided that each field of study or body of knowledge should have its own internally defined notion of truth or reality. Thus the truth of metatheory could be excluded from physics or electrodynamics. As a consequence, the truth or validity of different fields of study can no longer be compared. Each field is supervised by a group of experts in order to police the borders of that field with criteria for inclusion and exclusion. In the wake of Karl Popper's [23] influential work, falsifiability of theory predictions is often put forward as the criterion for distinguishing between good scientific theories and bad scientific theories in each field. However, falsifiability of theory predictions is a much weaker criterion than what the axiomatic method offered in the past. It is a necessary but not sufficient requirement for good science theories. **The more sufficient requirement is that the structure and foundations (axioms) of scientific theories as well as the predictions of scientific theories cannot be falsified. Thus no idealizations in the axioms or logic of proper scientific theories should be allowed when empirical data is available!** Unfortunately the search for axioms that were as true as possible, and the employment of rigorous logic to develop the theories of science have been downplayed in existential and postmodern philosophies. Only the falsifiability of the predictions of theories is used to discredit theories.

Under postmodernism each field of human study now has experts that define their criteria for truth and meaning. Each field is now a somewhat independent silo of knowledge and truth. The "uni-versity", which in the past tried to integrate all of man's knowledge into a unified whole, is now a "multi-versity" with each discipline being a somewhat independent field of knowledge.

3.8   Universal Force Approach Based on Structural Philosophy

The approach of this book builds on the ancient atomistic notions, the law of cause and effect, the deductive logic of the Axiomatic method, the inductive logic of Newton's empirical method, Poincaré's scientific realism based on structure, the Bourbaki notion of structure being the source of all reality and meaning, and the concept of universal truth across all fields of investigation. Here the underlying structures in science are based on a derived axiomatic universal electrodynamic contact force law, with the axioms of science being the minimal set of independent empirical electrodynamic laws (Ampere's Circuital Law, Faraday's Law of Induction, Gauss's Electric Flux Law, Gauss's Law for Magnetism, Lenz's Law, and the empirical fact that the smallest entities in nature (elementary particles) have finite-size and internal structures which give rise to their quantum properties. Lenz's Law allows the extension of the law of conservation of energy to the non-conservative forces in electromagnetic induction. **Thus the approach of this book is based on causal determinism and realism.** This is in stark contrast to the non-causal Copenhagen interpretation of quantum mechanics based on probabilities and unreal point particles and Einstein's relativity theory based on imaginary photons, the unreal idealistic notion of point particles and an unreal homogeneous and isotropic universe.

In order for the electrodynamic force to be a local contact force the electric and magnetic fields of charges must remain permanently attached to the charges. This was predicted by Cullwick [24] and experimentally confirmed by Hooper. [25] **Thus light is just a ripple or wave of energy in the electric and magnetic fields of charges and not a particle called the photon! Radio waves are light and perfectly support this conclusion!**

3.9 References

1. Aczel, Amir D., **The Artist and the Mathematician** (Thunder's Mouth Press, New York, 2006) pp. 130-131.

2. Aristotle, **Metaphysics, Vol. I**, 4, 985 b, 10-15.

3. Newton, Isaac. **Philosophiae. Naturalis Principia Mathematica**, 1687.

4. Newton, Isaac, **Mathematical Principles of Natural Philosophy** (1687) **Great Books of the Western World Vol. 34** (Encyclopedia Britannica Inc., Chicago, 1952).

5. Poincaré, H. **Science and Hypothesis,** 1902 version reproduced (Dover, New York) p. xxiv, (1952)

6. Poincaré, H. **The Value of Science,** 1905 version reproduced (Dover, New York) pp. 120-121 ((1958)

7. Poincare, H. **Last Essays,** 1913 version reproduced (Dover, New York) p. 44 (1963)

8. R. W. McAllister & Robert Hofstadter, "Elastic Scattering of 188 MeV Electrons from Proton and the Alpha Particle,*"* **Physical Review, V102,** p. 851 (1956)

9. Littauer, Schopper, and Wilson, "Structure of the Proton and Neutron", **Phys. Rev. Letters, Vol. 7,** pp. 144-147 (1961).

10. David L. Bergman and J. Paul Wesley, "Spinning Charged Ring Model of electron Yielding anomalous Magnetic Moment", **Galilean electrodynamics, Vol. 1,** pp. 63-67 (1990).

11. Aczel, Amir D., **The Artist and the Mathematician** (Thunder's Mouth Press, New York, 2006).

12. Mashaal, Maurice, **Bourbaki: A Secret Society of Mathematicians** (American Mathematical Society, 2006).

13. Sturrock, John, **Structuralism – Second Edition** (Blackwell Publishing, London, 2003).

14. Jakobson, Roman, **Essays on General Linguistics**, (University of Indiana Press, Bloomington, 1952).

15. Oulipo, **La littèrature potentielle** (Gallimard, Paris, 1973).

16. Lacan, Jacques, **The Seminar, Book IV "La relation d'objet"** (W.W. Norton & Co., New York, 1993).

17. Lévi-Strauss, Claude, **Elementary Structures of Kinship** (Les Structures Élémentaires del la Parenté). (Mouton, The Hague, 1947)

18. Lévi-Strauss, Claude, **Anthropologie structural** (Plon, Paris, 1958).

19. Leontief, Wassily, et al., **Studies in the Structure of the American Economy: Theoretical and Empirical Explorations in Input-Output Analysis** (Oxford University Press, New York, 1953).

20. Jean-Francois Lyotard, extracts from **The Postmodern Condition: a Report on Knowledge** (University of Minnesota Press, 1979).

21. Andrew Ross, editor, **Science Wars**, (Duck University Press, 1996).

22. Rorty, Richard, **Introduction, Objectivity, Relativism, and Truth**, (Cambridge University Press, 1991).

23. Karl Popper, **The Logic of Scientific Discovery**, (Basic Books, New York, 1959).

24. E. G. Cullwick, **Electromagnetism and Relativity** (Longsman Green and Co., 1957), p. 245.

25. Hooper, W. J., **New Horizons in Electric, Magnetic and Gravitational Field Theory** (Electrodynamic Gravity, Inc., 543 Broad Blvd, Cuyahoga Falls, Ohio, 44221, 1974), pp. 9-15. **http://www.rexresearch.com/hooper/horizon.htm**

Chapter 4 Derivation of an Improved Electrodynamic Force Law for Constant Velocity

It appears to me, that the study of electromagnetism in all its aspects has now become of the first importance as a means of promoting the progress of science. James Clerk Maxwell [1]

4.1 Proper Axioms of Electrodynamics

This chapter presents a new derivation of the electrodynamic force law for finite-size charged particles moving at constant velocity. In the derivation that follows [2, 3, 4] the approach is taken that a proper electrodynamic force law should be compatible with the following set of axioms which is more complete than that of Maxwell and corresponds better with reality:

1. Coulomb's law for the force between static charges
2. Ampere's generalized law for the force between current elements of closed loops
3. Faraday's law of electromagnetic induction
4. Gauss's laws
5. Only contact forces exist in nature
6. Lenz's law for induction due to motion
7. Charged particles have finite-size with interior structure
8. Fields of charges remain attached when charges move and have tensile strength
9. Galilean invariance
10. Newton's third law – for every action there is an equal and opposite reaction
11. Conservation of kinetic and radiation energy
12. Conservation of momentum (radiation reaction, etc.)
13. Mach's principle that local physical laws are determined by the masses and charges of the large-scale structure of the universe
14. Nonlinear electrodynamic processes occur in lasers and other phenomena

Note that axioms 5-14 are missing in the relativistic version of electrodynamics based on Maxwell's equations. Conservation of energy and momentum for dynamic magnetic fields is only contained in Lenz's law which is missing from Maxwell's electrodynamics. Also Maxwell's equations assume all interactions in electrodynamics are only linear, but lasers are based on non-linear interactions. Only Lenz's law describes nonlinear interactions. Furthermore Maxwell's equations do not satisfy Newton's third law for dynamic magnetic fields. Only Lenz's law satisfies Newton's third law for dynamic magnetic fields. Axiom 8 that the fields of charges remain attached when charges move and have tensile strength, precludes the customary use of photon particles in electrodynamics. It also prohibits the use of vector field and scalar fields in electrodynamics, because the volume of space is not unbounded and the vector does not vanish at large distances as required by Helmholtz's Theorem. See **Appendix B** for details on

Helmholtz's Theorem. Furthermore the discovery of magnetic monopoles also prohibits the use of vector potentials in electrodynamics according to vector calculus. (See **Appendix D**) Also the non-uniqueness of the vector potential makes is a non-physical quantity not to be compared with the physical energy potential.

The fundamental equations of electrodynamics are based upon six empirical laws that are valid for constant velocity, i.e.

1. **Generalized Ampere's Law**

$$\vec{B}_i(\vec{r}, \vec{v}) = \frac{\vec{v}}{c} \times \vec{E}_0(\vec{r}') \quad (4-1)$$

2. **Faraday's Law**

$$\int \vec{E}_i(\vec{r}', t') d\vec{l}' = -\frac{1}{c}\frac{d}{dt} \int \vec{B}_i(\vec{r}, t) \cdot \hat{n} da \quad (4-2)$$

3. **Gauss's Electrostatic Law**

$$\int \vec{E}_0(\vec{r}, t) \cdot \hat{n} da = 4\pi q \quad (4-3)$$

4. **Gauss's Magnetostatic Law**

$$\nabla \cdot \vec{B}_0 = 0 \quad (4-4)$$

5. **Lenz's Induction Law**

$$\vec{E}_i(\vec{r}, \vec{v}) \propto -\vec{E}_0(\vec{r})$$
$$= -\lambda(\vec{v})\vec{E}_0(\vec{r}) \quad (4-5)$$
$$where\ \vec{E}(\vec{r}, \vec{v}, t) = \vec{E}_0(\vec{r}) + \vec{E}_i(\vec{r}, \vec{v}) + \vec{E}_i(\vec{r}, t)$$

6. **Lorentz's Force Law**

$$\vec{F}(\vec{r}, \vec{v}, t) = q\vec{E}(\vec{r}, \vec{v}, t) - \frac{\vec{v}}{c} \times \vec{B}(\vec{r}, \vec{v}, t) \quad (4-6)$$

Note that both Ampere's law and Faraday's law involve the observer's reference frame and the primed moving frame of reference that are described by the Galilean transformation.

$$\vec{r}' = \vec{r} - \vec{v}t\ and\ t' = t \quad (4-7)$$

In modern science physicists have willfully discarded the Galilean transformation in favor of the relativistic Lorentz transformation to relate electromagnetic fields in the two frames. This derivation will show that this illogical procedure was totally unnecessary and resulted in the creation of the superfluous Theory of Special Relativity.

In this derivation the complete set of the fundamental empirical equations of electrodynamics are solved simultaneously by the method of substitution using the Galilean transformation. The resulting electric and magnetic fields in the observer's frame of reference will be derived for an elementary charged particle with an arbitrary finite-size elastic charge distribution of total charge q moving with relative velocity **v**. The resulting electric and magnetic fields will be found to be in agreement with both the experimentally observed fields of very high velocity charged particles in accelerator experiments and the relativistic version of Maxwell's equations derived using vector potentials.

4.2   Derivation of Electrodynamic Force Law

Light itself (including radiant heat and other radiations if any) is an electromagnetic disturbance in the form of waves propagated through the electromagnetic field according to electromagnetic laws.   James Clerk Maxwell [5]

In **Appendix A** of this paper is a derivation of Ampere's force law between elements of current loops, the generalized version of Ampere's force law and the proof that it satisfies Newton's 3rd law. Also included is a derivation of the Biot-Savart law and the Grassmann force law from the generalized Ampere force law. This derivation shows that both the Biot-Savart law and the Grassmann force law satisfy Newton's 3rd law, when one notes that the additional term in the generalized Ampere force law that is missing in the Grassmann force law gives no contribution for closed current loops. **Of course, Ampere's force law only applies to closed current loops!!**

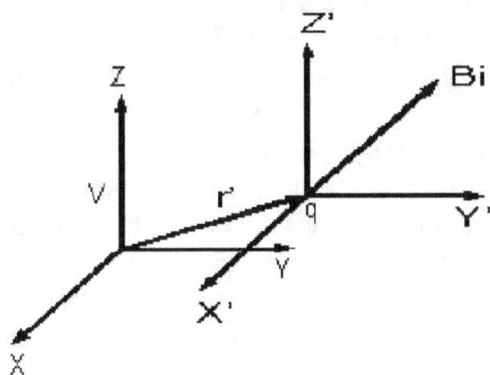

Figure 4-1 Induction Field B$_i$(r', t) Due to Charge q Moving with Velocity V

Using the Grassmann form of the generalized Ampere force law of equation (A-19) for a single point element of charge q moving with a relative velocity v, the induced flux density **B$_i$(r,v)** will be

$$\vec{B}_i(\vec{r}, \vec{v}) = \frac{q}{c} \frac{\vec{v} \times \vec{r}'}{|\vec{r}'|^3} = \frac{\vec{v}}{c} \times \vec{E}_0(\vec{r}') \quad (4-8)$$

where the familiar relativistic type notation of Figure 4-1 has been used for clarity. Note that equation (4-8) for a point element of charge gives the transformation of the **E$_0$(r')** field in the moving frame of reference to the induced field **B$_i$(r,v)** in the observer's frame of reference. This is assumed valid for all velocities **v** whether constant in time or changing, and will be used to obtain the fields of a charged particle with internal charge distribution.

Note that if Ampere's law, as represented by equation (4-8), is cast into its usual Maxwell form of equation (4-9)

$$\nabla \times \vec{B}(\vec{r},\vec{v}) = \frac{4\pi}{c}\vec{J}(\vec{r},\vec{v}) + \frac{1}{c}\nabla\int\frac{\nabla'\cdot\vec{J}(\vec{r}')}{|\vec{r}-\vec{r}'|}dr \approx \frac{4\pi}{c}\vec{J}_t(\vec{r},\vec{v}) \quad (4-9)$$

the reference frame transformation information of equation (4-2) is lost. Note that it is claimed that this equation can be separated into a transverse wave equation and a longitudinal wave equation where the second term on the right is claimed to be part of only the equation for the longitudinal field and can be ignored for the transverse wave equation. [6, p.138, 7, p. 174, 8, p.179]. Also the second term is claimed to be zero for steady state magnetic phenomena, but in general it is not zero for finite-size particles and induced self-field effects. For these reasons it appears that the covariant form of electrodynamics, based on Maxwell's equations, is technically incorrect. It is certainly not as fundamental as the empirical laws upon which electrodynamics is based.

From equation (4-2) the motion of the elementary charge q will produce an induced magnetic field $B_i(r, v, t)$. In order to obtain the $B_i(r, v, t)$ field that an observer would see in his reference frame, a coordinate substitution, known as the Galilean transformation, is used to obtain the expression for $B_i(r, v, t)$ in terms of the unprimed coordinates, i.e.

$$\vec{r}' = \vec{r} - \vec{v}t \ \text{ and } \ t' = t \ \ c' \ne c \qquad (4-10)$$

where v is assumed to be constant in time. Note that classical electrodynamics was originally based on Galilean relativity which assumed instant action-at-a-distance for longitudinal forces due to the Coulomb gauge potential [6 pp. 181-183, 7 pp. 220-223, 8 pp. 240-242] and absolute time as defined above such that the retarded time $\tau = t' - t = 0$. Thus it was logically inconsistent to introduce retardation effects into classical electrodynamics for non-radiation situations.

Note that $B_i(r, v, t)$ will be time varying. According to Faraday's law [6 pp. 170-173 or 7 pp. 210-213 or 8 pp. 208-211] $dB_i(r, v, t)/dt$ introduces an additional electric field $E_i(r, v, t)$. The distribution of charge within the finite-size elastic elementary charged particle rearranges itself to be consistent with the induced fields. This rearrangement causes a change in the internal binding energy of the particle and the kinetic energy stored in the particle's external electromagnetic fields. Using the fundamental empirical equations of electrodynamics one can calculate the induced fields in order to obtain the total fields of the moving charged particle, i.e.

$$\vec{E}(\vec{r},\vec{v},t) = \vec{E}_0(\vec{r}) + \vec{E}_i(\vec{r},\vec{v}) + \vec{E}_i(\vec{r},t)$$
$$\vec{B}(\vec{r},\vec{v},t) = \vec{B}_0(\vec{r}) + \vec{B}_i(\vec{r},\vec{v}) + \vec{B}_i(\vec{r},t) \qquad (4-11)$$

where $E_0(r)$ is the electrostatic field and $E_i(r, v)$ is the induced field due to motion v. Hooper [11] measured the properties of each of the three types of electric and magnetic fields and found that they were unique with different properties. Thus Maxwell and his followers made a mistake assuming that all the induced electric fields had idential properties and all the induced magnetic fields had identical properties.

In 1957 Cullwick [9] published results of his research indicating that the magnetic flux loops discovered by Oersted [10] were actually in motion along the linear conductor in the direction of the electron current giving rise to it. The loops moved with the electron drift velocity. This result was confirmed experimentally by Hooper. [11] This means that when motion is involved E(r', t') must be written in terms of the coordinates of its rest frame, i.e. E(r', t') becomes

$$\vec{E'}(\vec{r'}, t') = \vec{E}(\vec{r} - \vec{v}t, t) \qquad (4-12)$$

Faraday believed that the fields were an extension of the charge to infinity, that the field lines had tensile strength to attract and repel as shown in Figures 4-2 and 4-3, and that light was merely a ripple of angular momentum in the fields of the charge. [12]

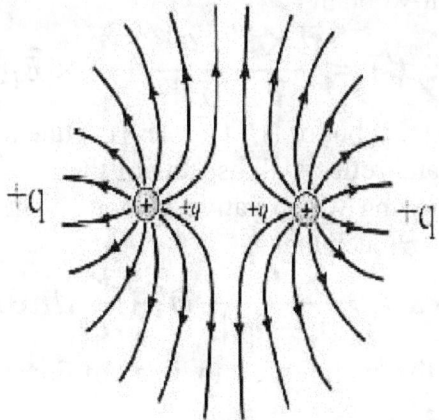

Figure 4-2 Repulsion of Like Charges

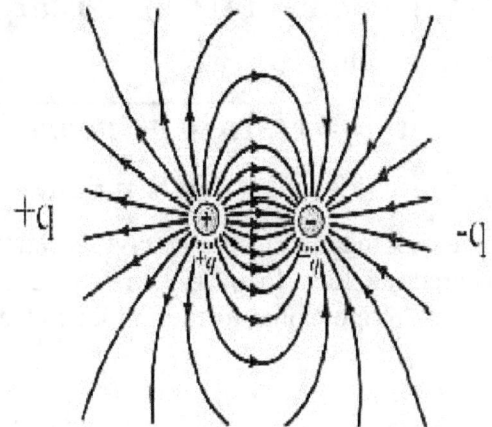

Figure 4-3 Attraction of Opposite Charges

Another significant aspect of this work is that the induced **B** x **v** field is not electrostatic in nature. According to Cullwick [9], Hooper [11], and Moon and Spencer [13], this means that the linear superposition principle as applied to electrical fields does not hold for the **B** x **v** generated fields. Thus in electrodynamics one must explicitly keep track of both electrostatic fields and the induced fields. Using the basic equations of electrodynamics one must calculate explicitly the induced fields in order to obtain the total fields of the moving charged particle. (Note that the covariant form of electrodynamics based upon Maxwell's equations assumes that the superposition principle holds and does not treat electrostatic and induced fields separately in disagreement with the experimental results cited. [11])

From the discussion above, equation (4-8) for a charged particle with finite-size and internal charge structure with total charge q must be written in terms of the position of the charge in its current rest frame. (See Figure 4-1 and note the origin of B_i in the primed coordinate system)

$$\vec{B}(\vec{r}',t') = \vec{B}_i(\vec{r} - \vec{v}t,t) = \frac{\vec{v}}{c} \times \left(\vec{E}_0(\vec{r}',t') + \vec{E}_i(\vec{r}',t')\right) \qquad (4-13)$$

Writing E_0 and E_i explicitly obtain

$$\vec{B}(\vec{r}',t') = \frac{\vec{v} \times \hat{r}'}{c} \frac{q}{|\vec{r}'|^2} + \frac{\vec{v}}{c} \times \vec{E}_i(\vec{r}',t') \qquad (4-14)$$

Note that $\mathbf{B}_i$ is not a function of $(\mathbf{r}, t)$ but of $(\mathbf{r}', t') = (\mathbf{r}-\mathbf{v}t,t)$ due to the fact that the induced field remains attached to the charge and reflects all aspects of the motion of the charge. Assume that the particle with charge q is moving with relative velocity v in the z-direction and that from symmetry $\mathbf{E}_i(\mathbf{r}', t') = E_i(\mathbf{r}',t') \, \mathbf{r}'/|\mathbf{r}'|$ such that

$$\vec{B}_i(\vec{r} - \vec{v}t,t) = \frac{v}{c} r \sin\theta \frac{q}{|\vec{r} - \vec{v}t|^3} \hat{\varphi}' + \frac{v}{c} \sin\theta E_i(\vec{r} - \vec{v}t,t)\hat{\varphi}' \qquad (4-15)$$

Taking the partial derivative with respect to time t of Equation (4-15) for constant relative velocity one obtains

$$\frac{\partial \vec{B}_i(\vec{r} - \vec{v}t,t)}{\partial t} = \frac{v}{c} r \sin\theta \hat{\varphi}' \left[\frac{3qv(r\cos\theta - vt)}{c|r - vt|^5} + \frac{\partial}{rc\partial t}E_i(\vec{r} - \vec{v}t,t)\right] \qquad (4-16)$$

where spherical coordinates are used with

$$|\vec{r}'| = |\vec{r} - \vec{v}t| = \sqrt{r^2\sin^2\theta + (r\cos\theta - vt)^2} \qquad (4-17)$$

$$\vec{v} \times \vec{r}' = \vec{v} \times (\vec{r} - \vec{v}t) = \vec{v} \times \vec{r} = vr\sin\theta\hat{\varphi}'$$

Now the induced time varying magnetic field $\mathbf{B}_i(\mathbf{r}-\mathbf{v}t, t)$ causes an electric field $\mathbf{E}_i(\mathbf{r}-\mathbf{v}t, t)$ to be induced. According to Faraday's Law [6 p. 173 or 7 p. 213 or 8 p. 211] the changing magnetic flux linked by a circuit is proportional to the induced electric field around the circuit, i.e.

$$\oint \vec{E}'_i(\vec{r}',t') \cdot d\vec{l}' = -\frac{1}{c}\frac{d}{dt}\oint \vec{B}_i(\vec{r} - \vec{v}t,t) \cdot \hat{n} da' \qquad (4-18)$$

where the circuits and fields are defined in Figure 4-4.

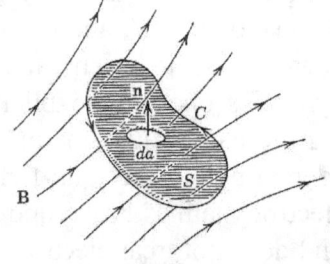

Figure 4-4 Definition of Fields for Faraday's Law

Note that Faraday's law gives the relationship between $\mathbf{E}_i(\mathbf{r'}, t')$ in the moving frame with respect to $\mathbf{B}_i(\mathbf{r\text{-}v}t, t)$ in the observer's frame. This empirical relationship is lost if one casts Faraday's law into its usual covariant Maxwell equation form as in equation (4-9).

Faraday's law can be used as a second equation to go with equation (4-15) to solve for the induced $\mathbf{B}_i$ and $\mathbf{E}_i$ fields. Eventually all the fundamental equations of electrodynamics will be required to obtain the solution. Faraday's law can be rewritten using Stokes theorem as

$$\oint \left[\vec{E}'_i(\vec{r}', t') - \frac{1}{c}\vec{v} \times \vec{B}_i(\vec{r}', t') \right] \cdot d\vec{l}' = -\frac{1}{c} \oint \frac{\partial \vec{B}_i(\vec{r} - \vec{v}t, t)}{\partial t} \cdot \hat{n} da' \quad (4-19)$$

where use of the convective derivative

$$\frac{d}{dt} = \frac{\partial}{\partial t} + \vec{v} \cdot \nabla' \quad (4-20)$$

gives

$$\frac{d\vec{B}_i(\vec{r} - \vec{v}t, t)}{dt} = \frac{\partial \vec{B}_i(\vec{r} - \vec{v}t, t)}{\partial t} + (\vec{v} \cdot \nabla')\vec{B}_i(\vec{r} - \vec{v}t, t)$$

$$= \frac{\partial \vec{B}_i(\vec{r} - \vec{v}t, t)}{\partial t} + \nabla' \times \left[\vec{B}_i(\vec{r} - \vec{v}t, t) \times \vec{v} \right] + \vec{v}\left[\nabla' \cdot \vec{B}_i(\vec{r} - \vec{v}t, t) \right]$$

$$= \frac{\partial \vec{B}_i(\vec{r} - \vec{v}t, t)}{\partial t} + \nabla' \times \left[\vec{B}_i(\vec{r} - \vec{v}t, t) \times \vec{v} \right] \; since \; \nabla' \cdot \vec{B} = 0 \quad (4-21)$$

and the Kelvin-Stokes theorem gives

$$\frac{1}{c} \oint \nabla' \times \left[\vec{B}_i(\vec{r} - \vec{v}t, t) \times \vec{v} \right] \cdot \hat{n} da' = \frac{1}{c} \oint \left[\vec{B}_i(\vec{r} - \vec{v}t, t) \times \vec{v} \right] \cdot d\vec{l}' \quad (4-22)$$

This is a statement of Faraday's law applied to a moving circuit C. We could also apply Faraday's law to a circuit C instantaneously at rest with v = 0. In this case one obtains

$$\oint \vec{E}'_i(\vec{r} - \vec{v}t, t) \cdot d\vec{l}' = -\frac{1}{c} \oint \frac{\partial \vec{B}_i(\vec{r} - \vec{v}t, t)}{\partial t} \cdot \hat{n}' da' \quad (4-23)$$

Galilean invariance requires that in this case $\mathbf{E'} = \mathbf{E}$ and $\mathbf{F'} = \mathbf{F}$. Thus the left-hand sides of equations (4-12) and (4-16) are equal giving

$$\vec{E}'_i(\vec{r}, \vec{v}, t) = \vec{E}_i(\vec{r} - \vec{v}t, t) + \frac{1}{c}\left[\vec{v} \times \vec{B}_i(\vec{r} - \vec{v}t, t) \right] \quad (4-24)$$

or

$$\vec{F}'(\vec{r}', t') = \vec{F}_0{}'(\vec{r}', t') + \vec{F}_i{}'(\vec{r}', t') = q\vec{E}_0{}'(\vec{r}', t') + q\vec{E}'_i(\vec{r}', t')$$

$$= q\vec{E}_0(\vec{r} - \vec{v}t, t) + q\vec{E}_i(\vec{r} - \vec{v}t, t) + \frac{q}{c}\left[\vec{v} \times \vec{B}_i(\vec{r} - \vec{v}t, t) \right] \quad (4-25)$$

or

$$\vec{F}(\vec{r} - \vec{v}t, t) = q\vec{E}(\vec{r} - \vec{v}t, t) + \frac{q}{c}\left[\vec{v} \times \vec{B}_i(\vec{r} - \vec{v}t, t)\right] \; Lorentz \; Force \quad (4-26)$$

where static fields are identical in all frames of reference. **Note that the Lorentz force law has been derived from Galilean invariance and the experimental fact that fields are a physical extension of the charge making the electromagnetic force a contact type force! The Lorentz force law should no longer be considered a separate independent fundamental axiom of electrodynamics!**

Faraday's law of equation (4-23) can be put into differential form by use of Stoke's theorem. The transformation of the line integral for the electric field into a surface integral leads to

$$\boldsymbol{\nabla}' \times \vec{E}_i(\vec{r} - \vec{v}t, t) = -\frac{1}{c}\frac{\partial}{\partial t}\vec{B}_i(\vec{r} - \vec{v}t, t) \quad (4-27)$$

Note that for self-consistency the charge distribution within the finite-size elementary particle has been altered to produce the field $E(r-vt, t) = E_0(r-vt, t) + E_i(r-vt, t)$, because the charge distribution is assumed to be in equilibrium with all forces or fields. Using equation (4-21) one may write Faraday's law in spherical coordinates as

$$\boldsymbol{\nabla}' \times \vec{E}_i(\vec{r}', t) = \frac{\hat{r}'}{rsin\theta}\left[\frac{\partial}{\partial\theta}\left(sin\theta\vec{E}_i(\vec{r}', t) \cdot \varphi'\right) - \frac{\partial}{\partial\varphi}\left(\vec{E}_i(\vec{r}', t) \cdot \theta'\right)\right] \quad (4-28)$$

$$+\hat{\theta}'\left[\frac{1}{rsin\theta}\frac{\partial}{\partial\varphi}\{\vec{E}_i(\vec{r}', t) \cdot \hat{r}'\} - \frac{1}{r}\frac{\partial}{\partial r}\{r\vec{E}_i(\vec{r}', t) \cdot \hat{\varphi}'\}\right]$$

$$+\frac{\hat{\varphi}'}{r}\left[\frac{\partial}{\partial r}\{r\vec{E}_i(\vec{r}', t) \cdot \hat{\theta}'\} - \frac{\partial}{\partial\theta}\{\vec{E}_i(\vec{r}', t) \cdot \hat{r}'\}\right]$$

$$= -\frac{1}{c}\frac{\partial}{\partial t}\vec{B}_i(\vec{r}', t) = -\frac{v}{c}rsin\theta\left[\frac{3qv(rcos\theta - vt)}{c|\vec{r} - \vec{v}t|^5} + \frac{\partial}{rc\partial t}E_i(\vec{r} - \vec{v}t, t)\right]\hat{\varphi}'$$

The symmetry of $\partial B_i/\partial t \propto \varphi'/|\varphi'|$ and $E_i(r-vt, t) \propto r'/|r'|$ allows equation (4-28) to be reduced to

$$-\frac{1}{r}\frac{\partial}{\partial\theta'}E_i(\vec{r} - \vec{v}t, t)\hat{\varphi}'$$

$$= -\frac{v}{c}rsin\theta\left[\frac{3qv(rcos\theta - vt)}{c|\vec{r} - \vec{v}t|^5} + \frac{\partial}{rc\partial t}E_i(\vec{r} - \vec{v}t, t)\right]\hat{\varphi}' \quad (4-29)$$

and

$$\frac{1}{rsin\theta}\frac{\partial}{\partial\varphi'}E_i(\vec{r} - \vec{v}t, t)\hat{\theta}' = 0$$

Thus Faraday's Law in differential form may be rearranged and integrated over θ to obtain

$$\int_0^{\theta'} dE_i(\vec{r} - \vec{v}t, t) = E_i(\vec{r} - \vec{v}t, t) - E_i(\vec{r} - \vec{v}t, t)|_{\theta'=0} \qquad (4-30)$$

$$= \frac{3v^2 r^2 q}{c^2} \int_0^{\theta'} \frac{(rcos\theta' - vt)}{|\vec{r} - \vec{v}t|^5} sin\theta' d\theta' + \frac{vr^2}{c} \int_0^{\theta'} \frac{\partial}{rc\partial t} E_i(\vec{r} - \vec{v}t, t) sin\theta' d\theta'$$

where the constant $E_i(r- vt, t)|_{\theta=0}$ diminishes the original electrostatic field.

From Lenz's law and symmetry of local forces, $E_i(r-vt, t)|_{\theta=0}$ should oppose the induced field $E_i(r-vt, t)r'/|r'|$, which is proportional to the moving static field $E_0(r-vt, t)r'/|r'|$, i.e.

$$E_i(\vec{r} - \vec{v}t, t)|_{\theta'=0}\hat{r}' = -\lambda(\vec{v})E_0(\vec{r} - \vec{v}t, t)\hat{r}' \qquad (4-31)$$

The use of Lenz's law allows one to satisfy Mach's principle. [14] According to Mach any correct physical theory must take into account all the charges and masses in the universe in a consistent way, because the electromagnetic and gravitational forces have infinite range. Since all massive particles seem to have charge structure, the use of Lenz's law appears to satisfy Mach's principle by taking into account the electromagnetic interaction of all other particles in the universe. **Einstein's Special Relativity Theory (SRT) incorporating the Lorentz transformation does not satisfy Mach's principle. The covariant form of electrodynamics based on Maxwell's equations also fails to satisfy Mach's principle.**

Equation (4-30) can be solved by iterative substitution for $E_i(r-vt, t)$ given by equation (4-31), where each successive iteration takes into account the next order of the induced field effect. In order to obtain the results observed in the laboratory frame, it is necessary to evaluate $E_i(r-vt, t)$ at t = 0 when the moving frame coincides with the laboratory frame after all terms for a given iteration have been evaluated. In order to simplify the iteration of successive terms the r' = r - vt terms are left in place in order to keep track of the correct power of r' for the derivative in the iterative term. The vt $(cos\theta'-1)$ terms are explicitly dropped, because they integrate to zero. Substituting equation (4-23) into equation (4-22) and using β = v/c obtain

$$E_i(\vec{r} - \vec{v}t, t)|_{t=0} + \lambda(\vec{v})E(\vec{r} - \vec{v}t, t)|_{t=0}$$

$$= 3r\beta \frac{q}{|\vec{r} - \vec{v}t|^5} \left\{ \frac{r}{2} sin^2\theta' + vt(cos\theta' - 1) \right\}|_{t=0}$$

$$+ \beta r^2 \int_0^{\theta'} \frac{\partial}{rc\partial t} E_i(\vec{r} - \vec{v}t, t) sin\theta' d\theta'|_{t=0} \qquad (4-32)$$

Iterating equation (4-32) obtain

$$E_i(\vec{r} - \vec{v}t, t)|_{t=0} = \frac{3}{2}\beta^2 r^3 \frac{q}{|\vec{r} - \vec{v}t|^5} sin^2\theta'|_{t=0} - \frac{\lambda(\vec{v})rq}{|\vec{r} - \vec{v}t|^3}|_{t=0}$$

$$+ \beta r^2 \int_0^{\theta'} \frac{\partial}{rc\partial t} \left[\frac{3}{2}\beta^2 r^3 \frac{q}{|\vec{r} - \vec{v}t|^5} sin^2\theta' - \frac{\lambda(\vec{v})rq}{|\vec{r} - \vec{v}t|^3} \right] sin\theta' d\theta'|_{t=0}$$

$$+\beta r^2 \int_0^{\theta'} \frac{\partial}{rc\partial t}\left[\beta r^2 \int_0^{\theta'} \frac{\partial}{rc\partial t} E_i(\vec{r}-\vec{v}t,t)sin\theta'd\theta'\right]sin\theta'd\theta'\Big|_{t=0} \quad (4-33)$$

After performing the partial derivatives with respect to time on all but the last term obtain

$$E_i(\vec{r}-\vec{v}t,t)\big|_{t=0} = \frac{3}{2}\beta^2 r^3 \frac{q}{|\vec{r}-\vec{v}t|^5}sin^2\theta'\big|_{t=0} - \frac{\lambda(\vec{v})rq}{|\vec{r}-\vec{v}t|^3}\big|_{t=0}$$

$$+\beta r^2 \int_0^{\theta'}\left[\frac{3}{2}\beta^2 r^3 \frac{q}{|\vec{r}-\vec{v}t|^7}5\beta rcos\theta'sin^2\theta' - \frac{\lambda(\vec{v})rq}{|\vec{r}-\vec{v}t|^5}3\beta rcos\theta'\right]sin\theta'd\theta'$$

$$+\beta^2 r^4 \int_0^{\theta'}\frac{\partial}{rc\partial t}\left[\int_0^{\theta'}\frac{\partial}{rc\partial t}E_i(\vec{r}-\vec{v}t,t)sin\theta d\theta\right]sin\theta'd\theta'\big|_{t=0} \quad (4-34)$$

In order to obtain the terms from the first iteration, one can finish taking the limit t=0 and θ'=θ to obtain the pattern emerging for all the iterative terms. Solving for **E(r) = E₀(r) + Eᵢ(r)** obtain

$$E_i(r) = \frac{q}{r^2}\frac{3\beta^2}{2}sin^2\theta - \frac{\lambda(v)q}{r^2}$$

$$+\frac{q}{r^2}\frac{15\beta^4}{2}\int_0^\theta sin^3\theta dsin\theta - \frac{\lambda(v)q3\beta^2}{r^2}\int_0^\theta sin\theta dsin\theta$$

$$=\frac{q}{r^2}\frac{3\beta^2}{2}sin^2\theta - \frac{\lambda(v)q}{r^2} + \frac{q}{r^2}\frac{15\beta^4}{8}sin^4\theta - \frac{\lambda(v)q}{r^2}\frac{3\beta^2}{2}sin^2\theta$$

$$=\frac{q}{r^2}\left[\frac{3}{2}\beta^2 sin^2\theta + \frac{15}{8}\beta^4 sin^4\theta\right] - \frac{\lambda(v)q}{r^2}\left[1+\frac{3}{2}\beta^2 sin^2\theta\right]$$

$$= E_0(r)\left[\frac{3}{2}\beta^2 sin^2\theta + \frac{15}{8}\beta^4 sin^4\theta\right] - \lambda(v)E_0(r)\left[1+\frac{3}{2}\beta^2 sin^2\theta\right] \quad (4-35)$$

Thus

$$\vec{E}(\vec{r},\vec{v}) = \vec{E}_0(\vec{r}) + \vec{E}_i(\vec{r},\vec{v}) \quad\quad (4-36)$$

$$= \vec{E}_0(r)\left[1+\frac{3}{2}\beta^2 sin^2\theta + \frac{15}{8}\beta^4 sin^4\theta\right] - \lambda(\vec{v})\vec{E}_0(r)\left[1+\frac{3}{2}\beta^2 sin^2\theta\right]$$

In the same manner the second iteration of the equation gives

$$\vec{E}(r) = \vec{E}_0(r)\left[1+\frac{3}{2}\beta^2 sin^2\theta + \frac{15}{8}\beta^4 sin^4\theta + \frac{35}{16}\beta^6 sin^6\theta + \cdots\right]$$

$$-\lambda(\vec{v})\vec{E}_0(r)\left[1+\frac{3}{2}\beta^2 sin^2\theta + \frac{15}{8}\beta^4 sin^4\theta + \cdots\right] \quad (4-37)$$

Using the binomial expansion below

$$\frac{1}{(1 - \beta^2 sin^2\theta)^{3/2}} = 1 + \frac{3}{2}\beta^2 sin^2\theta + \frac{15}{8}\beta^4 sin^4\theta + \frac{35}{16}\beta^6 sin^6\theta + \quad (4-38)$$

to sum the resultant self-consistent electric field of an elastic finite-size moving charged particle with total charge q as seen by an observer in his frame of reference obtain

$$\vec{E}(\vec{r}, \vec{v}) = \vec{E}_0(\vec{r}) + \vec{E}_i(\vec{r}, \vec{v}) = \frac{\{1 - \lambda(\vec{v})\}\vec{E}_0(\vec{r})}{(1 - \beta^2 sin^2\theta)^{3/2}} \quad (4-39)$$

In order to evaluate the constant $\lambda(v)$ one uses Gauss's law for electric charge. [6 p. 4 or 7 p. 30 or 8 p. 27]

$$\oint \vec{E}(r) \cdot \hat{n} da = 4\pi q \quad (4-40)$$

From Figure 4-5

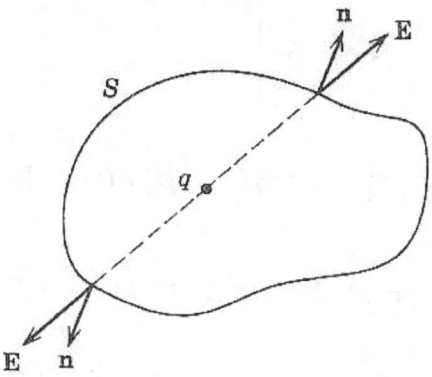

Figure 4-5 Geometry of Gauss's Law

using a spherical surface centered about the charge distribution q with spherical coordinates, and noting that E and n will then be in the same direction, equation (4-40) becomes

$$4\pi q = \int_0^{2\pi}\int_0^{\pi} \frac{[1 - \lambda(\vec{v})]E_0(r)}{(1 - \beta^2 sin^2\theta)^{3/2}} r^2 sin\theta d\theta d\varphi = 4\pi q\frac{1 - \lambda(\vec{v})}{1 - \beta^2} \quad (4-41)$$

where

$$\int_0^{2\pi} \frac{sin\theta d\theta}{(1 - \beta^2 sin^2\theta)^{3/2}} = \frac{2}{1 - \beta^2} \quad and \; E_0(r) = \frac{q}{r^2} \quad (4-42)$$

So $\lambda(v)$ must be equal to β^2 in order for the total flux of E(r) to be conserved. Thus the expression for the self-consistent fields may be written as

$$\vec{E}(\vec{r},\vec{v}) = \vec{E}_0(\vec{r})\frac{1-\beta^2}{(1-\beta^2 sin^2\theta)^{3/2}} \; and \; \vec{B}_i(\vec{r},\vec{v}) = \frac{\vec{v}}{c}\times\vec{E}(\vec{r},\vec{v}) \qquad (4-43)$$

Equation (4-43) is the derived version of the E and B fields in the lab frame for constant velocity v according to the Galilean transformation. **This equation is precisely the same as one would obtain from Special Relativity Theory using the Lorentz transformation for the electric and magnetic fields observed in the observer's frame of reference for a point charge q passing by with relative uniform velocity v. [6 p. 555 or 7 p. 560] Although the mathematical expressions are the same, only this derivation is based upon the complete set of the fundamental empirical laws of electrodynamics using real particles with finite-size and an internal charge structure and contact forces satisfying Mach's principle. Only this derivation takes into account electrical feedback effects on finite-size charge particles due to motion. Only this derivation is based on conservation of energy and momentum. Only this derivation was based on the classical axioms 5-14 supporting causal determinism and realism. Finally this derivation is simpler involving just electrodynamics and not electrodynamics plus Special Relativity Theory.**

Now the total electromagnetic force F exerted by the moving charge distribution on a test charge q' is using equation (4-26) for the Lorentz force

$$\vec{F}(\vec{r},\vec{v}) = q\left\{\vec{E}(\vec{r},\vec{v}) + \frac{\vec{v}}{c}\times\vec{B}_i(\vec{r},\vec{v})\right\}$$

$$= q\vec{E}_0(\vec{r})\frac{1-\beta^2}{(1-\beta^2 sin^2\theta)^{3/2}}\left[(1-\beta^2+\beta^2 cos^2\theta)\hat{r} - (\vec{\beta}\cdot\hat{r})\hat{r}\times(\hat{r}\times\vec{\beta})\right]$$

$$= q\vec{E}_0(\vec{r})\frac{1-\beta^2}{(1-\beta^2 sin^2\theta)^{3/2}}\left[(1-\beta^2 sin^2\theta)\hat{r} - (\vec{\beta}\cdot\hat{r})\hat{r}\times(\hat{r}\times\vec{\beta})\right]$$

$$= q\vec{E}_0(\vec{r})\frac{1-\beta^2}{(1-\beta^2 sin^2\theta)^{1/2}} - q|\vec{E}_0(\vec{r})|\frac{(1-\beta^2)(\vec{\beta}\cdot\hat{r})\hat{r}\times(\hat{r}\times\vec{\beta})}{(1-\beta^2 sin^2\theta)^{\frac{3}{2}}} \qquad (4-44)$$

where the identities (4-45) and (4-46) below were used

$$\frac{\vec{v}}{c}\times\left[\frac{\vec{v}}{c}\times\vec{E}_0(\vec{r})\right] = \frac{\vec{v}}{c}\cdot\vec{E}_0(\vec{r})\frac{\vec{v}}{c} - \frac{v^2}{c^2}\vec{E}_0(\vec{r})$$

$$= \frac{\vec{v}}{c}\cdot\vec{E}_0(\vec{r})\left[\frac{(\vec{v}\cdot\hat{r})\hat{r}}{c} - \frac{\hat{r}\times(\hat{r}\times\vec{v})}{c}\right] - \frac{v^2}{c^2}\vec{E}_0(\vec{r}) \qquad (4-45)$$

and

$$\vec{A}\times(\vec{B}\times\vec{C}) = (\vec{A}\cdot\vec{C})\vec{B} - (\vec{A}\cdot\vec{B})\vec{C}$$

$$\vec{v} = \vec{v} - (\vec{v}\cdot\hat{r})\hat{r} + (\vec{v}\cdot\hat{r})\hat{r} = (\vec{v}\cdot\hat{r})\hat{r} - \hat{r}\times(\hat{r}\times\vec{v}) \qquad (4-46)$$

Equation (4-44) is the derived version of the electrodynamic force **F** for an elastic finite-size charged particle to order v in the Galilean transformation. **The v/c factors typically identified with Special Relativity Theory are found to originate from finite-size electrical feedback effects, nonlinear effects, and conservation of energy and momentum.**

4.3 References

1. Clerk Maxwell, **Treatise on Electricity and Magnetism, Vol. 1**, Preface, p. vii.

2. Thomas G. Barnes, **Foundations of Electricity and Magnetism - Third Edition**(Thomas G. Barnes publisher, 2115 North Kansas, El Paso, TX 79902, 1977), pp. 368-384.

3. Joseph C. Lucas and Charles W. Lucas, Jr., "Electrodynamics of Real Particles vs. Maxwell's Equations, Relativity Theory and Quantum Mechanics", **Proceedings of the 1992 Twin-Cities Creation Conference** July 29- August 1, 1992 at Northwestern College, pp. 243-252.

4. Charles W. Lucas, Jr. and Joseph C. Lucas, "Weber's Force Law for Finite-Size Elastic Particles", **Galilean Electrodynamics 14 (1):** 3-10 (2003).

5. James Clerk Maxwell "A Dynamical Theory of the Electromagnetic Field", **Phil. Trans. R. Soc. Lond. Vol. 155**, p. 466 (1865).

6. J. D. Jackson, **Classical Electrodynamics** (John Wiley and Sons, Inc., New York, 1962).

7. J. D. Jackson, **Classical Electrodynamics-Second Edition** (John Wiley and Sons, Inc., New York, 1975).

8. J. D. Jackson, **Classical Electrodynamics-Third Edition** (John Wiley and Sons, Inc., New York, 1999).

9. E. G. Cullwick, **Electromagnetism and Relativity** (Longsman Green and Co., 1957), p. 245.

10. Hans Christian Oersted, **Experimenta circa effectum conflictus electrici in acum magneticam** (Copenhagen, 1820); English translation in Thompson's **Annals of Philosophy** *xvi:* 273 (1820).

11. W. J. Hooper, **New Horizons in Electric, Magnetic, and Gravitational Field Theory** (Electrodynamic Gravity, Inc. 543 Broad Blvd., Cuyahoga Falls, OH 44221, 1974), preface. http://www.rexresearch.com/hooper/horizon.htm

12. Faraday, M., **Philosophical Magazine Series 3 XXVIII (188)** May 1846.

13. P. Moon and D. E. Spencer, "A New Electrodynamics", **Journal of Franklin Institute 257,** pp. 369-382 (1954).

14. Mach, E., **Die Mechanik in Lihren Entwicklung, 8th ed.** (F. A. Brockhaus, Leipzig, 1921).

Chapter 5 Extension of Improved Electrodynamic Force Law to Include Acceleration

All electromagnetic phenomena ... have, as their fundamental basis, the mutual forces experienced by electric charges, and we have seen that these arise in three ways:

E_C - Two charges experience mutual forces in virtue of their positions. This is the electrostatic force of attraction or repulsion.

E_m - They experience additional forces in virtue of their velocities. Thence arise the forces experienced by a conductor carrying a steady current in a constant magnetic field, the forces between current-carrying conductors, and the induction of an emf in a conductor moving relatively to the source of a magnetic field.

E_t - They also experience additional forces by virtue of their accelerations, from which arise the induction of an emf by transformer actions, and electromagnetic radiation of energy. Geoffrey Cullwick [1]

5.1   Generalized Electromagnetic Potential U(r, v)

In the previous chapter the electrodynamic force for constant velocity β=v/c was derived to be

$$\vec{F}(\vec{r},\vec{v}) = \frac{q'\vec{E}_0(\vec{r})(1-\beta^2)}{(1-\beta^2 sin^2\theta)^{\frac{3}{2}}} - q'|\vec{E}_0(\vec{r})|\frac{(1-\beta^2)(\vec{\beta}\cdot\hat{r})\hat{r}\times(\hat{r}\times\vec{\beta})}{(1-\beta^2 sin^2\theta)^{\frac{3}{2}}} \quad (5-1)$$

For conservative systems the forces are derivable from a generalized potential U(r, v).

$$U(\vec{r},\vec{v}) = -\vec{F}(\vec{r},\vec{v})\cdot\vec{r}$$

$$\frac{dU}{dt} = -\vec{F}(\vec{r},\vec{v})\cdot\frac{d\vec{r}}{dt} - \frac{d\vec{F}(\vec{r},\vec{v})}{dt}\cdot\vec{r} = -\vec{F}(\vec{r},\vec{v})\cdot\vec{v} - \frac{d\vec{F}(\vec{r},\vec{v})}{dt}\cdot\vec{r} \quad (5-2)$$

For stability under a constant force d**F**/dt = 0, then

$$\frac{dU(\vec{r}, \vec{v})}{dt} = -\vec{F}(\vec{r}, \vec{v}) \cdot \vec{v} \qquad (5-3)$$

The generalized electromagnetic potential energy U corresponding to equation (5-1) for the electrodynamic force that is accurate to order v in the Galilean transformation is

$$U(r, v) = \frac{qq'}{r} \frac{(1 - \beta^2)}{(1 - \beta^2 sin^2\theta)^{1/2}} = \frac{qq'(1 - \vec{\beta}^2)}{\left[\vec{r}^2 - \dfrac{\{\vec{r} \times (\vec{r} \times \vec{\beta})\}^2}{\vec{r}^2} \right]^{\frac{1}{2}}} \qquad (5-4)$$

where β=v/c.

Although the expression derived for the electrodynamic force of equation (5-1) is identical to that obtained from the relativistic Maxwell equations based on using the vector potential A assuming no magnetic monopoles, only the approach taken in this book is allowed to define the energy potential U(r, v), because only this approach utilizes Lenz's Law in order to conserve energy for dynamic magnetic fields. **Also note in Appendix D that magnetic monopoles were discovered in 2009 potentially invalidating the vector potential approach.** Also the vector potential A is a non-unique and therefore non-physical entity of vector calculus which has nothing to do with the energy potential of classical physics. In summary, without energy being conserved by the force law, a generalized potential energy U cannot be legitimately defined.

Assuming that the generalized electromagnetic potential is a regular function and well-behaved, one can obtain the acceleration terms of the force by treating the velocity as a function of time.

$$\vec{v} \cdot \vec{F} = -\frac{dU(\vec{r}, \vec{v})}{dt} = -\frac{d}{dt} \frac{qq'(1 - \vec{\beta}^2)}{\left[\vec{r}^2 - \dfrac{\{\vec{r} \times (\vec{r} \times \vec{\beta})\}^2}{\vec{r}^2} \right]^{\frac{1}{2}}} \qquad (5-5)$$

$$= -qq' \left[\frac{\frac{d}{dt}(1 - \vec{\beta}^2)}{\left[\vec{r}^2 - \dfrac{\{\vec{r} \times (\vec{r} \times \vec{\beta})\}^2}{\vec{r}^2} \right]^{\frac{1}{2}}} + (1 - \vec{\beta}^2) \frac{\left(\dfrac{-1}{2} \right) \dfrac{d}{dt} \left[\vec{r}^2 - \dfrac{\{\vec{r} \times (\vec{r} \times \vec{\beta})\}^2}{\vec{r}^2} \right]}{\left[\vec{r}^2 - \dfrac{\{\vec{r} \times (\vec{r} \times \vec{\beta})\}^2}{\vec{r}^2} \right]^{3/2}} \right]$$

$$= qq' \left[\frac{2\frac{\vec{v}}{c}\cdot\frac{\vec{a}}{c}}{\left[\vec{r}^2 - \frac{\{\vec{r}\times(\vec{r}\times\vec{\beta})\}^2}{\vec{r}^2}\right]^{\frac{1}{2}}} + \frac{(1-\vec{\beta}^2)}{2}\frac{\left[2\vec{r}\cdot\vec{v} - \{\vec{r}\times(\vec{r}\times\vec{\beta})\}^2\frac{2\vec{r}\cdot\vec{v}}{r^4}\right]}{\left[\vec{r}^2 - \frac{\{\vec{r}\times(\vec{r}\times\vec{\beta})\}^2}{\vec{r}^2}\right]^{3/2}} \right]$$

$$+ qq' \left[\frac{\frac{(1-\vec{\beta}^2)}{2}\frac{2\vec{r}\times(\vec{r}\times\vec{\beta})}{r^2}\cdot\frac{d}{dt}\{\vec{r}(\vec{r}\cdot\vec{\beta}) - \vec{\beta}(\vec{r}\cdot\vec{r})\}}{\left[\vec{r}^2 - \frac{\{\vec{r}\times(\vec{r}\times\vec{\beta})\}^2}{\vec{r}^2}\right]^{3/2}} \right]$$

$$= qq' \left[\frac{2\frac{\vec{v}}{c}\cdot\frac{\vec{a}}{c}}{\left[\vec{r}^2 - \frac{\{\vec{r}\times(\vec{r}\times\vec{\beta})\}^2}{\vec{r}^2}\right]^{\frac{1}{2}}} + \frac{(1-\vec{\beta}^2)\vec{r}\cdot\vec{v}\left[1 - \frac{\{\vec{r}\times(\vec{r}\times\vec{\beta})\}^2}{r^4}\right]}{\left[\vec{r}^2 - \frac{\{\vec{r}\times(\vec{r}\times\vec{\beta})\}^2}{\vec{r}^2}\right]^{3/2}} \right]$$

$$+ qq'(1-\vec{\beta}^2)\frac{\frac{\vec{r}\times(\vec{r}\times\vec{\beta})}{r^2}\cdot\left\{\vec{v}(\vec{r}\cdot\vec{\beta}) + \vec{r}\frac{\vec{v}^2}{c} + \vec{r}\left(\vec{r}\cdot\frac{\vec{a}}{c}\right) - \frac{\vec{a}}{c}\vec{r}^2 - 2\vec{\beta}(\vec{v}\cdot\vec{r})\right\}}{\left[\vec{r}^2 - \frac{\{\vec{r}\times(\vec{r}\times\vec{\beta})\}^2}{\vec{r}^2}\right]^{3/2}}$$

$$= \frac{qq'}{\vec{r}^2}\frac{(1-\vec{\beta}^2)\vec{v}\cdot\vec{r} + \frac{2\vec{r}^2}{c^2}\vec{v}\cdot\vec{a}}{\left[\vec{r}^2 - \frac{\{\vec{r}\times(\vec{r}\times\vec{\beta})\}^2}{\vec{r}^2}\right]^{\frac{1}{2}}}$$

$$+ \frac{qq'}{\vec{r}^2}(1-\vec{\beta}^2)\frac{\vec{r}\times(\vec{r}\times\vec{\beta})\cdot\left\{-\frac{\vec{v}}{c}(\vec{v}\cdot\vec{r}) + \frac{\vec{r}}{c}\vec{v}^2 + \vec{r}\times\left(\vec{r}\times\frac{\vec{a}}{c}\right)\right\}}{\left[\vec{r}^2 - \frac{\{\vec{r}\times(\vec{r}\times\vec{\beta})\}^2}{\vec{r}^2}\right]^{3/2}}$$

$$= \frac{qq'}{\vec{r}^2} \frac{(1-\vec{\beta}^2)\vec{v}\cdot\vec{r} + \frac{2\vec{r}^2}{c^2}\vec{v}\cdot\vec{a}}{\left[\vec{r}^2 - \frac{\{\vec{r}\times(\vec{r}\times\vec{\beta})\}^2}{\vec{r}^2}\right]^{\frac{1}{2}}} + \frac{qq'}{\vec{r}^2}(1-\vec{\beta}^2)$$

$$x \frac{\vec{r}\times(\vec{r}\times\vec{\beta})\cdot\left\{\frac{-\vec{v}}{c}(\vec{v}\cdot\vec{r}) + \frac{\vec{r}}{c}\vec{v}^2\right\} + \{\vec{r}(\vec{r}\cdot\vec{\beta}) - \vec{\beta}\vec{r}^2\}\cdot\vec{r}\times\left(\vec{r}\times\frac{\vec{a}}{c}\right)}{\left[\vec{r}^2 - \frac{\{\vec{r}\times(\vec{r}\times\vec{\beta})\}^2}{\vec{r}^2}\right]^{3/2}}$$

$$= \frac{qq'}{\vec{r}^2} \frac{(1-\vec{\beta}^2)\vec{v}\cdot\vec{r} + \frac{2\vec{r}^2}{c^2}\vec{v}\cdot\vec{a}}{\left[\vec{r}^2 - \frac{\{\vec{r}\times(\vec{r}\times\vec{\beta})\}^2}{\vec{r}^2}\right]^{\frac{1}{2}}}$$

$$- \frac{qq'}{\vec{r}^2}(1-\vec{\beta}^2) \frac{\vec{r}\times(\vec{r}\times\vec{\beta})\cdot\frac{\vec{v}}{c}(\vec{v}\cdot\vec{r}) + \vec{r}^2\vec{r}\times\left(\vec{r}\times\frac{\vec{a}}{c}\right)\cdot\vec{\beta}}{\left[\vec{r}^2 - \frac{\{\vec{r}\times(\vec{r}\times\vec{\beta})\}^2}{\vec{r}^2}\right]^{3/2}}$$

where the following vector identities were used

$$\vec{A}\times(\vec{B}\times\vec{C}) = (\vec{A}\cdot\vec{C})\vec{B} - (\vec{A}\cdot\vec{B})\vec{C}$$

$$\vec{r}\cdot\vec{r}\times(\vec{r}\times\vec{\beta}) = (\vec{r}\cdot\vec{\beta})(\vec{r}\cdot\vec{r}) - (\vec{r}\cdot\vec{r})(\vec{r}\cdot\vec{\beta}) = 0 \qquad (5-6)$$

$$\vec{r}\cdot\vec{r}\times(\vec{r}\times\vec{a}) = (\vec{r}\cdot\vec{a})(\vec{r}\cdot\vec{r}) - (\vec{r}\cdot\vec{r})(\vec{r}\cdot\vec{a}) = 0$$

Thus the electrodynamic force including acceleration terms in Gaussian units is

$$\vec{F}(\vec{r},\vec{v},\vec{a}) = \frac{qq'}{\vec{r}^2}\left[\frac{(1-\vec{\beta}^2)\vec{r} + \frac{2\vec{r}^2}{c^2}\vec{a}}{\left[\vec{r}^2 - \frac{\{\vec{r}\times(\vec{r}\times\vec{\beta})\}^2}{\vec{r}^2}\right]^{\frac{1}{2}}}\right.$$

$$\left. - (1-\vec{\beta}^2)\frac{(\vec{\beta}\cdot\vec{r})\vec{r}\times(\vec{r}\times\vec{\beta}) + (\vec{r}\cdot\vec{r})\vec{r}\times\left(\vec{r}\times\frac{\vec{a}}{c^2}\right)}{\left[\vec{r}^2 - \frac{\{\vec{r}\times(\vec{r}\times\vec{\beta})\}^2}{\vec{r}^2}\right]^{3/2}}\right] \qquad (5-7)$$

This force in the limit of constant velocity, i.e. **a** = 0, corresponds to the covariant relativistic electrodynamic force of equation (5-1) based on Maxwell's equations. [2 p. 555 or 3 p. 560] Note that the second group of terms in the equation for the force is responsible for the curling corkscrew type of motion of charges in plasma currents, in fluorescent lights, and plasma lightning balls sold at many gift shops. Figure 5-1 shows the curling corkscrew motion of charge due to the electrodynamic force for constant velocity. In general there is a second smaller scale curling motion superimposed upon the first due to acceleration.

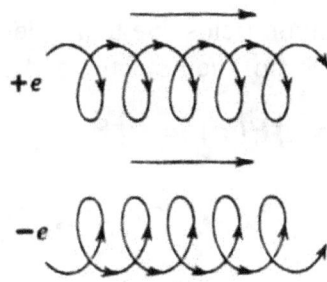

Figure 5-1 Curling of Corkscrew Motion of Charge

According to Ampere [4] the effect of a current flowing in a circuit twisted into small sinuosities is the same as if the circuit were smoothed out. For $v^2/c^2 \ll 1$ and $2a/rc^2 \ll 1$ this is true, but not for relativistic velocities $v/c \approx 1$. Thus the force between two moving charges is increased due to the reduction of the effective distance r between the charges due to the spiraling motion of each charge caused by the r x (r x v) term in the electromagnetic force.

This derived version of the electromagnetic force law, which includes acceleration terms, is equivalent to Phipps's [5, 6] and Wesley's [7] proposed relativistic versions of the force in the limiting case of transverse or circular motion ($\theta = \pi/2$ and v' = 0) to order v^2/c^2, i.e.

$$\vec{F}(\vec{r}, \vec{v}, \vec{a})_{\vec{r} \perp \vec{v}, \vec{a} = 0} = \frac{qq'\hat{r}}{\vec{r}^2} \left[\frac{(1 - \beta^2)}{(1 - \beta^2 sin^2\theta)^{1/2}} \right]_{\theta = \pi/2} = \frac{qq'\hat{r}}{\vec{r}^2}(1 - \beta^2)^{1/2} \quad (5-8)$$

$$\approx \vec{F}_{Phipps} \quad and \quad \approx \vec{F}_{Wesley}$$

It is also equivalent to Weber's version of the force law for linear parallel motion, i.e. v‖r‖a such that sin θ = 0

$$\vec{F}(\vec{r}, \vec{v}, \vec{a})_{\vec{v}\|\vec{r}\|\vec{a}} = \frac{qq'}{|\vec{r}|^2} \left[(1 - \beta^2)\hat{r} + \frac{2\vec{r}^2}{c^2}\vec{a} + \cdots \right] \approx \vec{F}_{Weber} \quad (5-9)$$

In the non-relativistic limit v << c, the magnetic field **B** ∝ v/c x **E** ≈ 0 may be neglected such that

$$F_{v<<c} \approx qE_{v<<c} \approx qE_0 + q(E_a + E_{Rad}) \approx \frac{qq'}{c^2r}[2\vec{a} + \vec{r}(\vec{r} \cdot \vec{a}) - \vec{a}] \quad (5-10)$$

$$= \frac{qq'\hat{r}}{|\vec{r}|^2} + \frac{qq'}{c^2r}[\vec{a} + \hat{r}(\hat{r} \cdot \vec{a})]$$

This is identical to the result for the Taylor series expansion of the Lienard-Wichert electric field for **v** = 0 and **a** > 0 [3, p. 698 problem 14.2]

5.2   Acceleration Fields and Radiation

In order to determine the radiation fields due to acceleration, it is convenient to define the electromagnetic fields and force in terms of their electric and magnetic components.

$$\vec{F}_E = q\vec{E}(\vec{r}) = q[\vec{E}_0(\vec{r}) + \vec{E}_i(\vec{r})] \quad (5-11)$$

$$\vec{F}_M = \frac{\vec{v}}{c} \times [\vec{B}_0(\vec{r}) + \vec{B}_i(\vec{r})]$$

The induced fields may also be defined in terms of the type of motion that produced them, i.e.

$$\vec{E}_i(\vec{r}) = \vec{E}_v(\vec{r}) + \vec{E}_a(\vec{r}) + \vec{E}_{Rad}(\vec{r}) + \cdots \quad 5-12)$$

$$\vec{B}_i(\vec{r}) = \vec{B}_v(\vec{r}) + \vec{B}_a(\vec{r}) + \vec{B}_{Rad}(\vec{r}) + \cdots$$

Thus the force may be written in terms of the type of motion

$$\vec{F} = \vec{F}_{E_0} + \vec{F}_{E_v} + \vec{F}_{E_a} + \cdots + \vec{F}_{B_0} + \vec{F}_{B_v} + \vec{F}_{B_a} + \cdots + \vec{F}_{Rad} \quad (5-13)$$

Since the velocity term in the induced fields is known experimentally to not obey the superposition principle, it is assumed that the acceleration and higher order time derivative terms also do not obey the superposition principle. Thus each of these fields should have unique properties and may correspond to the unique static, velocity and acceleration electric and magnetic fields reported by Hooper [8]. These fields are prime candidates for an O (3) [9] type of electrodynamics that can be used to develop a string theory of elementary particles. One O (3) electrodynamic theory of elementary particles has already been published. [10]

Comparing equation (5-13) with equation (5-7) one obtains

$$\vec{F}_{E_0} + \vec{F}_{E_v} = \frac{qq'\vec{r}}{|\vec{r}|^3} \frac{(1 - \beta^2)}{(1 - \beta^2 sin^2\theta)^{1/2}} = q'(\vec{E}_0 + \vec{E}_v) \qquad (5 - 14)$$

$$\vec{F}_{B_0} + \vec{F}_{B_v} = \frac{qq'\vec{r}}{|\vec{r}|^3} \frac{(1 - \beta^2)\beta^2 sin^2\theta}{(1 - \beta^2 sin^2\theta)^{1/2}} + \frac{qq'}{|\vec{r}|^4} \frac{(1 - \beta^2)(\vec{\beta} \cdot \vec{r})\vec{r} \times (\vec{r} \times \vec{\beta})}{(1 - \beta^2 sin^2\theta)^{3/2}}$$

where the explicit vector dependence has been removed from the denominator.

5.3   Radiation from Real Finite-Size Particles

If a charged particle is accelerated, it is observed to give off radiation. This radiation is due to the curling motion of the electric and magnetic fields produced by the acceleration. The instantaneous energy flux of the radiation is given by the Poynting vector S [3, p. 665]

$$\vec{S} = \frac{c}{4\pi} \vec{E}_{Rad} \times \vec{B}_{Rad} = \frac{c}{4\pi} \left|\vec{E}_{Rad}\right|^2 \hat{r}$$

$$= \frac{q^2}{4\pi c^3 r^2} \frac{(1 - \beta^2)^2}{(1 - \beta^2 sin^2\theta)^3} |\hat{r} \times (\hat{r} \times \vec{a})|^2 \qquad (5 - 15)$$

where θ is the angle between r and v and all observed radiation is due to acceleration a. Thus the power radiated per unit solid angle is

$$\frac{dP}{d\Omega} = r^2|\vec{S}| = \frac{q^2}{4\pi c^3} \frac{(1 - \beta^2)^2}{(1 - \beta^2 sin^2\theta)^3} |\hat{r} \times (\hat{r} \times \vec{a})|^2 \qquad (5 - 16)$$

In the non-relativistic limit v/c << 1 is

$$\frac{dP}{d\Omega} = \frac{q^2}{4\pi c^3} \vec{a}^2 sin^2\theta' \qquad (5 - 17)$$

where θ' is the angle between r and a. The total instantaneous power radiated is obtained by integrating dP/dΩ over all solid angles

$$P = \int_0^{2\pi} \int_{-1}^{1} \frac{q^2}{4\pi c^3} \vec{a}^2 (1 - cos^2\theta') dcos\theta' d\varphi' = \frac{2q^2\vec{a}^2}{3c^3} \qquad (5 - 18)$$

This is the standard Larmor formula for total power radiated from a non-relativistic accelerated charge. [1 p. 469 or 2 p. 659 or 3 p. 665]

For the relativistic case of a circular accelerator where v is perpendicular to r such that sin θ = 1 or r·v = 0 and a is perpendicular to β = v/c

$$\frac{dP}{d\Omega} = \frac{q^2}{4\pi c^3} \frac{(1-\beta^2)^2}{(1-\beta^2 sin^2\theta)^3} \vec{a}^2 sin^2\theta' = \frac{q^2}{4\pi c^3}\gamma^4(1-\beta^2)\vec{a}^2 sin^2\theta' \quad (5-19)$$

$$P_{Cir} = \frac{2q^2\gamma^4}{3c^3}\vec{a}^2(1-\beta^2) = \frac{2q^2\gamma^4}{3c^3}\left[\vec{a}^2 - (\vec{\beta}\times\vec{a})^2\right]$$

where the integral of equation (5-18) was used.

This is the fully relativistic Lienard result first obtained in 1898. [1 p. 470 or 2 p.660 or 3 p. 666] Note that there is a relationship between θ and θ', i.e. as sin θ → 1 sin θ' → 0. This causes all the radiation to be emitted at very small angles θ' producing the search light effect observed in particle accelerators and shown in Figure 5-2.

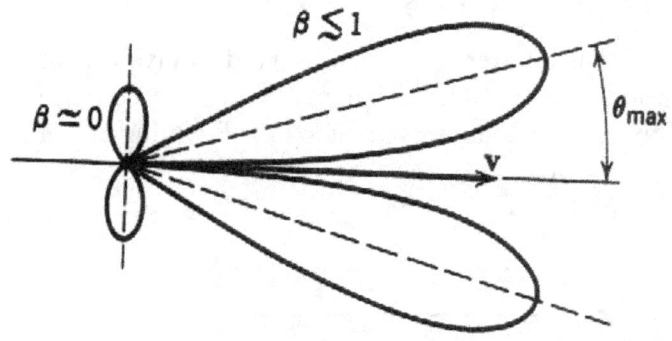

Figure 5-2 Searchlight Effect in Radiation Pattern for Charge in Motion β≈1

5.4   References

1. E. Geoffrey Cullwick: **The Fundamentals of Electromagnetism**, (Cambridge Univ. Press), p. 285 (1949).
2. J. D. Jackson, **Classical Electrodynamics-Second Edition** (John Wiley and Sons, Inc., New York, 1975).
3. J. D. Jackson, **Classical Electrodynamics-Third Edition** (John Wiley and Sons, Inc., New York, 1999).
4. Ampere, **Mem. De l'Acad. VI**, p. 175 (1825).
5. Phipps, T. E. and T. E. Phipps, Jr., **Physics Letters A 146:** p. 6 (1990).
6. T. E. Phipps, Jr. **Physics Essays 3:** p. 198 (1990).
7. J. P. Wesley, **Classical Quantum Theory** (Benjamin Wesley, Germany, 1996), p. 296.
8. Hooper, W. J., **New Horizons in Electric, Magnetic and Gravitational Field Theory** (Electrodynamic Gravity, Inc., 543 Broad Blvd, Cuyahoga Falls, Ohio, 44221, 1974), pp. 9-15. http://www.rexresearch.com/hooper/horizon.htm
9. Evans, Myron W. "The Enigmatic Photon-Volume 5: O(3) Electrodynamics" published in **Fundamental Theories of Physics Volume 106** (Kluwer Academic Publishers, 1999).
10. Lucas, Jr., C. W., "A Classical Electromagnetic Theory of Elementary Particles" **J. of New Energy 6 (4)** 80-108 (2003).

Chapter 6 Extension of Improved Force Law to Include Radiation Reaction da/dt

6.1 Radiation Reaction and Self-fields

In the field of classical electrodynamics two approaches are typically used for describing nature.

1. **The charges and current sources of fields are specified and the fields are calculated**
2. **The fields are specified and the motions of particles are calculated**

However, in some fields of physics, such as plasma physics, this approach is not sufficient. There the fields generated by the sources and the motions of the sources must be described in a self-consistent manner. Usually the motion of a selected source is calculated in response to fields generated by all other sources. Rarely is the motion of a source particle due to the fields generated by that same particle calculated. There are two reasons for this.

1. **Neglect of the "self-fields" usually leads to results that are accurate enough for most applications**
2. **Inclusion of "self-fields" leads to problems with infinities in physics requiring renormalization due to the use of the point particle approximation**

6.2 Outstanding Problems Using Self fields

According to Jackson [1, 2, 3] there are still outstanding conceptual problems created by the "self-fields" of point particles. The difficulties presented by this problem touch one of the most fundamental aspects of physics, the nature of the elementary particle. Although partial solutions, workable within limited areas, can be given, the basic problem remains unsolved. One might hope that the transition from classical to quantum-mechanical treatments would remove the difficulties. While the scientific community still hopes that this may eventually occur, the present quantum-mechanical discussions are beset with even more elaborate troubles than the classical ones. It is one of the triumphs of comparatively recent years (~ 1948 - 1950) that the concepts of Lorentz covariance and gauge invariance were exploited sufficiently cleverly to circumvent these difficulties in quantum electrodynamics and so allow the calculation of very small radiative effects to extremely high precision, in full agreement with experiment. From a fundamental point of view, however, the difficulties remain.

Radiation reaction is the effect on a radiation-emitting particle due to the fact that energy, momentum, and angular momentum are carried away by the radiation. The basic approach to calculating the radiation reaction effect, both classically and quantum mechanically, has been to calculate the effect on "self-generated" fields. The approach presented in this work does not suffer from the use of the point-particle approximation. There are no embarrassing infinities to renormalize away.

6.3 Derivation of Non-Relativistic Radiation Reaction Force

By defining an electrodynamic potential energy U in Gaussian units for the electrodynamic force for constant velocity, we were able to extend the electrodynamic force to include acceleration a effects in **Chapter 5**. (See equations 6-1 and 6-2 below)

$$U(r,v) = \frac{qq'}{r}\frac{(1-\beta^2)}{(1-\beta^2 sin^2\theta)^{1/2}} = \frac{qq'(1-\vec{\beta}^2)}{\left[\vec{r}^2 - \frac{\{\vec{r}\times(\vec{r}\times\vec{\beta})\}^2}{\vec{r}^2}\right]^{\frac{1}{2}}} \quad (6-1)$$

where β=v/c and

$$\frac{dU(\vec{r},\vec{v})}{dt} = -\vec{v}\cdot\vec{F}(\vec{r},\vec{v},\vec{a})$$

$$\vec{F}(\vec{r},\vec{v},\vec{a}) = \frac{qq'}{\vec{r}^2}\left[\frac{(1-\vec{\beta}^2)\vec{r}+\frac{2\vec{r}^2}{c^2}\vec{a}}{\left[\vec{r}^2 - \frac{\{\vec{r}\times(\vec{r}\times\vec{\beta})\}^2}{\vec{r}^2}\right]^{\frac{1}{2}}}\right.$$

$$\left. - (1-\vec{\beta}^2)\frac{(\vec{\beta}\cdot\vec{r})\vec{r}\times(\vec{r}\times\vec{\beta}) + (\vec{r}\cdot\vec{r})\vec{r}\times\left(\vec{r}\times\frac{\vec{a}}{c^2}\right)}{\left[\vec{r}^2 - \frac{\{\vec{r}\times(\vec{r}\times\vec{\beta})\}^2}{\vec{r}^2}\right]^{3/2}}\right] \quad (6-2)$$

Thus we were able to predict radiation directly from the electrodynamic force law which was not possible to do directly from the relativistic covariant version of the electrodynamic force law, because Maxwell's electrodynamics did not conserve energy for magnetic fields allowing the definition of an electrodynamic potential. Thus for Maxwell's electrodynamics and the covariant version of electrodynamics built upon it, one cannot legitimately define a proper potential energy. Instead the relativistic version of Maxwell's electrodynamics is built upon the vector potential of vector calculus which has nothing to do with conservation of energy. In vector calculus, a **vector potential** is a vector field whose curl is a given vector field. In the relativistic version of Maxwell's electrodynamics the magnetic field B is the curl of the vector potential A, since the divergence of the magnetic field B is zero due to the lack of magnetic monopoles.

Assuming that the charge in a finite-size elementary particle flows in closed loops[4-6], then from conservation of energy the average work done on the particle by the electrodynamic force in emitting radiation in the non-relativistic limit $v \ll c$ is the negative of the power integrated over one period from τ_1 to τ_2

$$\int_{\tau_1}^{\tau_2} \vec{F}_{Rad} \cdot \vec{v} dt = -\int_{\tau_1}^{\tau_2} P dt = -\int_{\tau_1}^{\tau_2} \frac{2q^2}{3c^3} \vec{a}^2 dt = -\int_{\tau_1}^{\tau_2} \frac{2q^2}{3c^3} \frac{d\vec{v}}{dt} \cdot \frac{d\vec{v}}{dt} dt \quad (6-3)$$

Notice that we can integrate the above expression by parts. If we assume that there is periodicity in the charged particle structure, the boundary term in the integral by parts disappears to give

$$\int_{\tau_1}^{\tau_2} \vec{F}_{Rad} \cdot \vec{v} dt = -\frac{2q^2}{3c^3} \frac{d\vec{v}}{dt} \cdot \vec{v} \bigg|_{\tau_2} + \frac{2q^2}{3c^3} \frac{d\vec{v}}{dt} \cdot \vec{v} \bigg|_{\tau_1} + \frac{2q^2}{3c^3} \int_{\tau_1}^{\tau_2} \frac{d^2\vec{v}}{dt^2} \cdot \vec{v} dt$$

$$= 0 + \frac{2q^2}{3c^3} \int_{\tau_1}^{\tau_2} \frac{d^2\vec{v}}{dt^2} \cdot \vec{v} dt \quad (6-4)$$

Comparing the integrands of the left and right hand sides of the equation, we can identify the experimentally confirmed non-relativistic radiation reaction force [3, p. 748] as

$$\vec{F}_{Rad} = \frac{2q^2}{3c^3} \frac{d^2\vec{v}}{dt^2} = \frac{2q^2}{3c^3} \frac{d\vec{a}}{dt} \quad (6-5)$$

6.4   Derivation of Relativistic Radiation Reaction Force

For the relativistic case of a circular accelerator where v is perpendicular to r such that sin $\theta = 1$ or $\mathbf{r} \cdot \mathbf{v} = 0$

$$\int_{\tau_1}^{\tau_2} \vec{F}_{Rad} \cdot \vec{v} dt = -\int_{\tau_1}^{\tau_2} P dt = -\int_{\tau_1}^{\tau_2} \frac{2q^2 \gamma^2}{3c^3} \vec{a}^2 dt = -\int_{\tau_1}^{\tau_2} \frac{2q^2 \gamma^2}{3c^3} \frac{d\vec{v}}{dt} \cdot \frac{d\vec{v}}{dt} dt$$

$$= 0 + \frac{2q^2 \gamma^2}{3c^3} \int_{\tau_1}^{\tau_2} \frac{d^2\vec{v}}{dt^2} \cdot \vec{v} dt \quad (6-6)$$

Again assuming that there is periodicity in the charged particle structure, the boundary terms in the integral by parts disappears to give

$$\vec{F}_{Rad} = \frac{2q^2 \gamma^2}{3c^3} \frac{d\vec{a}}{dt} \quad (6-7)$$

6.5 Significance of the Radiation Reaction Boundary Condition

During the times of the ancient Greeks, the smallest thing in the universe was an atom. The term atom means uncutable or indivisible. Essentially its properties were unknowable. Thousands of years later the electron was identified as a particle in 1897 by J. J. Thomson and his team of British physicists. [7] The electron was conceived as an indivisible quantity of electric charge by the British natural philosopher Richard Laming. [8] Earnest Rutherford discovered the proton to be the hydrogen nucleus in 1920. In 1932 James Chadwick discovered the neutron as a subatomic particle. At this point the ancient Greek indivisible thing, called the atom, consisted of electrons, protons, and neutrons. In theories of the atom these particles were idealized as points, since their internal structure and size were unknown. After that the electron, proton and neutron were discovered to have a magnetic moment indicative of current flowing in some sort of finite-size structure.

Then in 1956 Robert Hofstadter [9] began a series of high energy electron scattering experiments in which he measured the size and internal structure of the proton and neutron as shown in Figure 6-1. This data shows not only the finite size but the internal structure of the proton and neutron. Both the proton and neutron consist of at least three sub-structures.

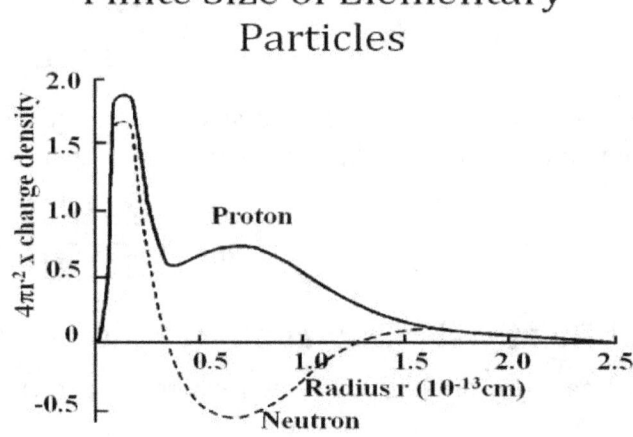

Figure 6-1 Scattering Data Showing Proton and Neutron Internal Structure and Finite –Size [9]

The significance of this detail is greatly enhanced in Chapters 9 and 10 where improved versions of the force of gravity and the force of inertia are derived that appear to be more complete than all such theories in the past in that they describe many new phenomena. Also it is significant to note that the only other remaining so-called "fundamental forces" in nature, i.e. the weak and strong interaction forces, have a range that is on the order of the size of an elementary particle suggesting that they are really just mocking up the finite-size electrodynamic effects of elementary particles. Chapter 11 uses the authoritative NIST atomic and nuclear data to show that there is no atomic or nuclear data to support the existence of the strong or weak nuclear force within the nucleus. Only the electrodynamic force exists within the nucleus. All this lends support to the notion that the improved electrodynamic force is actually the universal force.

Thus it appears that the only force existing within an elementary particle is the electrodynamic force. This combined with the information that all elementary particles consist of closed charge loop structures enables us to determine the structural relations between the closed charge loops within elementary particles. This is the subject of the next book in the series **The Universal Force Volume 2 - An Electrodynamic Model of Elementary Particles**. Thus the unknowable "things" of Poincare are reduced in size from elementary particles to the closed charge loops within elementary particles. Now the question is "What is charge?"

6.6   The Demise of Inertial Reference Frames

The more complete definition of the electrodynamic force as a function of the relative coordinate variables r, v, a, and da/dt instead of just r and v of the idealized inertial reference frame signals the end of the temporary use of inertial reference frames. One can now use the more complete or true reference frame that takes into account all types of interactions with all the matter of the universe.

This includes the effects of acceleration. In the past scientists have not been able to make any measurements or experiments in a true inertial reference frame. This is due to the fact that all experiments performed on the earth are subject to the acceleration effects due to the rotation of the earth on its axis, the acceleration due to the earth's orbit around the sun, the acceleration due to the sun's orbit around the center of the Milky Way galaxy, and the acceleration due to the Milky Way galaxy's orbit around the center of the universe as defined by the Hubble redshifts. Even experiments performed in the outer space of our solar system are subject to three of these accelerations.

Furthermore inertial reference frames were never able to take into account the effect of absorption and emission of radiation and the accompanying radiation reaction or recoil due to the absorption and emission of radiation. This recoil produces a Doppler shift in the radiation. Since all matter is continually absorbing and emitting radiation, the idealized inertial reference frame was an incomplete approximate way of looking at the interactions of matter. Thus the time to discard the idealized inertial reference frame has arrived.

6.7   The Demise of Relativity Theory and Covariance

Einstein's special relativity theory was the result of attempts by Einstein, Lorentz and Poincarè to explain how the speed of light was observed to be independent of idealized inertial reference frames and to understand the covariant symmetries of the laws of electromagnetism. Einstein's theory of special relativity was designed as an upgrade of Lorentz's relativity theory while still supporting the covariant symmetry of the equations of electrodynamics.

In this work the results of Einstein's special relativity theory were derived from the complete set of the empirical laws of electrodynamics by removing the point particle idealization and using Lenz's Law to conserve energy and momentum which was omitted in Maxwell's equations. The point particle idealization was used by Heaviside and others to obtain Maxwell's equations in covariant form by changing the cross gradient of the magnetic field to be consistent with covariance.

$$\nabla \times \vec{B}(\vec{r}, t) = \frac{4\pi}{c}\vec{J}(\vec{r}, t) + \frac{1}{c}\nabla \int \frac{\nabla' \cdot \vec{J}(\vec{r}', t')}{|\vec{r} - \vec{r}'|} d^3r' \approx \frac{4\pi}{c}\vec{J}(\vec{r}, t) \quad (6-8)$$

$$\|$$

$$0$$

POINT PARTICLE IDEALIZATION TO SUPPORT COVARIANCE

Once the point particle idealization is removed, the notion of covariance in electrodynamics based on the vector potential is lost. Since the unique physical energy potential is to be preferred above the non-unique mathematical vector potential anyway, **it appears that the removal of the point particle idealization from electrodynamics invalidates both special relativity theory and the notion of covariance.**

6.8 References

1. J. D. Jackson, **Classical Electrodynamics** (John Wiley and Sons, Inc., New York, 1962), p. 578.

2. J. D. Jackson, **Classical Electrodynamics-Second Edition** (John Wiley and Sons, Inc., New York, 1975), p. 781.

3. J. D. Jackson, **Classical Electrodynamics-Third Edition** (John Wiley and Sons, Inc., New York, 1999), p. 745.

4. Lucas Jr., Charles W., "Derivation of a Universal Electromagnetic Force Law for Finite-Size Elastic Charged Particles" **Proceedings of the Natural Philosophy Alliance Vol. 2, No. 1, 12th Annual Conference of the NPA** at the University of Connecticut, at Storrs, CT May 23-27, pp. 85-108 (2005).

5. Lucas Jr., Charles W., "Derivation of the Classical Universal Electrodynamic Force Law", "The Electrodynamic Origin of the Force of Inertia", "The Electrodynamic Origin of the Force of Gravity", "A Classical Electrodynamic Theory of the Atom", "A Classical Electrodynamic Theory of the Nucleus", "A Classical Electrodynamic String Theory of Elementary Particles", "The Electrodynamic Origin of Life in Organic Molecules Such as DNA and Proteins" **Proceedings of the Natural Philosophy Alliance (NPA) Volume 3, 13th Annual Conference of the NPA,** 3-7 April 2006 at the University of Tulsa at Tulsa, OK USA (2006).

6. Lucas Jr., Charles W., "A Classical Electromagnetic Theory of Elementary Particles" **Journal of New Energy, Vol. 6, No. 4**, **Proceedings of "Physics as a Science" International Workshop**, Arrecife, Lanzarote, Canary Islands July 1-5, 2002, pp. 81-109 (2002).

7. Thomson, J.J., "Cathode Rays". **Philosophical Magazine 44,** pp. 269-293 (1897). doi:10.1080/14786449708621070

8. Arabatzis, T. (2006). **Representing Electrons: A Biographical Approach to Theoretical Entities**. (University of Chicago Press, Chicago). pp. 70–74 (2006). ISBN 0-226-02421-0.

9. Littauer, Schopper, and Wilson, "Structure of the Proton and Neutron", **Phys. Rev. Letters, Vol. 7,** pp. 144-147 (1961).

Chapter 7 Logical Arguments from Metatheory

In questions of science, the authority of a thousand is not worth the humble reasoning of a single individual. Galileo Galilei [1]

7.1 Metatheory

Metatheory, the theory of theories, is a branch of metamathematics. It is the study of the principles, conceptual elements, consistency and other aspects of logical systems. From the days of the earliest natural philosophers science or natural philosophy has been developed as a logical system derived from postulates or axioms. As such scientific theories are subject to various logical principles based upon inductive and deductive logic and consistency.

Henri Poincarè is generally credited as founding the field of metatheory or metamathematics. Being one of the last of the true natural philosophers, he was concerned about the structure of scientific theories and the logical basis of truth. Poincarè was the co-discoverer of relativity theory with Einstein and he actually published one year before Einstein. However, neither he nor Einstein ever received the Nobel Prize for this work, because of Poincarè's own arguments from metatheory on the subject.

7.2 Argument that Special Relativity is of Electrodynamic Origin

Poincarè made logical arguments from metatheory [2] that no two fundamental theories in nature could employ the same fundamental constants such as c the velocity of light. This was then combined with another logical argument that only fundamental theories could be true theories.

- **Electrodynamics uses c in wave equation** $\lambda f = c$ (7 - 1)

- **Special Relativity uses c in space-time interval**

$$ds^2 = dx^2 + dy2 + dz^2 - c^2dt^2 \quad (7 - 2)$$

- **Quantum Mechanics uses c in energy quantum**

$$E = h\nu = h(2\pi c/\lambda) \quad (7 - 3)$$

- **General Relativity uses c in Einstein's field equation**

$$G_{\mu\nu} + \Lambda g_{\mu\nu} = \frac{8\pi G}{c^4} T_{\mu\nu} \quad (7 - 4)$$

Poincare noticed that four "so-called" fundamental theories of modern science used the same fundamental constant c for the velocity of light, i.e. electrodynamics, special relativity, quantum mechanics, and general relativity. According to his logical criterion only one of these four theories could be fundamental. Poincare suggested that the fundamental theory was electrodynamics and that eventually it would explain all of the data explained by these other theories. As a result Einstein never received a Nobel Prize for his development of Special Relativity Theory and General Relativity Theory. He did receive a Nobel Prize for his wife Mileva's work on the photoelectric effect. He gave her the entire amount of the Nobel Prize when he divorced her.

7.3   Argument that Gravity is of Electrodynamic Origin

Poincarè also published another interesting logical argument from metatheory. [3] In this logical argument he showed that no two fundamental force laws could have the same mathematical form such as $1/R^2$. Now the electrodynamic force law and the force of gravity both have a $1/R^2$ form. Since Einstein's General Theory of Relativity involves the fundamental constant c, Poincarè reasoned that gravity must also be of electrodynamic origin.

$$F_G = -G\frac{m_1 m_2}{R^2} \qquad F_{EM} = \frac{1}{4\pi\epsilon_0}\frac{q_1 q_2}{R^2} \qquad (7-5)$$

7.4   Argument Based on the Superposition Principle

The Superposition Principle [4] describes the properties of linear systems needed for coherence and stability. Systems may be a collection of moving charges in electrodynamics and a combination of theories to describe these charges such as electrodynamics, relativity theory, and quantum mechanics.

A linear system is one that satisfies the homogeneity and additivity properties required by the Superposition Principle for coherence and stability shown below.

$$F(x_1+x_2+...) = F(x_1) + F(x_2) + ... \qquad \textbf{Additivity (7 - 6)}$$

$$F(ax) = a\,F(x) \text{ for scalar a} \qquad \textbf{Homogeneity (7 - 7)}$$

If one wants to describe matter in the form of an elementary particle, an atom or a molecule by combining electrodynamics, relativity theory and quantum mechanics, each of these theories must separately satisfy the Superposition Principle.

$$F(x) = F_{EM}(x) + F_{SR}(x) + F_{QM}(x) \qquad (7-8)$$

Maxwellian electrodynamics cannot be combined with Special Relativity, because the electrodynamic field and force is nonlinear in r.

$$\vec{E}(\vec{r}) = \frac{q\hat{r}}{4\pi\epsilon_0 r^2} \qquad (7-9)$$

Special Relativity is not a proper theory to add to electrodynamics, since it modifies electrodynamics further making it more nonlinear by giving rise to the expression for the electric field that is nonlinear in v due to the $\beta^2 = (v/c)^2$ terms.

$$\vec{E}(\vec{r},\vec{v}) = \frac{q}{4\pi\epsilon_0 r^2} \frac{\left(1 - \frac{v^2}{c^2}\right)\hat{r}}{\left[1 - \left(\frac{v}{c}\right)^2 sin^2\,\varphi\right]^{3/2}} \quad (7-10)$$

$$\text{where} \quad \mathbf{sin\varphi = \hat{r} \times \hat{v}} \quad (7-11)$$

In quantum mechanics the principal task is to compute how a certain type of wave propagates. The wave is called a wave function. The equation governing the behavior of the wave is the Schrödinger wave equation. The primary approach to computing the behavior of a wave function is to write the wave function as a linear superposition of special quantum wave functions known as stationary states. Since the non-relativistic time-dependent Schrödinger's wave equation is *claimed to be* linear, the behavior of the original wave function can be computed through the Superposition Principle.

$$i\hbar\frac{\partial}{\partial t}\Psi(r,t) = \left[\frac{-\hbar^2}{2m}\nabla^2 + V(r,t)\right]\Psi(r,t) \quad (7-12)$$

7.5   Conclusions from Metatheory

Thus we see that electrodynamics is a nonlinear theory, Special Relativity is a nonlinear theory and quantum mechanics is a linear theory. According to the Superposition Principle no two of these theories may be combined to describe matter in the form of an elementary particle, atom or molecule, because a nonlinear theory cannot be combined with another nonlinear or linear theory. Only linear theories can be combined. **Only one nonlinear theory is possible. If a nonlinear theory is valid, it has to be the one and only theory, i.e. the universal theory.**

In order for the newly derived version of electrodynamics to be a proper theory, it must be the universal theory. It must explain all the forces in nature, relativistic effects, and quantum effects. The following chapters in this book will attempt to show just that.

7.6   References

1. http://www.goodreads.com/author/quotes/14190.Galileo_Galilei
2. Poincaré, Henri, Oeuvres, Vol. 9, p. 497 (1954).
3. Poincaré, Henri, Oeuvres, vol. 9, pp. 489-93
4. http://en.wikipedia.org/wiki/Superposition_principle

Chapter 8 Hierarchy of Electrodynamic Interactions

For a successful technology, reality must take precedence over public relations, for Nature cannot be fooled. Richard P. Feynman [1]

In fluid dynamics and chemistry there has long been an acknowledgement that there is a hierarchy of electrodynamic interactions that need to be taken into account in order to understand the basic interactions in those fields. Since electrical charges can form finite size structures such as neutral electric dipoles and neutral electric quadrupoles, etc. there appears to be a hierarchy of interactions in electrodynamics with decreasing strength.

8.1 Charge to Charge – Coulomb Force

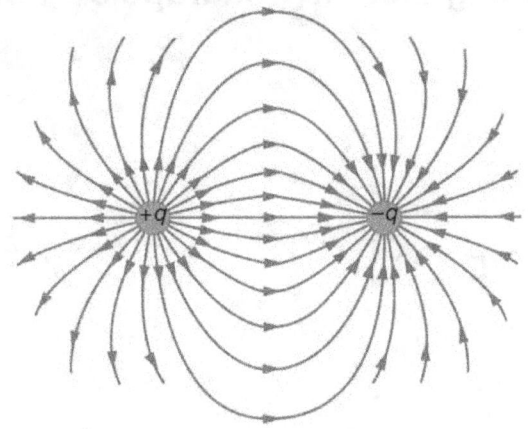

Figure 8-1 Electric Charge to Electric Charge Force

8.2 Charge to Vibrating Electric Dipole – Force of Inertia

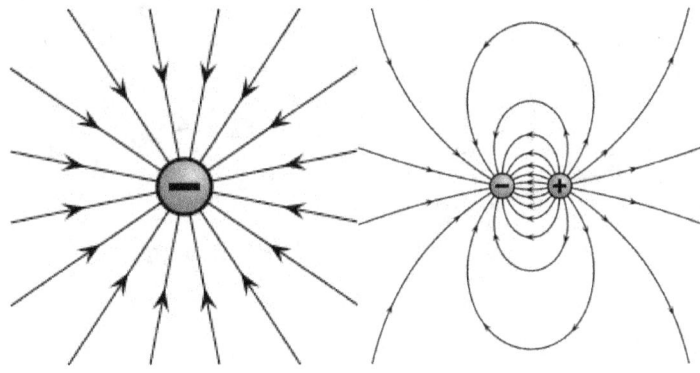

Figure 8-2 Electric Charge to Electric Dipole Force

8.3 Vibrating Electric Dipole to Vibrating Electric Dipole – Force of Gravity

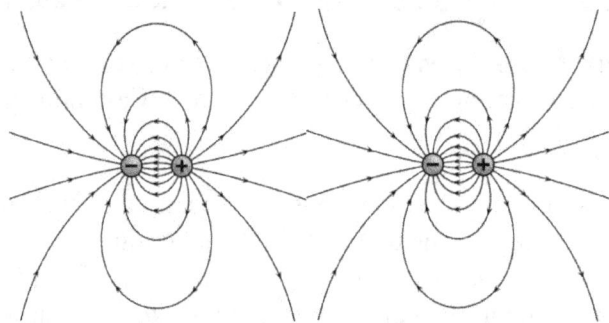

Figure 8-3 Electric Dipole to Electric Dipole Force

8.4 Charge to Vibrating Electric Quadrupole – Force of Inertia

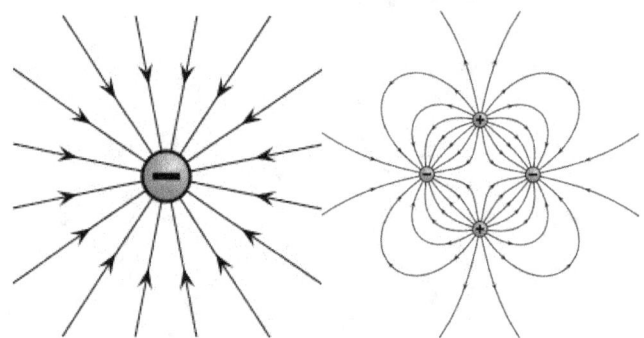

Figure 8-4 Electric Charge to Electric Quadrupole Force

8.4 Vibrating Electric Dipole to Vibrating Electric Quadrupole – Force of Gravity

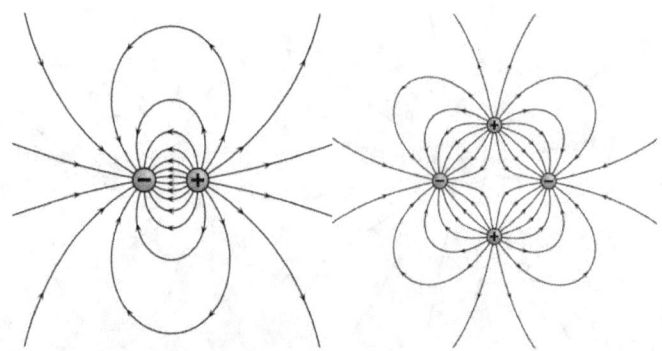

Figure 8-5 Electric Dipole to Electric Quadrupole Force

8.5   Vibrating Electric Quadrupole to Vibrating Electric Quadrupole – Force of Gravity

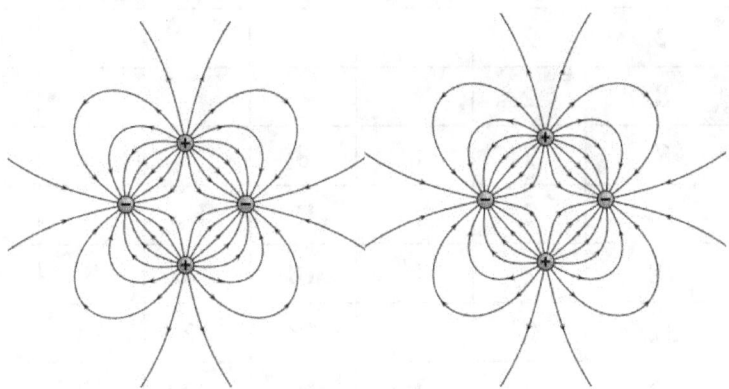

Figure 8-6 Electric Quadrupole to Electric Quadrupole Force

8.6   Vibrating Charge Structures in Atoms

The most natural large scale vibrating charge structures in atoms consist of atomic shells of electrons and nuclear shells of protons. These structures can support vibrations and rotations which result in a net force when averaged over a large collection of particles. Note that quadrupole moments are responsible for non-negligible electric field and torque contributions to aggregation of like-charged dust grains and charged plasma particles. The quadrupole torque mostly determines how these aggregates will rotate. [2]

Chapters 9 and 10 will investigate **force 3** the vibrating electric dipole to vibrating electric dipole giving rise to the force of gravity and **force 2** the charge to vibrating electric dipole giving rise to the force of inertia. Both of these forces average to zero, if there is no periodic vibration of the neutral electric dipoles. Thus these forces are periodic motion dependent.

If one examines the structure of the Periodic Table of the Elements or the Table of Nuclides, one sees that the outer electrons of the atom and the nuclear protons are arranged in geometrical shells of size 2, 8, 18, 32, and 50 charged particles. In **The Universal Force Volume 3** the role of combinatorial geometry subject to particular electrodynamic constraints determines these physical shell structures. **Table 1** below gives the distribution of electrons in packing shells. **Table 2** gives the distribution of nucleons in packing shells. Note that the buildup of nuclear shells is different from the buildup of atomic shells, because the nucleus has nothing at its center to keep the inner shells from coming apart to form the more highly bound larger shells. **Table 3** gives the step by step buildup of the 4th packing shell of atomic electrons. From the atomic electron data of Tables 1 and 3, one can see that not only are the dipole and quadrupole interactions important, but one should take into account the groupings of 8, 18, and 32 electrons. Then there is the question of what to do about the incomplete or partially filled shells as shown in Table 3. Also the shells in the nucleus do not perfectly match the atomic electron shells. A proper physical approach must deal with all of these issues.

Table 1 Distribution of Atomic Electrons in Packing Shells

Total Electrons	Shell Electrons	K Shell	L Shell	M shell	N Shell	O Shell	P Shell	Q Shell
2	2	2						
10	8	2	8					
18	8	2	8	8				
36	18	2	8	18	8			
54	18	2	8	18	18	8		
86	32	2	8	18	32	18	8	
118	32	2	8	18	32	32	18	8

Table 2 Distribution of Nucleons in Packing Shells

Total Nucleons	2 Shell	2 Shell	8 Shell	8 Shell	18 Shell	18 Shell	32 Shell	32 Shell	50 Shell	50 Shell
2	2									
8			8							
20	2				18					
28	2		8		18					
50					18		32			
82							32		50	
126			8		18	18	32		50	

Table 3 Step by Step Buildup of the 4ᵗʰ Packing Shell of Atomic Electrons

Atomic Symbol	Atomic Number	1ˢᵗ Shell	2ⁿᵈ Shell	3ʳᵈ Shell	4ᵗʰ Shell
Ar	18	2	8	8	
K	19	2	8	8	1
Ca	20	2	8	8	2
Sc	21	2	8	9	2
Ti	22	2	8	10	2
V	23	2	8	11	2
Cr	24	2	8	13	1
Mn	25	2	8	13	2
Fe	26	2	8	14	2
Co	27	2	8	15	2
Ni	28	2	8	16	2
Cu	29	2	8	18	1
Zn	30	2	8	18	2
Ga	31	2	8	18	3
Ge	32	2	8	18	4
As	33	2	8	18	5
Se	34	2	8	18	6
Br	35	2	8	18	7
Kr	36	2	8	18	8

8.7   References

1. http://www.brainyquote.com/quotes/authors/r/richard_p_feynman.html
2. Coleman, D. A., Matthews, L. S., and Hyde, T. W., "Effects of Electric Quadrupole Moments on Dust Aggregation Dynamics and Morphology of Like charged Grains in a Plasma"
    http://www.baylor.edu/content/services/document.php/124346.pdf

Chapter 9 Electrodynamic Origin of Gravitational Forces

No more causes of natural things should be admitted than are both true, and sufficient to explain their phenomena. Isaac Newton [1]

9.1 Introduction

In the past natural philosophers and astronomers thought that we lived in a disconnected universe consisting of the Milky Way Galaxy with a vacuum or void between the stars and other objects. In recent times astronomers have discovered that space is filled with charged particles forming dynamic plasmas. From Figure 9-1 we can see that there are magnetic fields due to electrical currents stretching across many light years in these plasmas.

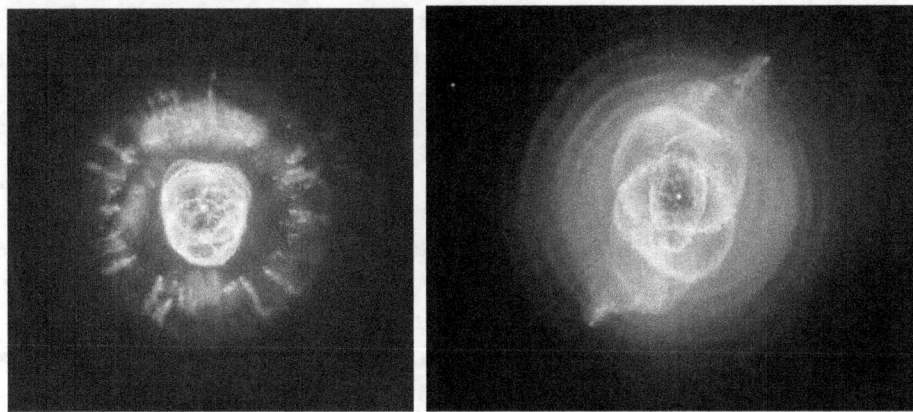

Figure 9-1 Eskimo Nebula [2] and Cat's Eye Nebula [3]

Plasma Structure of Nebulas Controlled by Magnetic Fields

These magnetic fields appear as current filaments or threads extending for light years connecting things together. Galaxies appear as something like streetlights along the current filaments of an electric universe. Within our own galaxy, stars appear aligned in the same way along filaments formed by Birkeland currents that pervade plasma space. This is very different from the universe of Isaac Newton's day which consisted primarily of empty space with astronomical bodies orbiting one another due to the action-at-a-distance gravitational force. The electrodynamic plasma force is approximately 10^{40} times as strong as the gravitational force and appears to play a more dominant role in the universe than the gravitational force as it connects things together.

Einstein's General Relativity Theory, the replacement theory for Newton's theory of gravity, has run into great difficulty in these days. It cannot explain the role of electrodynamic plasmas in the organization of the universe. Furthermore, it cannot explain the observed structure of spiral galaxies without inventing dark matter which cannot be observed in the laboratory like other types of matter. In addition, in order to explain the expansion of the universe using

119

relativity theory, it has been necessary to invent dark energy which cannot be observed directly either. Dark matter and dark energy are thought to make up 95% of the universe.

Instead of pursuing these problematic approaches to a disconnected universe, it would seem better to consider an electrodynamic approach for a connected universe in agreement with Mach's Principle. Ever since the discovery of Coulomb's electrostatic force law (Coulomb, 1785) between charges q_1 and q_2 and Newton's universal gravitational force law (Newton, 1687) between masses m_{g1} and m_{g2},

$$F_{Coulomb} = \frac{q_1 q_2}{r^2} \text{ and } F_{Gravity} = G \frac{m_{g1} m_{g2}}{r^2} \text{ where } r = |\vec{r}_2 - \vec{r}_1| \quad (9-1)$$

scientists have imagined that there might be a relationship between electrodynamics and gravity, because the $1/r^2$ form of the force laws is identical.

The philosopher Poincaré [41] argued from the logic of meta-theory (theory of theories) that any two fundamental theories that employed the same fundamental constants or the same mathematical form were not both fundamental. The most advanced theory of gravity, Einstein's General Relativity Theory as represented in his field equation (9-2), is expressed in terms of c the velocity of light which is the fundamental constant of electrodynamics.

$$G_{\mu v} = -\frac{8\pi}{c} G T_{\mu v} \quad (9-2)$$

Einstein also said in his Principle of Equivalence that gravity and inertia are intimately related. Poincaré suggested that electrodynamics might be the origin of both the relativistic gravitational force as well as the related relativistic inertial force, because both force laws involve mass and the velocity of light which is the fundamental constant of electrodynamics.

According to the Lorentz force law combined with Ampere's law of induction for a moving charge ($B_i = qv/cE_0$),

$$\vec{F} = q\vec{E}_0 + \frac{\vec{v}}{c} \times \vec{B}_i = q\vec{E}_0 + \frac{\vec{v}}{c} \times \left(\frac{q\vec{v}}{c} \times \vec{E}_0\right) = q\vec{E}_0 \left(1 + \frac{v^2}{c^2}\right) \quad (9-3)$$

the induced B_i field adds a term to the static Coulomb field E_0 proportional to v^2/c^2. The free electron drift velocity in conductors is typically on the order of 0.03 m/sec such that

$$\frac{v^2}{c^2} \approx \left(\frac{3 \times 10^{-2} m/sec}{3 \times 10^8 m/sec}\right)^2 \approx 10^{-20} \quad (9-4)$$

Now the gravitational force is approximately 10^{-40} times smaller than the static Coulomb electric force, suggesting that the gravitational force might be a higher order v^4/c^4 term multiplying the static Coulomb force.

The derived universal electrodynamic contact force [4], shown in equation (5-7) and (9-5) below, gives terms of this sort. In the past such higher order terms in the electrodynamic force were never fully investigated.

$$\vec{F}(\vec{r},\vec{v},\vec{a}) = \frac{qq'}{\vec{r}^2} \frac{(1-\vec{\beta}^2)\vec{r} + \frac{2\vec{r}^2}{c^2}\vec{a}}{\left[\vec{r}^2 - \{\vec{r}\times(\vec{r}\times\vec{\beta})\}^2/\vec{r}^2\right]^{\frac{1}{2}}}$$

$$-\frac{qq'}{\vec{r}^2}(1-\vec{\beta}^2)\frac{(\vec{\beta}\cdot\vec{r})\vec{r}\times(\vec{r}\times\vec{\beta}) + (\vec{r}\cdot\vec{r})\vec{r}\times\left(\vec{r}\times\frac{\vec{a}}{c^2}\right)}{\left[\vec{r}^2 - \{\vec{r}\times(\vec{r}\times\vec{\beta})\}^2/\vec{r}^2\right]^{\frac{3}{2}}} \quad (9-5)$$

In the electrodynamic force equation (9-5) the first acceleration "**a**" term gives rise to Newton's Second Law (**F**=m**a**) for the force of inertia and the second acceleration term "**r x a**" gives rise to absorption and emission of electromagnetic radiation or light.

For acceleration **a**=0 and writing out β=v/c terms obtain

$$\vec{F}(\vec{r},\vec{v},\vec{a}) = \frac{qq'}{\vec{r}^2}\frac{(1-\vec{\beta}^2)\hat{r}}{[1-\beta^2 sin^2\theta]^{1/2}} - \frac{qq'}{\vec{r}^2}\frac{(1-\vec{\beta}^2)\left(\hat{r}\cdot\frac{\vec{v}}{c}\right)\hat{r}\times\left(\hat{r}\times\frac{\vec{v}}{c}\right)}{[1-\beta^2 sin^2\theta]^{\frac{3}{2}}} \quad (9-6)$$

Equation (9-6) is identical, mathematically speaking, to the action-at-a-distance covariant relativistic electrodynamic force based on Maxwell's equations. [5 p. 555 or 5 p. 560] However, it gives rise to a very different interpretation of nature.

9.2   Origin of Gravitational Forces

In questions of science, the authority of a thousand is not worth the humble reasoning of a single individual. Galileo Galilei 1564-1642

The force of gravity is normally measured between neutral charge bodies. Since the time of Newton, scientists have learned that the neutral atom consists of electrons, protons, and neutrons. Even elementary charged particles such as the neutron consist of smaller components of positive and negative charge such as quarks or charge fibers. This work explores the possibility that the attractive force of gravity is due to a small residual electrodynamic force between vibrating neutral electric dipoles. These electric dipoles could be inside an elementary particle, such as the neutron, or in an atom involving the atomic electron and nuclear proton.

In the Gaussian system of units assuming constant velocity **a**=0 and **v/c=β** is very small, one may expand the terms in equation (9-6) for the electrodynamic force for the radial term using the binomial expansion and keeping only terms to order β^4 and substituting $sin^2\theta = 1 - cos^2\theta$ to obtain

$$F(r,v) = \frac{qq'\hat{r}}{\vec{r}^2}(1-\vec{\beta}^2)\left[1 + \frac{1}{2}\beta^2 sin^2\theta + \frac{\left(\frac{1}{2}\right)\left(\frac{3}{2}\right)}{2}\beta^4 sin^4\theta + \cdots\right] \quad (9-7)$$

$$-\frac{qq'}{\vec{r}^2}(\hat{r}\cdot\vec{\beta})\hat{r}\times(\hat{r}\times\vec{\beta})(1-\vec{\beta}^2)\left[1+\frac{3}{2}\beta^2 sin^2\theta+\frac{\left(\frac{3}{2}\right)\left(\frac{5}{2}\right)}{2}\beta^4 sin^4\theta+\cdots\right]$$

$$=\frac{qq'\hat{r}}{\vec{r}^2}\left[1-\frac{1}{2}\beta^2-\frac{1}{2}\beta^2 cos^2\theta-\frac{1}{8}\beta^4-\frac{1}{4}\beta^4 cos^2\theta+\frac{3}{8}\beta^4 cos^4\theta+\cdots\right]$$

$$-\frac{qq'}{\vec{r}^2}(\hat{r}\cdot\vec{\beta})\hat{r}\times(\hat{r}\times\vec{\beta})\left[1+\frac{1}{2}\beta^2-\frac{3}{2}\beta^2 cos^2\theta-\frac{3}{8}\beta^4-\frac{9}{4}\beta^4 cos^2\theta\right.$$

$$\left.+\frac{15}{8}\beta^4 cos^4\theta+\cdots\right]$$

Consider the force between two vibrating neutral electric dipoles each consisting of positive protons and negative electrons. If we label the charges q_{1+}, q_{1-}, q_{2+}, and q_{2-}, the total force between the two neutral dipoles is given by equation (9-8)

$$F = F_{2+,1+} + F_{2+,1-} + F_{2-,1+} + F_{2-,1-} \quad (9-8)$$

where the dipoles are defined in Figure 9-2 for hydrogen atoms where the larger toroid is the electron and the smaller toroid is the proton.

$$\vec{r}_{1+2+} = \vec{r}_{2+} - \vec{r}_{1+} \quad \omega_1 = 2\pi f_1 \quad \omega_2 = 2\pi f_2$$

$$\vec{r}_{1-2+} = \vec{r}_{2+} - \vec{r}_{1-} - \vec{A}_1 \cos(\omega_1 t + \phi_1) \quad A_1 f_1 = v_1$$

$$\vec{r}_{1+2-} = \vec{r}_{1+} - \vec{r}_{2-} - \vec{A}_2 \cos(\omega_2 t + \phi_2) \quad A_2 f_2 = v_2$$

$$\vec{r}_{2-1-} = \vec{r}_{2-} - \vec{r}_{1-} - \vec{A}_2 \cos(\omega_2 t + \phi_2) - A_1 \cos(\omega_1 t + \phi_1)$$

$$|\leftarrow - r_{1+2+} - \rightarrow|$$

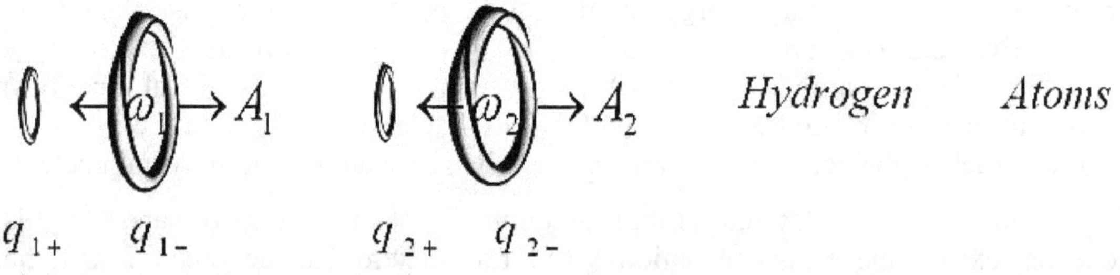

$$q_{1+} \quad q_{1-} \quad\quad q_{2+} \quad q_{2-}$$

Figure 9-2 Oscillations of Electrons in Neutral Dipoles

Note from **Chapter 6** that all elementary particles must be composed of closed toroidal rings in order for the derived radiation reaction force to be in agreement with the empirically observed radiation reaction force. The amplitudes of oscillation of vibration A_1 and A_2 should be on the order of the size of the atom 10^{-10} m or less. From the wave equation $\lambda f=v$ where λ is approximately equal to the amplitude A, the frequencies of oscillation ω_1 and ω_2 must be in the microwave range $\approx 10^{10}$ per second. At t=0 the amplitude of vibration is maximum as represented by cos (ωt+φ).

In order to simplify the calculations assume that the positively charged proton is much more massive than the negatively charged electron, such that the vibratory motion of the dipole can be considered as due primarily to the motion of the electron. Since the mass of the proton is 1836 times the mass of an electron, this is a reasonable approximation.

In order to calculate a quantity comparable to the force of gravity, it will be necessary to perform a number of averages. Each oscillating dipole has a different phase that must be averaged over. Each oscillating dipole may have a different physical orientation that needs to be averaged over for all the dipoles in the material body. In order to obtain a time independent value it will be necessary to perform a time average on each of the oscillating dipoles. Thus the force to be compared with gravity is

$$\vec{F}(\vec{r}, \vec{v}) = \frac{1}{\tau_1} \int_0^{\tau_1} dt_1 \frac{1}{\tau_2} \int_0^{\tau_2} dt_2 \frac{1}{2\pi} \int_0^{2\pi} d\varphi_1 \frac{1}{2\pi} \int_0^{2\pi} d\varphi_2 \frac{1}{2\pi} \int_0^{2\pi} d\varphi \frac{1}{\pi} \int_0^{\pi} sin\theta d\theta$$

$$\vec{F}(r, \theta, \varphi, A_1, \omega_1, \varphi_1, t_1, A_2, \omega_2, \varphi_2, t_2, v) \quad (9-9)$$

For simplicity consider that there are two collections of dipoles to average over and that the two collections have spherical symmetry. In this case the integral over φ becomes just 2π giving

$$\vec{F}(\vec{r}, \vec{v}) = \frac{1}{\tau_1} \int_0^{\tau_1} dt_1 \frac{1}{\tau_2} \int_0^{\tau_2} dt_2 \frac{1}{2\pi} \int_0^{2\pi} d\varphi_1 \frac{1}{2\pi} \int_0^{2\pi} d\varphi_2 \frac{1}{\pi} \int_0^{\pi} sin\theta d\theta$$

$$\vec{F}(r, \theta, \varphi, A_1, \omega_1, \varphi_1, t_1, A_2, \omega_2, \varphi_2, t_2, v) \quad (9-10)$$

Note that there are two fundamental type terms in the force given in equation (9-7). The first term is radial and is proportional to **r**. The second term is non-radial and proportional to (**β·r**) {**r** x (**r** x **β**)}. The v^4/c^4 part of the first term will lead to Newton's universal law of gravitation. The v^4/c^4 part of the second term will lead to a new (previously unknown?) gravitational force term which causes a corkscrew type motion and quantization of gravitational orbits. Since the first term can be used to identify with Newton's universal force of gravitation, it will be calculated first.

9.3   Computation of Radial Force Term

The four radial force terms of equation (9-8) are shown below where the coordinates are given explicitly.

$$\vec{F}_{2-,1+} = \frac{q_1 q_2 \hat{r}_{21}}{|\vec{r}_2 + \vec{A}_1 - \vec{r}_1|^2} \left[1 - \frac{1}{2}(\vec{\beta}_2 - \vec{\beta}_1)^2 - \frac{1}{2}(\vec{\beta}_2 - \vec{\beta}_1)^2 \cos^2\theta \right. \quad (9-11)$$

$$\left. - \frac{1}{8}(\vec{\beta}_2 - \vec{\beta}_1)^4 - \frac{1}{4}(\vec{\beta}_2 - \vec{\beta}_1)^4 \cos^2\theta + \frac{3}{8}(\vec{\beta}_2 - \vec{\beta}_1)^4 \cos^4\theta + \cdots \right]$$

$$\vec{F}_{2+,1+} = \frac{q_1 q_2 \hat{r}_{21}}{|\vec{r}_2 - \vec{r}_1|^2} \left[1 - \frac{1}{2}(\vec{\beta}_2 - \vec{\beta}_1)^2 - \frac{1}{2}(\vec{\beta}_2 - \vec{\beta}_1)^2 \cos^2\theta - \frac{1}{8}(\vec{\beta}_2 - \vec{\beta}_1)^4 \right.$$

$$\left. - \frac{1}{4}(\vec{\beta}_2 - \vec{\beta}_1)^4 \cos^2\theta + \frac{3}{8}(\vec{\beta}_2 - \vec{\beta}_1)^4 \cos^4\theta + \cdots \right]$$

$$\vec{F}_{2+,1-} = \frac{q_1 q_2 \hat{r}_{21}}{|\vec{r}_2 - \vec{r}_1 - \vec{A}_1|^2} \left[1 - \frac{1}{2}(\vec{\beta}_2 - \vec{\beta}_1)^2 - \frac{1}{2}(\vec{\beta}_2 - \vec{\beta}_1)^2 \cos^2\theta \right.$$

$$\left. - \frac{1}{8}(\vec{\beta}_2 - \vec{\beta}_1)^4 - \frac{1}{4}(\vec{\beta}_2 - \vec{\beta}_1)^4 \cos^2\theta + \frac{3}{8}(\vec{\beta}_2 - \vec{\beta}_1)^4 \cos^4\theta + \cdots \right]$$

$$\vec{F}_{2-,1-} = \frac{q_1 q_2 \hat{r}_{21}}{|\vec{r}_2 + \vec{A}_2 - \vec{r}_1 - \vec{A}_1|^2} \left[1 - \frac{1}{2}(\vec{\beta}_2 - \vec{\beta}_1)^2 - \frac{1}{2}(\vec{\beta}_2 - \vec{\beta}_1)^2 \cos^2\theta \right.$$

$$\left. - \frac{1}{8}(\vec{\beta}_2 - \vec{\beta}_1)^4 - \frac{1}{4}(\vec{\beta}_2 - \vec{\beta}_1)^4 \cos^2\theta + \frac{3}{8}(\vec{\beta}_2 - \vec{\beta}_1)^4 \cos^4\theta + \cdots \right]$$

Now the force of gravity is normally measured in lab experiments where $r_2 - r_1 \gg A_1$ and $r_2 - r_1 \gg A_2$. Thus to good approximation the A_1 and A_2 terms in the denominator may be dropped such that

$$\vec{F}_{q_2,q_1} = \frac{q_1 q_2 \hat{r}_{21}}{|\vec{r}_2 - \vec{r}_1|^2} \left[1 - \frac{1}{2}(\vec{\beta}_2 - \vec{\beta}_1)^2 - \frac{1}{2}(\vec{\beta}_2 - \vec{\beta}_1)^2 \cos^2\theta - \frac{1}{8}(\vec{\beta}_2 - \vec{\beta}_1)^4 \right.$$

$$\left. - \frac{1}{4}(\vec{\beta}_2 - \vec{\beta}_1)^4 \cos^2\theta + \frac{3}{8}(\vec{\beta}_2 - \vec{\beta}_1)^4 \cos^4\theta + \cdots \right] \quad (9-12)$$

For the velocity terms in the [ ] of the expression for the force above, $\beta_2 \approx \beta_1$ and $\beta_2 - \beta_1 \approx 0$ for most laboratory measurements of gravity. In this case only the $A_1\omega_1$ and $A_2\omega_2$ terms are left as shown below where $q_1 = q_2 = e$ is the charge of the proton and $-e$ is the charge of the electron.

$$\vec{F}_{2+,1+} = \frac{e^2 \hat{r}_{21}}{|\vec{r}_2 - \vec{r}_1|^2} [1]$$

$$\vec{F}_{2+,1-} = \frac{-e^2 \hat{r}_{21}}{|\vec{r}_2 - \vec{r}_1|^2} \left[1 - \left\{ \frac{A_1\omega_1}{c} sin(\omega_1 t_1 + \varphi_1) \right\}^2 \left(\frac{1 + cos^2\theta}{2} \right) \right.$$

$$\left. + \left\{ \frac{A_1\omega_1}{c} sin(\omega_1 t_1 + \varphi_1) \right\}^4 \left(-\frac{1}{8} - \frac{1}{4}cos^2\theta + \frac{3}{8}cos^4\theta \right) \right]$$

$$\vec{F}_{2-,1+} = \frac{-e^2 \hat{r}_{21}}{|\vec{r}_2 - \vec{r}_1|^2} \left[1 - \left\{ \frac{A_2\omega_2}{c} sin(\omega_2 t_2 + \varphi_2) \right\}^2 \left(\frac{1 + cos^2\theta}{2} \right) \right.$$

$$\left. + \left\{ \frac{A_2\omega_2}{c} sin(\omega_2 t_2 + \varphi_2) \right\}^4 \left(-\frac{1}{8} - \frac{1}{4}cos^2\theta + \frac{3}{8}cos^4\theta \right) \right]$$

$$\vec{F}_{2-,1-} = \frac{-e^2 \hat{r}_{21}}{|\vec{r}_2 - \vec{r}_1|^2} \left[1 \right.$$

$$- \left\{ \frac{A_1\omega_1}{c} sin(\omega_1 t_1 + \varphi_1) + \frac{A_2\omega_2}{c} sin(\omega_2 t_2 + \varphi_2) \right\}^2 \left(\frac{1 + cos^2\theta}{2} \right)$$

$$+ \left\{ \frac{A_1\omega_1}{c} sin(\omega_1 t_1 + \varphi_1) + \frac{A_2\omega_2}{c} sin(\omega_2 t_2 + \varphi_2) \right\}^4 \left(-\frac{1}{8} \right.$$

$$\left. \left. - \frac{1}{4}cos^2\theta + \frac{3}{8}cos^4\theta \right) \right] \tag{9-13}$$

One can see that the sum of the first terms, i.e. 1 terms, in the [ ] of the four forces is just 0. The sum of the second terms in the [ ] does not sum to 0, since the cross term remains. Two parts of the third term in the [ ] cancel leaving the cross terms. The resulting sum of the four forces gives

$$\vec{F} = \vec{F}_{2+,1+} + \vec{F}_{2+,1-} + \vec{F}_{2-,1+} + \vec{F}_{2-,1-} \tag{9-14}$$

$$= \frac{-e^2 \hat{r}_{21}}{|\vec{r}_2 - \vec{r}_1|^2} \left[2 \frac{A_1 \omega_1}{c} sin(\omega_1 t_1 + \varphi_1) \frac{A_2 \omega_2}{c} sin(\omega_2 t_2 + \varphi_2) \left(\frac{1 + cos^2\theta}{2} \right) \right.$$

$$+ 4 \frac{A_1^{\ 3} \omega_1^{\ 3}}{c} sin^3(\omega_1 t_1 + \varphi_1) \frac{A_2 \omega_2}{c} sin(\omega_2 t_2 + \varphi_2) \left(-\frac{1}{8} \right.$$

$$\left. -\frac{1}{4} cos^2\theta + \frac{3}{8} cos^4\theta \right)$$

$$+ 4 \frac{A_2^{\ 3} \omega_2^{\ 3}}{c} sin^3(\omega_2 t_2 + \varphi_2) \frac{A_1 \omega_1}{c} sin(\omega_1 t_1 + \varphi_1) \left(-\frac{1}{8} \right.$$

$$\left. -\frac{1}{4} cos^2\theta + \frac{3}{8} cos^4\theta \right)$$

$$- 6 \frac{A_1^{\ 2} \omega_1^{\ 2}}{c} sin^2(\omega_1 t_1 + \varphi_1) \frac{A_2^{\ 2} \omega_2^{\ 2}}{c} sin^2(\omega_2 t_2 + \varphi_2) \left(-\frac{1}{8} \right.$$

$$\left. \left. -\frac{1}{4} cos^2\theta + \frac{3}{8} cos^4\theta \right) \right]$$

Now the integrals in equation (9-10) can be evaluated. The symmetry of the integrals over φ_1 and φ_2 will cause the odd powers of sin $(\omega_1 t_1 + \varphi_1)$ and sin $(\omega_2 t_2 + \varphi_2)$ to average to zero as shown below in equation (9-15).

$$\frac{1}{\tau} \int_0^\tau \frac{1}{2\pi} \int_0^{2\pi} sin(\omega t + \varphi) d\varphi dt = \frac{\omega}{2\pi} \int_0^{\frac{2\pi}{\omega}} \frac{1}{2\pi} \int_0^{2\pi} sin(\omega t + \varphi) d\varphi dt$$

$$= \frac{1}{2\pi} \int_0^{2\pi} \frac{1}{2\pi} \int_0^{2\pi} sin(x + \varphi) d\varphi dx \quad (x = \omega t)$$

$$= \frac{1}{2\pi} \int_0^{2\pi} \frac{1}{2\pi} \int_0^{2\pi} (sinx cos\varphi - cosx sin\varphi) d\varphi dx$$

$$= \left(\frac{1}{2\pi} \right)^2 \int_0^{2\pi} (sinx cos\varphi - cosx sin\varphi) d\varphi \Big|_0^{2\pi}$$

$$= \left(\frac{1}{2\pi} \right)^2 (-2)(0 - 0) = 0 \qquad (9-15)$$

Thus from the symmetry of the integrals the average force of equation (9-14) may be reduced to

$$\vec{F}(\vec{r}, \vec{v}) = \frac{1}{\tau_1}\int_0^{\tau_1} dt_1 \frac{1}{\tau_2}\int_0^{\tau_2} dt_2 \frac{1}{2\pi}\int_0^{2\pi} d\varphi_1 \frac{1}{2\pi}\int_0^{2\pi} d\varphi_2 \frac{1}{\pi}\int_0^{\pi} sin\theta d\theta$$

$$\vec{F}(r, \theta, \varphi, A_1, \omega_1, \varphi_1, t_1, A_2, \omega_2, \varphi_2, t_2, v)$$

$$= \frac{\omega_1}{2\pi}\int_0^{\frac{2\pi}{\omega_1}} dt_1 \frac{\omega_2}{2\pi}\int_0^{\frac{2\pi}{\omega_2}} dt_2 \frac{1}{2\pi}\int_0^{2\pi} d\varphi_1 \frac{1}{2\pi}\int_0^{2\pi} d\varphi_2 \frac{1}{\pi}\int_0^{\pi} sin\theta d\theta$$

$$\left[\frac{e^2 \hat{r}_{21}}{|\vec{r}_2 - \vec{r}_1|^2} 6\frac{A_1{}^2\omega_1{}^2}{c^2} sin^2(\omega_1 t_1 + \varphi_1)\frac{A_2{}^2\omega_2{}^2}{c^2} sin^2(\omega_2 t_2 + \varphi_2)\left(-\frac{1}{8}\right.\right.$$
$$\left.\left. -\frac{1}{4}cos^2\theta + \frac{3}{8}cos^4\theta\right)\right]$$

$$= -\frac{\omega_1}{2\pi}\int_0^{\frac{2\pi}{\omega_1}} dt_1 \frac{\omega_2}{2\pi}\int_0^{\frac{2\pi}{\omega_2}} dt_2 \frac{1}{2\pi}\int_0^{2\pi} d\varphi_1 \frac{1}{2\pi}\int_0^{2\pi} d\varphi_2$$

$$\left[\frac{e^2 \hat{r}_{21}}{|\vec{r}_2 - \vec{r}_1|^2} 6\frac{A_1{}^2\omega_1{}^2}{c^2} sin^2(\omega_1 t_1 + \varphi_1)\frac{A_2{}^2\omega_2{}^2}{c^2} sin^2(\omega_2 t_2 + \varphi_2)\frac{4}{15\pi}\right]$$

$$= -\frac{\omega_1}{2\pi}\int_0^{\frac{2\pi}{\omega_1}} dt_1 \frac{\omega_2}{2\pi}\int_0^{\frac{2\pi}{\omega_2}} dt_2 \frac{e^2 \hat{r}_{21}}{|\vec{r}_2 - \vec{r}_1|^2} 6\frac{A_1{}^2\omega_1{}^2}{c}\frac{A_2{}^2\omega_2{}^2}{c}\left(\frac{1}{2}\right)^2\frac{4}{15\pi}$$

$$= -\frac{e^2 \hat{r}_{21}}{|\vec{r}_2 - \vec{r}_1|^2}\frac{A_1{}^2\omega_1{}^2}{c^2}\frac{A_2{}^2\omega_2{}^2}{c^2}\frac{2}{5\pi} \quad (attractive\ force\ only) \qquad (9-16)$$

where in the calculation of equation (9-16) above $-sin\theta\ d\theta = dcos\theta$, $x = \omega t$, $dx = \omega\ dt$, such that

$$\frac{1}{2\pi}\int_0^{2\pi} sin(\omega t + \varphi)d\varphi$$

$$= \frac{1}{2\pi}\int_0^{2\pi} (sin^2\omega t cos^2\varphi + 2cos\omega t sin\omega t sin\varphi cos\varphi + cos^2\omega t sin^2\varphi)d\varphi$$

127

$$= \frac{1}{2\pi}(\pi cos^2\omega t + 0 + \pi sin^2\omega t) = \frac{1}{2}(cos^2\omega t + sin^2\omega t = \frac{1}{2} \qquad (9-17)$$

and

$$\frac{1}{\pi}\int_0^{\pi} sin\theta d\theta \left(-\frac{1}{8}-\frac{1}{4}cos^2\theta+\frac{3}{8}cos^4\theta\right)$$

$$= -\frac{1}{\pi}\int_{-1}^{1} dcos\theta \left(-\frac{1}{8}-\frac{1}{4}cos^2\theta+\frac{3}{8}cos^4\theta\right)$$

$$= -\frac{1}{\pi}\left(-\frac{cos\theta}{8}-\frac{1}{4}\frac{cos^3\theta}{3}+\frac{3}{8}\frac{cos^5\theta}{5}\right)\Big|_{-1}^{1}$$

$$= -\frac{1}{\pi}\left(-\frac{2}{8}-\frac{1}{4}\frac{2}{3}+\frac{3}{8}\frac{2}{5}\right) = \frac{4}{15\pi} \qquad (9-18)$$

This attractive only force is to be compared with Newton's Universal Law of Gravitation

$$\vec{F}_G(\vec{r}) = -G\frac{m_{g1}m_{g2}\hat{r}_{21}}{|\vec{r}_2-\vec{r}_1|} \qquad (9-19)$$

We need to see if the following relationship is reasonable.

$$Gm_{g1}m_{g2} = \frac{A_1{}^2\omega_1{}^2}{c^2}\frac{A_2{}^2\omega_2{}^2}{c^2}\frac{2}{5\pi} \qquad (9-20)$$

Note that for two bodies with collections of N_1 and N_2 atoms of atomic number Z_1 and Z_2 this formula becomes

$$Gm_{g1}m_{g2} = \frac{2}{5\pi}\frac{N_1Z_1A_1{}^2\omega_1{}^2}{c^2}\frac{N_2Z_2A_2{}^2\omega_2{}^2}{c^2} \qquad (9-21)$$

There will be a range of combinations of amplitude A and frequency ω for which equation (9-20) above holds. Although equation (9-16) looks very similar to Newton's Universal Law of Gravitation, it is very different. First it is a local contact force. Second it says gravity is decaying over time, because according to the laws of electrodynamics all vibrating charge systems must radiate away energy. Third the second term of the gravitational force will turn out to be a non-radial term causing many effects including the quantization of gravity. All of these gravitational effects must be observed in order to claim that this force law is valid. If the properties of the derived force of gravity are supported by experimental data, then it can be claimed to be more complete than the previous theories of gravity such as Newton's Universal Law of Gravitation and Einstein's General Relativity Theory.

9.4   Corroborating Evidence for Radiative Decay of Gravity

If the conjecture that the source of gravity is due to a statistical residual electromagnetic force between collections of vibrating neutral electric dipoles originating from the $(v/c)^4$ terms of the electromagnetic force is correct, then there are consequences which can be used to verify the conjecture. According to electrodynamics these oscillating dipoles must radiate energy. Since the gravitational force dominates on the large scale in the physical universe, the energy radiated by these oscillating dipoles in every atom should be greatest in the vicinity of matter, and be easily observable in its microwave frequency range.

Now hydrogen is the most abundant element in the universe comprising 75% of all visible matter. [7] In order to test this conjecture on the origin of the force of gravity, let us calculate the wavelength λ for this dipole radiation assuming hydrogen atoms. For simplicity assume $N_1 = N_2 = 1$, $q_1 = q_2 = e$, $\omega_1 = \omega_2 = \omega$, $m_1 = m_2 = m$ of hydrogen, and $A_1 = A_2 = A \leq$ size of hydrogen atom. From the wave equation $\lambda f = c$, we have $\omega = 2\pi f = 2\pi c/\lambda$. Thus

$$Gm^2 \geq \frac{2e^2}{5\pi} \frac{A^4 \omega^4}{c^4} = \frac{2e^2}{5\pi} \frac{A^4}{c^4} \left(\frac{2\pi c}{\lambda}\right)^4 = \frac{2e^2}{5\pi} \frac{16\pi^4 A^4}{\lambda^4} \quad (9-22)$$

Solving for λ obtain

$$\lambda^4 \leq \frac{2e^2}{5\pi} \frac{16\pi^4 A^4}{Gm^2} \quad (9-23)$$

Using the following values for the hydrogen constants from the **CRC Handbook of Chemistry and Physics** [8] and the radius A of the hydrogen atom from Zumdahl [9] one obtains an upper limit for λ

$G = 6.67390 \times 10^{-8} \text{ cm}^3/\text{gs}^2$

$A \leq 0.37 \times 10^{-8} \text{ cm}$

$e = 4.803 \times 10^{-11} \text{ statC} = 4.803 \times 10^{-11} \text{g}^{1/2}\text{cm}^{3/2}\text{s}^{-1}$

$m_e = 1.6726 \times 10^{-24} \text{ g}$

$$\lambda^4 \leq \frac{2}{5\pi} \left(4.803 \times 10^{-11} g^{\frac{1}{2}} cm^{\frac{3}{2}} s^{-1}\right)^2$$

$$x \frac{16\pi^4 (0.37 \times 10^{-8} \ cm)^4}{6.67390 \times 10^{-8} \ cm^3/gs^2 (1.6726 \times 10^{-24} g)^2}$$

$$\lambda \leq 1.46 \ cm = 14.6 \ mm \quad (9-24)$$

Note that λ is in the microwave range. The less than relation comes from the assumption that the electron could not stay bound to the atom if it oscillated away from the nucleus beyond the size of the atom.

One of the most significant sources of radiation in the universe is known as the 2.735 ^{0}K cosmic background radiation as shown in Figure 9-3 as measured by NASA's COBE satellite.

Note that the peak in the radiation is at 1 mm wavelength which is much less than 14.6 mm. Our calculation shows that the derived force of gravity can be made to simultaneously predict the measured experimental strength of the force of gravity and the observed cosmic background radiation by making the current amplitude of vibration of the electron much smaller than the radius of the atom. This is a very reasonable value and what one might expect. Note that the cosmic background radiation is non-isotropic and shows variations reflecting the matter distribution in space as shown in additional COBE satellite data in Figure 9-4.

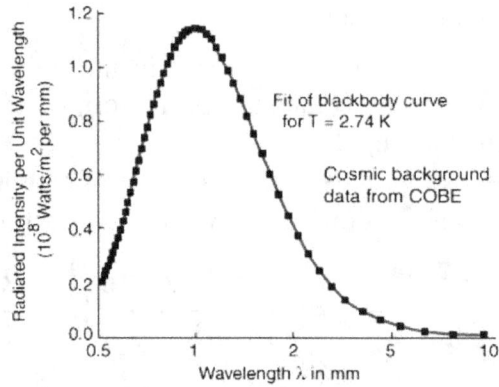

Figure 9-3 Cosmic Background Radiation from NASA's COBE1 Satellite

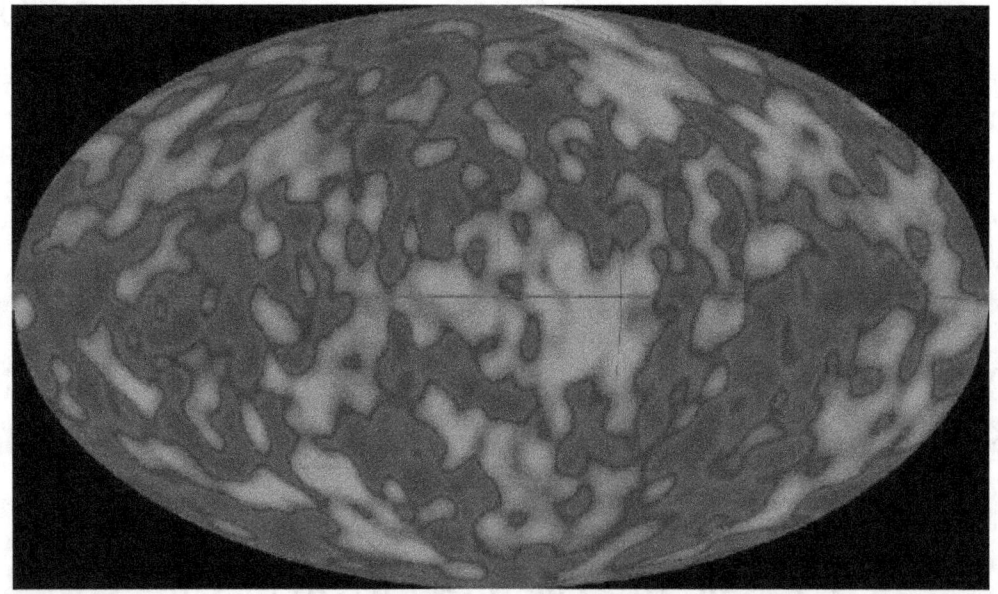

Figure 9-4 Spatial Distribution of Cosmic Background Radiation COBE2 NASA Skymap [10]

In summary the derived radial term of the electrodynamic force of gravity is not only able to predict the observed magnitude and radial direction of the force of gravity, but it also explains the origin of the cosmic background radiation at the same time. Thus it has an advantage over previous theories of gravity in that it explains more observed data. Note that the customary blackbody wavelength distribution as shown in Figure 9-3 can be shown to be completely classical in origin for finite-size electrons in the shape of a toroid without any need of an auxiliary theory such as quantum mechanics. [11] See **Appendix C Derivation of Blackbody Radiation Formula.**

9.5   Decay of the Force of Gravity

Another consequence of this electrodynamic theory of gravitation is that the force of gravity is decreasing over time. The emission of the radiation above causes a decay of the force of gravity due to a decrease in the value of the mass. The rate of decay depends on an atom's position in an astronomical body and the size of the astronomical body. Since the oscillating electrons in all atoms can both absorb and emit radiation, those atoms nearest the center of an astronomical body lose their oscillation energy the slowest while those atoms nearest the surface of the astronomical body lose their energy the fastest. The rate of decay of the gravitational force of an astronomical body will depend on the ratio of the volume of the body to its surface area. Thus, the larger the radius of an astronomical body, the slower its force of gravity decays. So the force of gravity within a planet would decay faster than the force of gravity within the sun.

Applying these notions to the universe as a whole, the rate of weakening of gravity depends on a body's position in the universe. Since the oscillating electrons in all atoms can both absorb and emit radiation, those atoms in large astronomical bodies nearest the center of the universe lose their oscillation energy the slowest while those atoms in astronomical bodies nearest the edge of the universe lose their energy the fastest. Similarly those atoms near the center of a galaxy lose their oscillation energy slower than those atoms near the edge of the galaxy. Thus the rate of decay of the gravitational force depends on position in the universe.

Is there any evidence that the force of gravity has decayed? Yes, the expansion of the earth and the resulting separation of the continents have been documented. Figures 9-5 and 9-6 show the three dimensional stretch marks under the oceans and through the continents that details the approximately 70% radial expansion of the earth since its surface solidified. The weakening of the force of gravity is the only reasonable explanation for the expansion of the Earth. Most cosmological models, such as the Big Bang model, have the Earth contracting over time as it cools, with gravity being constant, and cannot explain this data.

According to Hook's law of elasticity in three dimensions the elastic material of the crust of the earth expands very slowly due to the change in the strength of gravity, but it eventually reaches its elastic limits and starts to crack and come apart. The giant three-dimensional stretch marks below show that the origin of the bursting of a seam in the surface of the earth started at the position of the present Dead Sea in Palestine then proceeded down the Red Sea into the Indian Ocean where it forked to form the Pacific and Atlantic Oceans. Although the splitting up of the surface of the earth formed large pieces called plates, the continual movement of the plates apart from one another (See Figures 9-8 and 9-9) can only be explained by an expanding earth. Only an expanding earth model can conserve energy and angular momentum for the movement of the continental plates.

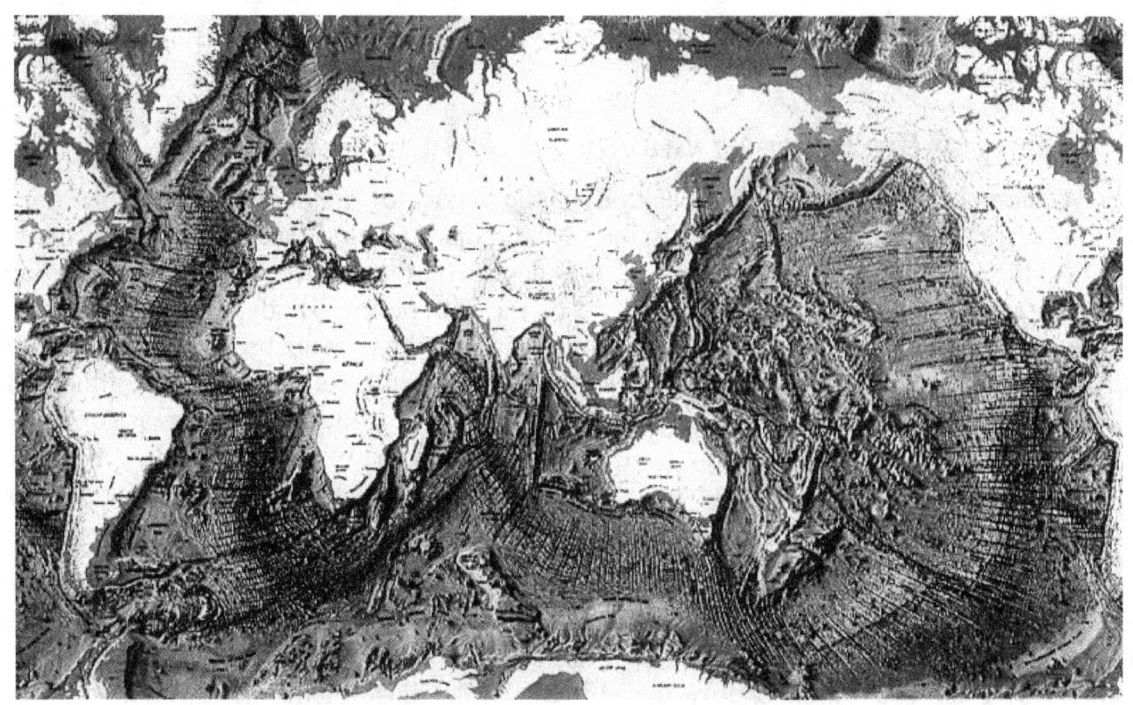

Figure 9-5 Stretch Marks of Earth's Expansion

The maps by Marie Tharp (Figures 9-5 and 9-6) were confirmed by the World Ocean Floor (1977) map of the U.S. Navy Office of Naval Research shown in Figure 9-7. The motion of the continental plates away from each other in Figures 9-8 also confirms the expansion of the earth of about 25 cm per year currently.

Figure 9-6 Close-up of Earth's Expansion Stretch Marks

Note that politically correct mainstream science in the United States does not allow this data showing the expansion of the earth to be presented in any private or public school textbooks. This is due to the absence of any politically correct theories than can explain the expansion of the earth and conserve energy and momentum. The mechanism by which this data is kept out of private and public school textbooks is the Federal and State education frameworks which must be adhered to by any schools receiving Federal aid or accreditation.

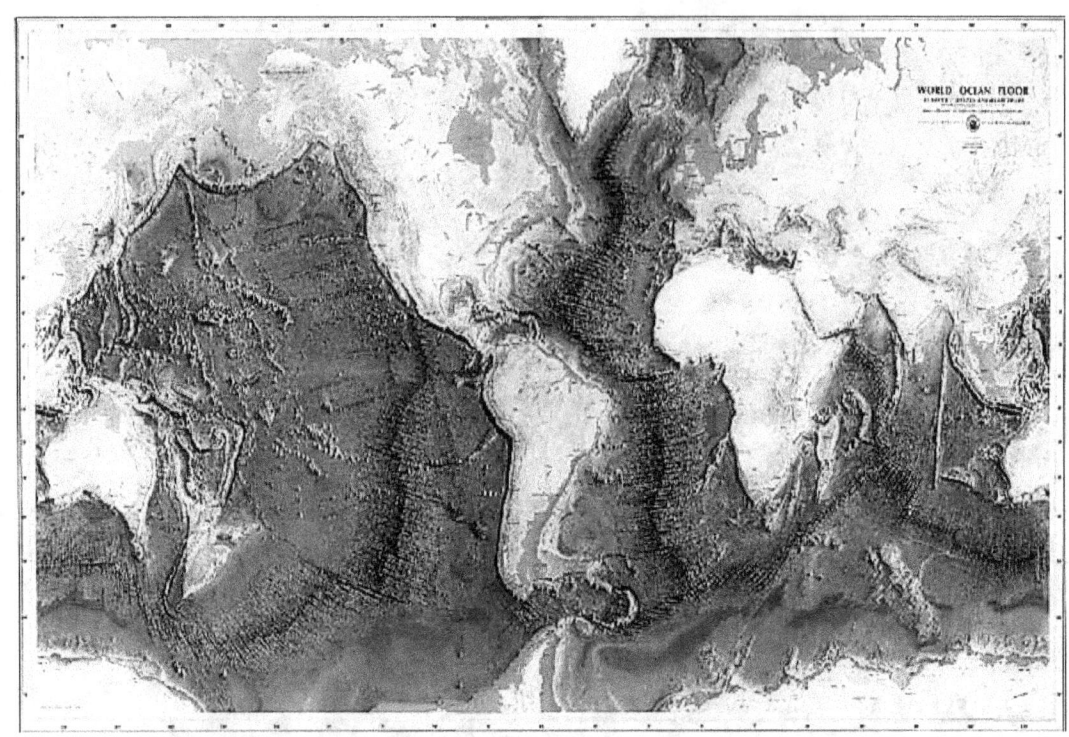

Figure 9-7 US Office of Naval Research World Ocean Floor Map 1977 [12]

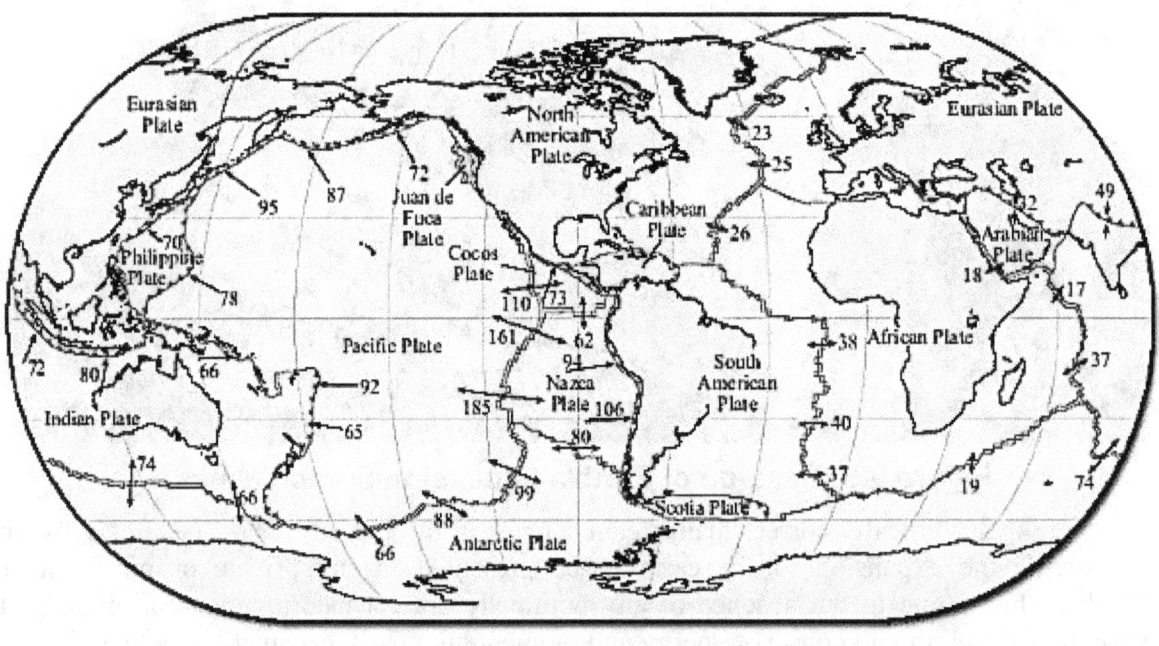

key   // constructive plate boundary   ⤢ destructive plate boundary   / transform fault plate boundary

Figure 9-8 Movement of Tectonic Plates [14]

Arrows Giving Direction in mm/yr Support Earth Expansion

The expansion of the earth caused the north pole of the earth to rotate with respect to the surface of the earth due to conservation of energy and angular momentum. This caused the newest stripes of matter being added to the ocean bottom along the mid-ocean ridges to be magnetized with varying degrees of magnetization and orientation. Scientists have measured the magnetization of the ocean bottom by measuring the magnetic field strength at a certain depth in the ocean using a cable dragged magnetometer and subtracting out the theoretically expected strength of the magnetic field of the earth as shown in Figure 9-9.

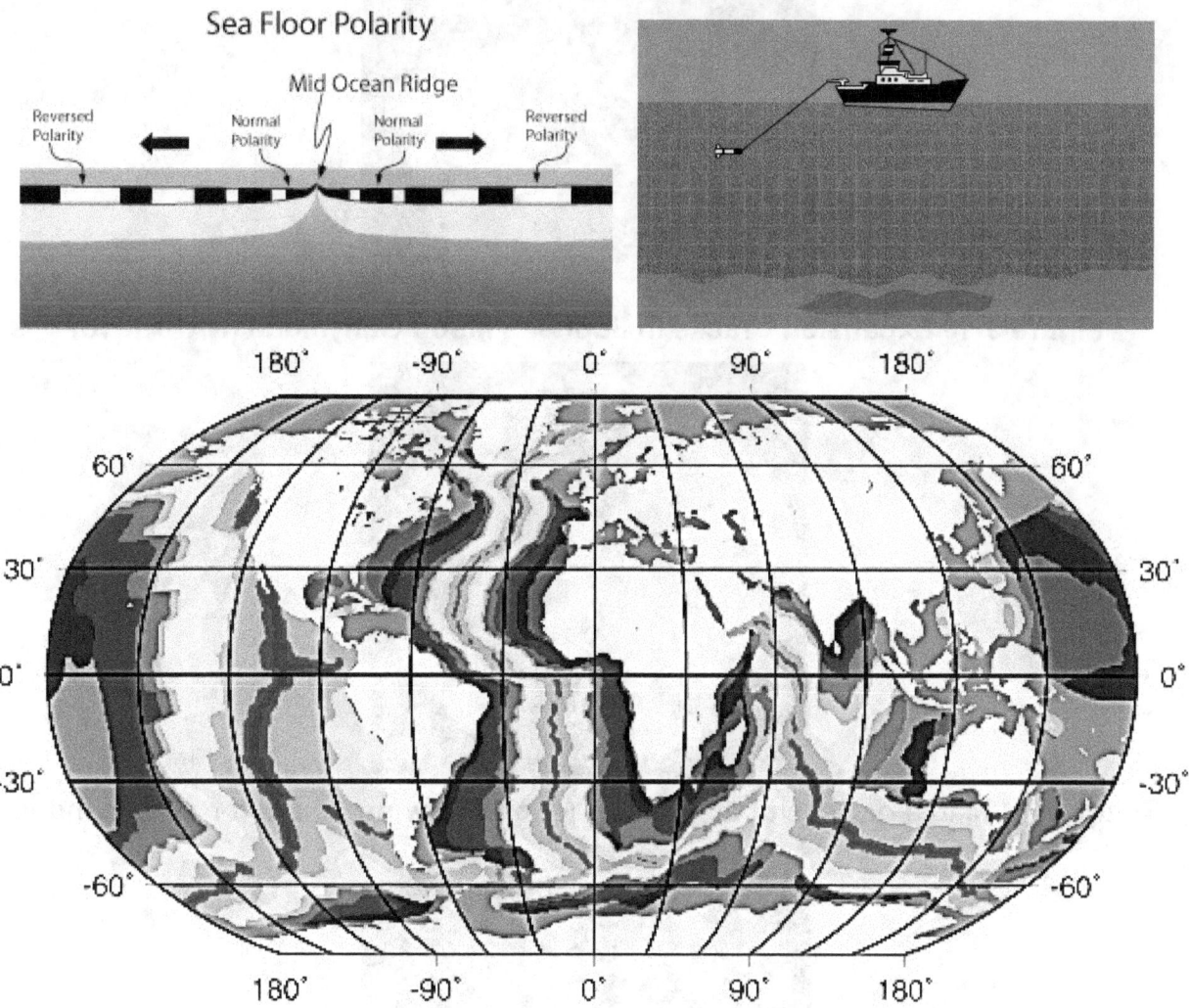

Figure 9-9 Parallel Magnetic Ocean Floor Stripes

On Both Sides of Mid Ocean Ridges [15]

This reveals that there are stripes of similar magnetization that are parallel to the mid ocean ridges indicating that a three-dimensional expansion has occurred. A closer examination of the sea floor polarity in the central upper part of Figure 9-9 reveals that rate of expansion of the ocean bottom was much greater in the past than it is now. This supports the notion of a very strong initial decay rate and a very weak decay rate at the present time.

The expansion of the earth should not be unique in our solar system. The pictures of the surface of Jupiter's moon Ganymede in Figure 9-10 shows clearly the expansion cracks without the presence of oceans. Figure 9-11 shows the mares or seas of the Earth's moon showing where it expanded. Figure 9-12 shows where the expansion is presently occurring on the planet Venus. Figure 9-13 shows an expansion crack in the surface of the planet Mars thousands of miles long as well as the division of continents.

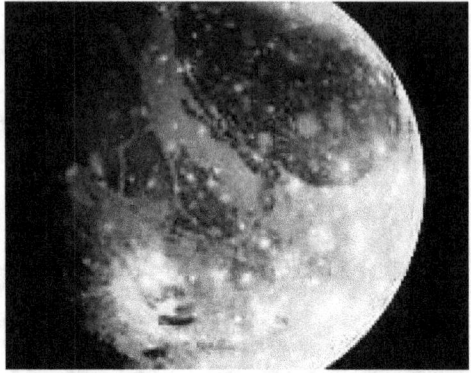

Figure 9-10 Expansion Cracks in Jupiter's Moon Ganymede (NASA) [16]

Figure 9-11 Mares or Seas of the Earth's Moon Showing Where It Has Expanded [17]

Figure 9-12 Radar Image Showing Expansion of Venus [18]

Figure 9-13 Expansion Crack in Surface of Planet Mars

Thus the electrodynamic derived force of gravity appears to be the only theory of gravity that describes an expanding earth, moon, planets, and stars as observed, due to the rapid decay of the strength of gravity producing the cosmic background radiation. The decay of gravity combined with conservation of energy explains the high velocity of stars in the outer arms of spiral galaxies. See **Chapter 13** for a more complete explanation.

9.6 Computation of Non-Radial Gravitational Force Term

In a manner similar to that for the radial term, the non-radial term of the gravitational force may be calculated from equation (9-6) to be

$$\vec{F}_{2+,1+} = \frac{e^2}{|\vec{r}_2 - \vec{r}_1|^2}(\hat{r}\cdot\vec{\beta})\hat{r}\times(\hat{r}\times\vec{\beta})[1] \qquad\qquad (9-25)$$

$$\vec{F}_{2+,1-} = \frac{-e^2(\hat{r}\cdot\vec{\beta})\hat{r}\times(\hat{r}\times\vec{\beta})}{|\vec{r}_2 - \vec{r}_1|^2}\left[1 - \left\{\frac{A_1\omega_1}{c}sin(\omega_1 t_1 + \varphi_1)\right\}^2\left(\frac{1+cos^2\theta}{2}\right)\right.$$
$$\left. + \left\{\frac{A_1\omega_1}{c}sin(\omega_1 t_1 + \varphi_1)\right\}^4\left(-\frac{1}{8} - \frac{1}{4}cos^2\theta + \frac{3}{8}cos^4\theta\right)\right]$$

$$\vec{F}_{2-,1+} = \frac{-e^2(\hat{r}\cdot\vec{\beta})\hat{r}\times(\hat{r}\times\vec{\beta})}{|\vec{r}_2 - \vec{r}_1|^2}\left[1 - \left\{\frac{A_2\omega_2}{c}sin(\omega_2 t_2 + \varphi_2)\right\}^2\left(\frac{1+cos^2\theta}{2}\right)\right.$$
$$\left. + \left\{\frac{A_2\omega_2}{c}sin(\omega_2 t_2 + \varphi_2)\right\}^4\left(-\frac{1}{8} - \frac{1}{4}cos^2\theta + \frac{3}{8}cos^4\theta\right)\right]$$

$$\vec{F}_{2-,1-} = \frac{e^2(\hat{r}\cdot\vec{\beta})\hat{r}\times(\hat{r}\times\vec{\beta})}{|\vec{r}_2-\vec{r}_1|^2}\Bigg[1$$

$$-\left\{\frac{A_1\omega_1}{c}sin(\omega_1 t_1+\varphi_1)+\frac{A_2\omega_2}{c}sin(\omega_2 t_2+\varphi_2)\right\}^2\left(\frac{1+cos^2\theta}{2}\right)$$

$$+\left\{\frac{A_1\omega_1}{c}sin(\omega_1 t_1+\varphi_1)+\frac{A_2\omega_2}{c}sin(\omega_2 t_2+\varphi_2)\right\}^4\left(-\frac{1}{8}\right.$$

$$\left.-\frac{1}{4}cos^2\theta+\frac{3}{8}cos^4\theta\right)\Bigg]$$

From the symmetry of the integrals the average force of equation (9-25) for the non-radial terms may be reduced to

$$\vec{F}(\vec{r},\vec{v}) = \frac{1}{\tau_1}\int_0^{\tau_1} dt_1 \frac{1}{\tau_2}\int_0^{\tau_2} dt_2 \frac{1}{2\pi}\int_0^{2\pi} d\varphi_1 \frac{1}{2\pi}\int_0^{2\pi} d\varphi_2 \frac{1}{\pi}\int_0^{\pi} sin\theta d\theta$$

$$\vec{F}(r,\theta,\varphi,A_1,\omega_1,\varphi_1,t_1,A_2,\omega_2,\varphi_2,t_2,v)$$

$$= \frac{\omega_1}{2\pi}\int_0^{\frac{2\pi}{\omega_1}} dt_1 \frac{\omega_2}{2\pi}\int_0^{\frac{2\pi}{\omega_2}} dt_2 \frac{1}{2\pi}\int_0^{2\pi} d\varphi_1 \frac{1}{2\pi}\int_0^{2\pi} d\varphi_2 \frac{1}{\pi}\int_0^{\pi} sin\theta d\theta$$

$$\left[\frac{e^2(\hat{r}\cdot\vec{\beta})\hat{r}\times(\hat{r}\times\vec{\beta})}{|\vec{r}_2-\vec{r}_1|^2}6\frac{A_1^2\omega_1^2}{c^2}sin^2(\omega_1 t_1+\varphi_1)\frac{A_2^2\omega_2^2}{c^2}sin^2(\omega_2 t_2\right.$$

$$\left.+\varphi_2)\left(-\frac{1}{8}-\frac{1}{4}cos^2\theta+\frac{3}{8}cos^4\theta\right)\right]$$

$$= \frac{\omega_1}{2\pi}\int_0^{\frac{2\pi}{\omega_1}} dt_1 \frac{\omega_2}{2\pi}\int_0^{\frac{2\pi}{\omega_2}} dt_2 \frac{1}{2\pi}\int_0^{2\pi} d\varphi_1 \frac{1}{2\pi}\int_0^{2\pi} d\varphi_2 \frac{e^2(\hat{r}\cdot\vec{\beta})\hat{r}\times(\hat{r}\times\vec{\beta})}{|\vec{r}_2-\vec{r}_1|^2}$$

$$6\frac{A_1^2\omega_1^2}{c^2}sin^2(\omega_1 t_1+\varphi_1)\frac{A_2^2\omega_2^2}{c^2}sin^2(\omega_2 t_2+\varphi_2)\left(\frac{-3}{2\pi}\right)$$

$$= \frac{\omega_1}{2\pi}\int_0^{\frac{2\pi}{\omega_1}} dt_1 \frac{\omega_2}{2\pi}\int_0^{\frac{2\pi}{\omega_2}} dt_2 \frac{e^2(\hat{r}\cdot\vec{\beta})\hat{r}\times(\hat{r}\times\vec{\beta})}{|\vec{r}_2-\vec{r}_1|^2}6\frac{A_1^2\omega_1^2}{c^2}\left(\frac{1}{2}\right)\frac{A_2^2\omega_2^2}{c^2}\left(\frac{1}{2}\right)\left(\frac{-3}{2\pi}\right)$$

$$= -\frac{e^2(\hat{r}\cdot\vec{\beta})\hat{r}\times(\hat{r}\times\vec{\beta})}{|\vec{r}_2-\vec{r}_1|^2}\frac{A_1{}^2\omega_1{}^2}{c^2}\frac{A_2{}^2\omega_2{}^2}{c^2}\left(\frac{9}{4\pi}\right) \qquad (9-26)$$

$$where \quad \frac{1}{\pi}\int_0^\pi sin\theta d\theta\left(-\frac{3}{8}-\frac{9}{4}cos^2\theta+\frac{15}{8}cos^4\theta\right)$$

$$= -\frac{1}{\pi}\int_{-1}^1 dcos\theta\left(-\frac{3}{8}-\frac{9}{4}cos^2\theta+\frac{15}{8}cos^4\theta\right)$$

$$= -\frac{1}{\pi}\left(-\frac{3cos\theta}{8}-\frac{9}{4}\frac{cos^3\theta}{3}+\frac{15}{8}\frac{cos^5\theta}{5}\right)\Big|_{-1}^1$$

$$= -\frac{1}{\pi}\left(-\frac{3}{8}2-\frac{9}{4}\frac{2}{3}+\frac{15}{8}\frac{2}{5}\right) = -\frac{3}{2\pi} \qquad (9-27)$$

From equations (9-16) and (9-26) the full gravitational force $\mathbf{F}_G$ may be rewritten using the definition of mass from the first term to show this second term in more familiar notation as

$$\vec{F}_G(\vec{r}) = -G\frac{m_{g1}m_{g2}}{|\vec{r}_2-\vec{r}_1|^2}\left[\hat{r}_{21}-\frac{45}{8}(\hat{r}_{21}\cdot\vec{\beta})\hat{r}_{21}\times(\hat{r}_{21}\times\vec{\beta})\right] \qquad (9-28)$$

The first term is Newton's universal gravitational force. The second term is a new term that gives rise to a corkscrew type of spiraling motion. The strength of the second term is much less than the first due to the β^2 factor. The first term causes planets to orbit the sun with an elliptical orbit in the equatorial plane of the sun. The second term modifies the orbit to lie on a spiral that is centered on an ellipse in the equatorial plane of the sun. Is this notion supported by observations?

9.7 Corroborating Evidence for Spiraling Orbits

Astronomers have found that the elliptical orbits of the various planets are tilted with respect to the equatorial plane of the sun as shown in Figure 9-14. [19] As a planet goes around the sun on a spiral, the effective orbit appears to be an elliptical orbit tilted with respect to the equatorial plane of the sun.

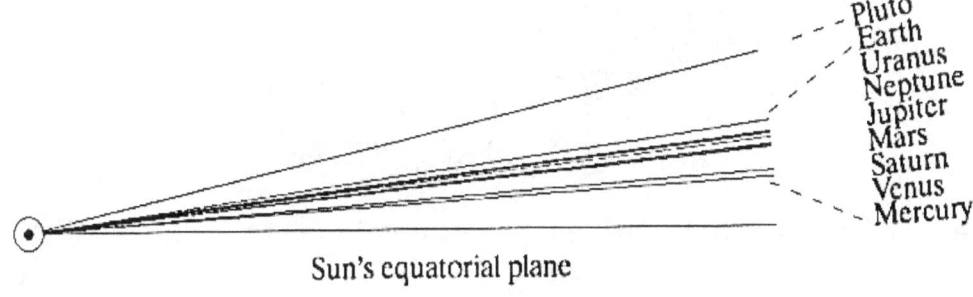

Figure 9-14 Tilt of Planetary Orbits with Respect to Equatorial Plane of Sun [19]

Figure 9-15 shows the motion of four of Jupiter's moons about the orbit of Jupiter about the sun. Note the spiral or corkscrew orbits of the moons about Jupiter's orbit. Also note the relative periods of the spirals are integer multiples of one another, i.e. Io = 2, Europa = 4, Ganymede = 8, Callisto=16. The quantization of the orbits of the moons of Jupiter is a necessary condition for the stability of the system. Jupiter and its moons must have spatial coherence for stability which causes the moons to return periodically to the same relative positions on the spiral of their orbits.

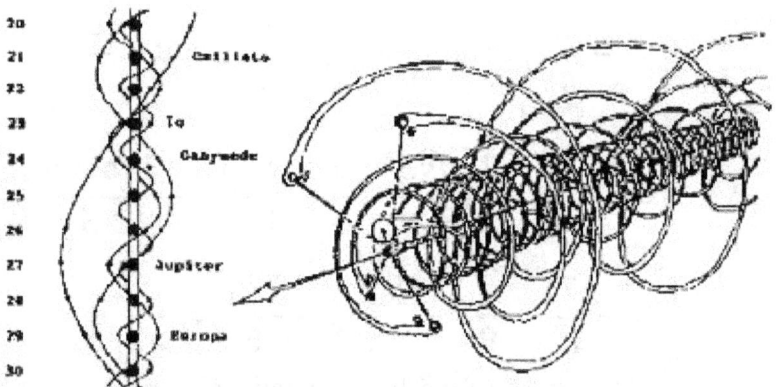

Figure 9-15 Spiral Orbits of Jupiter's Moons

Io, Callisto, Europa and Ganymede [20]

This electrodynamic theory of gravity appears to be the only theory of gravity able to explain the tilting of the orbits of the planets with respect to the equatorial plane of the sun and the quantization of these orbits. In contrast Newton's Universal Law of Gravitation and Einstein's General Theory of Relativity predict that the orbits of all the planets of the sun should lie in the equatorial plane of the sun like the rings of Saturn and that there is no quantization of orbits due to gravity.

9.8 Origin of Hubble's Law Due to Gravitational Redshifts

Edwin Hubble discovered that the light from distant stars is shifted in color toward the red part of the spectrum as shown in Figure 9-16. The farther away the star the greater the red shift.

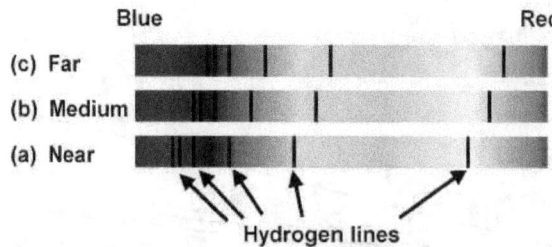

Figure 9-16 Red Shift of Hydrogen Absorption Lines for Near, Medium, Far Distance Stars [21]

The decrease in the force of gravity over time has a significant effect on the light that we see from distant stars. From conservation of energy light emitted from a stellar surface on a star of mass M and radius R is expected to have a red shift equal to the difference in gravitational potential. Using G for Newton's universal gravitation constant this potential at the stellar surface is -GM/R and zero at infinity, so the red shift z may be defined as

$$z = \frac{\Delta\lambda}{\lambda} = \frac{GM}{c^2 R} \quad (9-29)$$

This equation for the gravitational red shift was confirmed experimentally by Pound & Rebka in 1960. [21]

If the force of gravity is decreasing, then GM/R would have been greater in the past when gravity was stronger. Thus, in general, the gravitational red shift of light from stars should be larger the farther away the star is independent of the star's velocity or type as shown in Figure 9-17. The star's velocity can add to or reduce the red shift due to the Doppler Effect. Also stars or galaxies that are larger decay more slowly giving rise to a larger red shift at the same distance. Stars that are newer, such as quasars which are found near the center of active young galaxies, should have a significantly higher redshift than older galaxies around it, even if they are bound to the older galaxy.

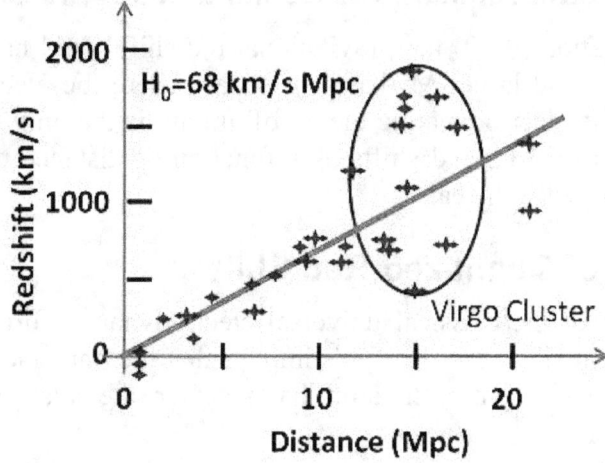

Figure 9-17 Hubble's Law Red Shifts vs. Distance or Brightness [23]

The data that Hubble used to formulate his famous law that red shifts are roughly proportional to distance is shown in Figure 9-17. Note that the Doppler red shift due to velocity and size effects cause deviations from a perfect straight line which are small in comparison with the main effect of the gravitational red shift from earlier times.

Halton Arp [22] discovered quasars that, according to gamma ray spectroscopy, are physically connected to galaxies, yet their respective cosmological redshifts are dramatically larger. For instance galaxy NGC 4319 and quasar Mark 205 are physically connected according to gamma spectroscopy. Mark 205 has a redshift of z = 0.07 while the associated galaxy NCG 4316 has a redshift of only z = 0.0056. (See Figure 9-18)

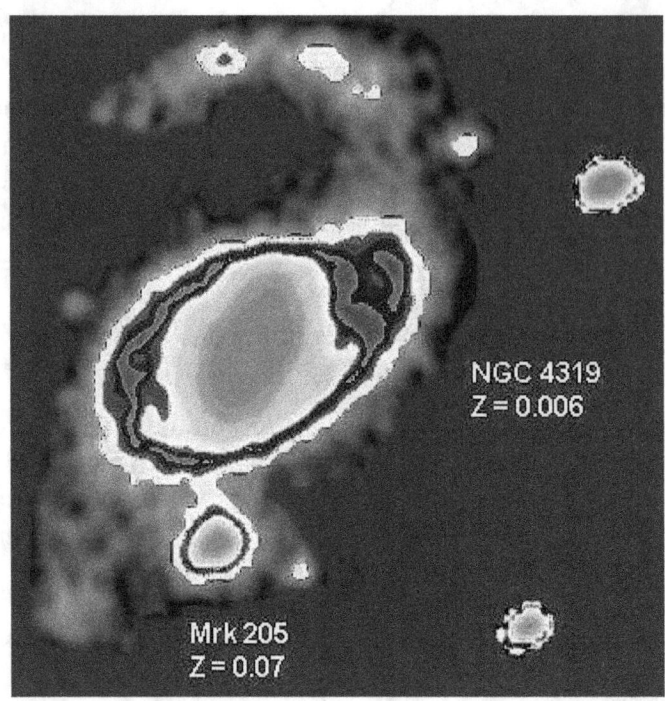

Figure 9-18 Quasar Mark 205 Bound to NGC4319 Galaxy [22]

According to equation (9-29) the gravitational red shift could have been much larger in the past due to smaller R and larger M than it is today where the electrons vibrate with very small amplitudes. Thus the electrodynamic theory of gravity is the only theory of gravity that is able to describe Hubble's Law for red shifts as a function of distance or equivalently time for light emitted in the past to reach the earth.

9.9   Significance of Quantized Red Shifts

One consequence of the classical universal electrodynamic force law is that all forces have a $1/R^2$ dependence on all size scales. This implies that the universe must have a center just as elementary particles have a center, the atom has a center, the solar system has a center, and galaxies have a center.

In the case of our solar system, matter in the form of planets only exists at particular quantized radii as represented by the modern version of Bode's Law by Stanley Dermott [24]. Also the matter of the moons about the planets such as Uranus only exists at particular quantized radii as represented by Bode's Law. Thus one might suspect that on a very large scale in the universe matter might also exist at particular quantized radii as represented by Bode's Law. (See Figures 9-19, 9-20, 9-21)

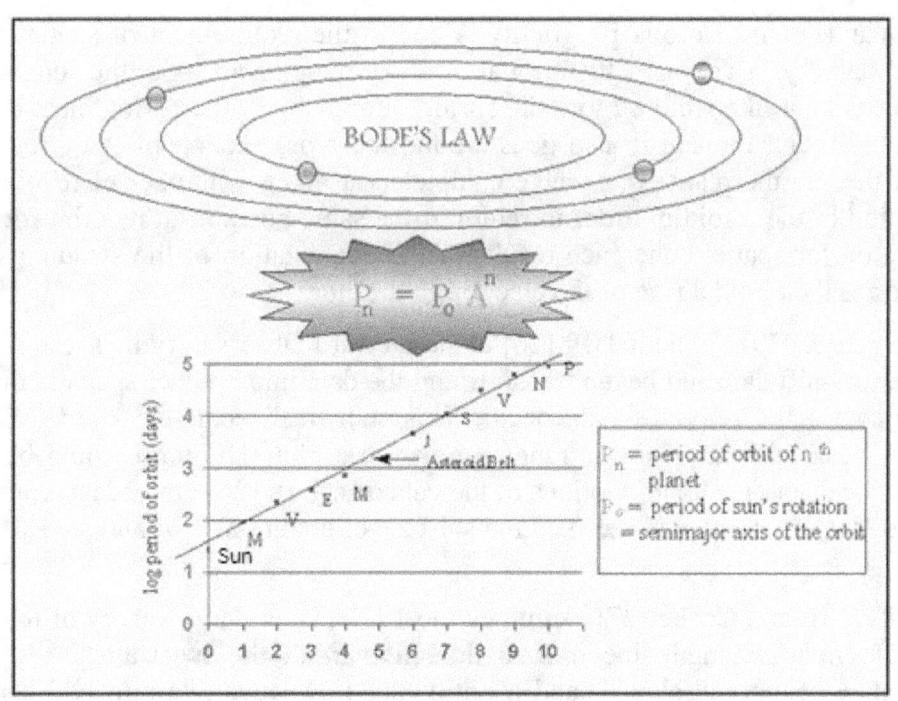

Figure 9-19 Planetary Data Supporting Modern Bode's Law [24]

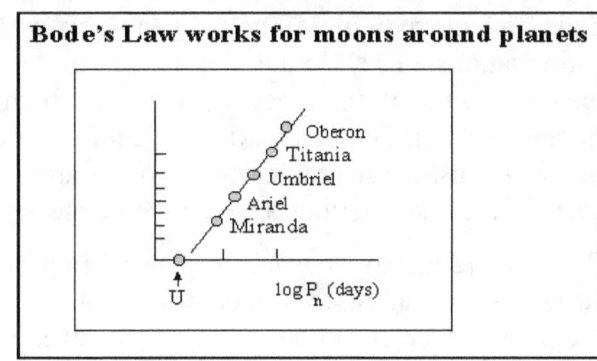

Figure 9-20 Uranus Moon Data Supporting Bode's Law [24]

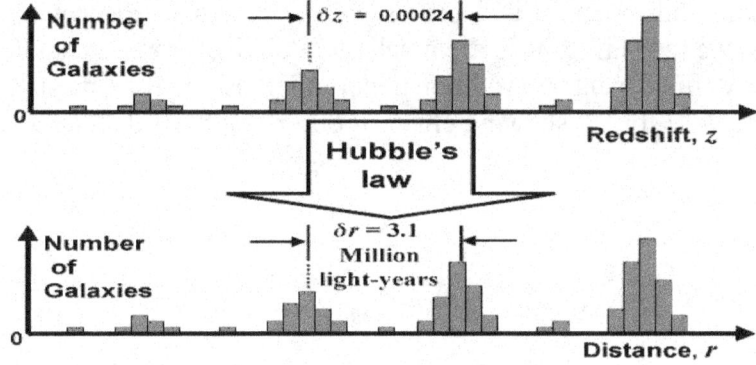

Figure 9-21 Tifft's Quantized Red Shifts Support Bode's Law on Universal Scale (Idealized format without background) [25]

The quantization of orbits by gravity is due to the requirement of spatial coherence in order to have stability. Basically, the toroidal orbits of the planets about the sun and the moons about the planets look like a closed toroidal spring. For stability the spring must be closed. As the planet goes around the sun it also goes around the cross section of the toroid. After one revolution of the sun the planet must have completed an integer number of revolutions around the cross section of the toroid in order to return to the same position in its orbit for stability. A similar condition for spatial coherence results in the quantization of the standing waves in the charge that forms the toroidal ring of the electron in the atom.

In the early 1970s William Tifft [26] at the Steward Observatory in Tucson, Arizona was analyzing the red shift data and began transforming the data into "power spectra" that show how various spacings in the red shift data occur. This statistical technique shows difficult-to-see regularities as peaks rising above the random noise in a plot. The noise could be due to such things as the "local" or "peculiar" motions of the galaxies. Tifft [26] noticed a surprisingly strong peak corresponding to an interval between red shift z's of about 0.00024 and a weak peak at ½ of 0.00024.

In 1984 Tifft and Cocke [27] examined the 1981 Fisher-Tully survey of red shifts in the radiowave (21 cm wavelength line from hydrogen) part of the spectrum. They found sharp periodicities at exact submultiples 1/3 and 1/2 of 0.00024. However, despite Tifft's steady stream of publications, astronomers remained skeptical about the notion of quantized red shifts.

Then in 1997, an independent study of 250 galaxy red shifts by Napier and Guthrie [28] confirmed Tifft's basic observations. They found the red shift distribution to be strongly quantized in the galactocentric frame of reference with a very high confidence level. The galactocentric frame of reference is the frame at rest with respect to the center of our own galaxy, the Milky Way. When they compensated for the earth's motion around the sun and the sun's motion around the galaxy center, the quantizations appeared more clearly.

In 1996 and 1997 Tifft [29, 30] showed that it is important to compensate the galactocentric red shifts further by accounting for our galaxy's motion with respect to the cosmic microwave background radiation. Doppler shifts of the microwaves show that our galaxy is moving about 560 km/s in a direction south of the constellation Hydra [31]. Accounting for this motion converts the galactocentric red shifts to a frame of reference which is at rest with respect to the cosmic background radiation and presumably at rest with respect to the universe as a whole. In this frame the red shift groups are much more distinct from one another suggesting that the universe has a defined center. Additional periodicities of 1/4 and 1/8 of 0.00024 were observed. See Figure 9-22 for the skewing effect of observing the red shifts away from the center of the universe.

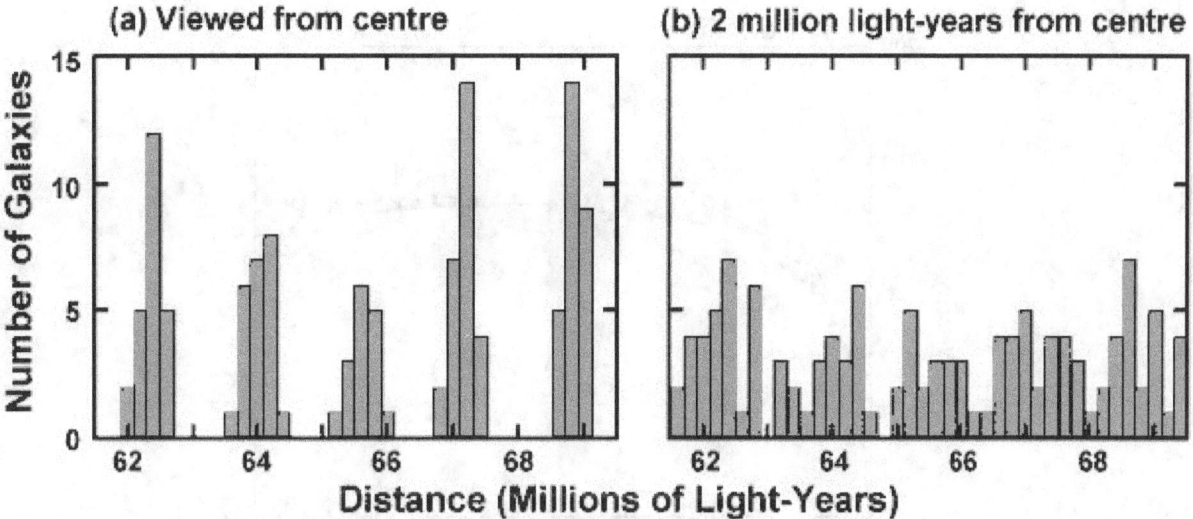

Figure 9-22 Effect of Observing Red Shifts Away From the Center of Universe [25]

In 1992 Tifft [32] in an anonymous paper claimed that galactic red shifts have actually decayed slightly in just a few years. This is consistent with red shifts being primarily intrinsic gravitational red shifts and the force of gravity declining rapidly far away near the edge of the universe. The electrodynamic theory of gravity is the only theory of gravity that predicts the general decay of all red shifts in the universe.

9.10 MOND vs. Dark Matter

The observed rotational speeds of objects in extragalactic systems exceed what can be explained by the visible mass of stars and gas. One approach to explain this discrepancy is to infer that there is more mass than meets the eye, i.e. dark matter and dark energy exist.[33] Another approach that appears to be less drastic is to assume that there is a MOdified Newtonian Dynamics (MOND) in these regions. Figure 9-23 shows the plot of the rotational velocity V data versus distance R from the center of a typical spiral galaxy NGC 2403 and compares that with the solid line MOND predictions and the dashed line predicted Newtonian values. [34] The dotted curve is Newtonian rotation curve due to the gaseous components (hydrogen plus primordial helium).

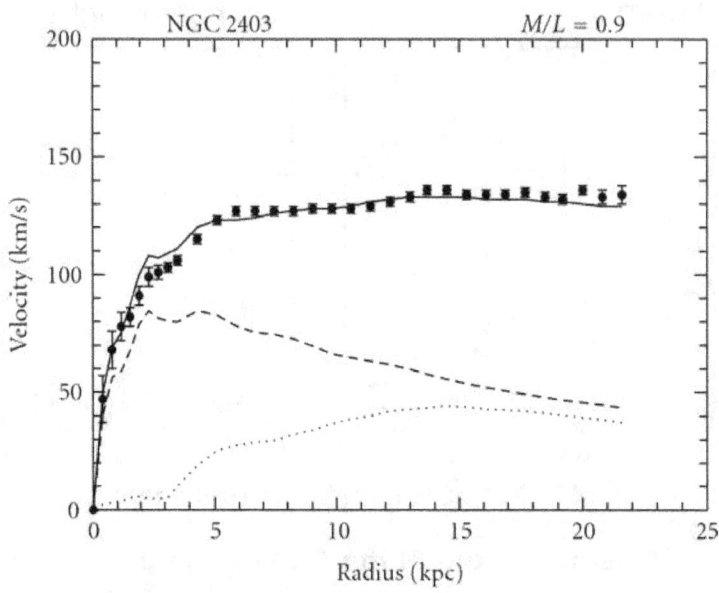

Figure 9-23 NGC 2403 Spiral Galaxy Graph of Rotational Velocity vs. Distance from Center [34] - - - Newtonian Prediction

The astronomical data indicates that the velocity of rotation of the outer spiral arms is significantly higher than would be predicted by Newtonian dynamics which is expected to be valid in this region. Milgrom [35] in 1983 was the first astronomer to suggest that many different types of data could be explained by assuming some sort of Modified Newtonian Dynamics (MOND). He documented that MOND correctly maps the observed mass to the observed dynamics. In his paper Milgrom arranged the typical acceleration of various physical processes in a two dimensional diagram. One parameter was the acceleration of the local process itself, the other parameter was the acceleration induced by the environment of the rest of the universe.

Tulley-Fisher [36] examined many spiral galaxies and found a linear relationship between the orbital speed of a galaxy's outskirt stars and the galaxy's brightness. In general brightness, or luminosity, is proportional to mass M. This implies a mass-velocity relationship of the form $M \propto V^4$. This result agrees with the MOND predictions. (See Figure 9-24 below) However, astronomers are still looking for a satisfactory explanation of why the velocity increases like it does. See **Chapter 13** for the answer.

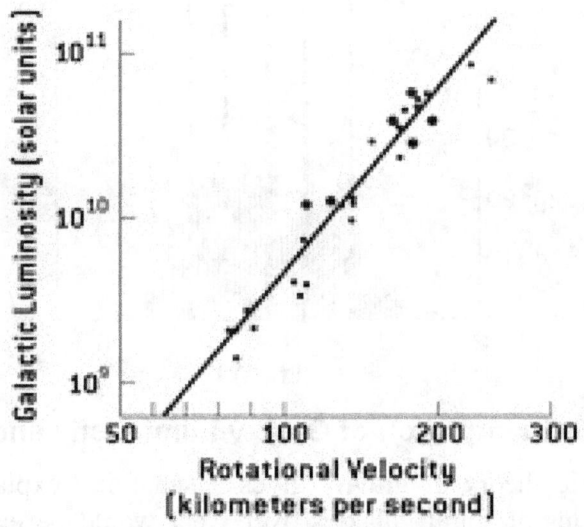

Figure 9-24 Relationship of Rotational Velocity to Galactic Luminosity [36]

In 1999 Roscoe [37] performed an extensive analysis of 900 Tully-Fisher rotation curves for spiral galaxies. He confirmed the Tully-Fisher relationship to a very high level of confidence >95% as shown in Figure 9-25.

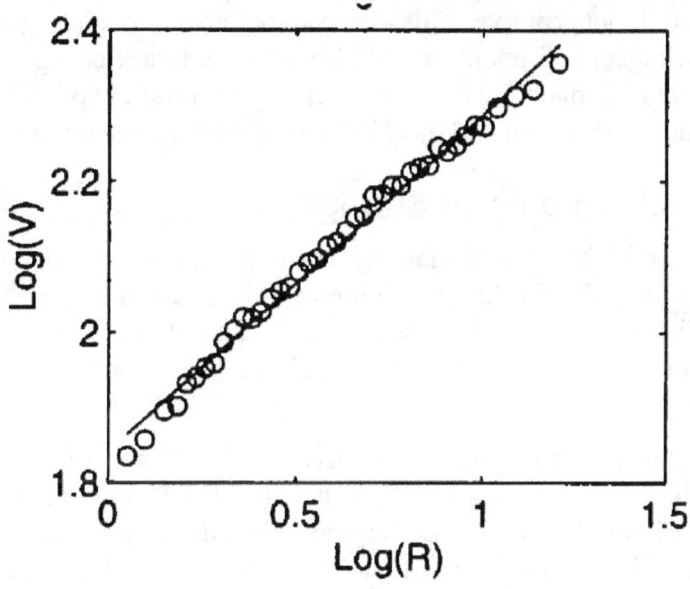

Figure 9-25 Analysis of 900 Optical Rotation Curves of Spiral Galaxies [37]

Furthermore, Roscoe [37] was able to confirm by analysis that the size of spiral galaxies and their luminosities were discrete or quantized as shown in Figure 9-26. Bode's law is reappearing here as it did for the red shifts. Thus there is a consistency.

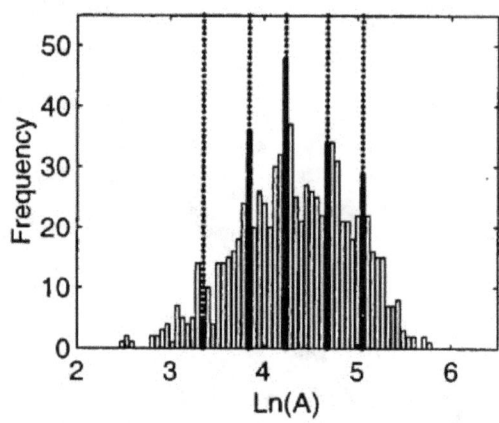

Figure 9-26 Quantization of Galaxy Luminosity and Size [37]

The electrodynamic theory of gravity gives a satisfying explanation of the increased velocities of the outer arms of spiral galaxies over what would be expected from Newtonian dynamics. According to the derived electrodynamic theory of gravity, the mass in the outer stars of the galaxy decays faster than the mass in the center of the galaxy. Conservation of kinetic energy for the outer stars indicates that at some time in the past the mass of the galaxy center was significantly greater, such as when it was originally formed. Thus the outer stars in the spiral galaxy are now in the process of escaping from the galaxy whose mass has decayed to the point that it can no longer hold them captive. This electrodynamic approach to gravity does not require the invention of the illogical and unphysical dark matter and dark energy. Also its second term (see section 9.6) is able to explain the discrete sizes of the spiral galaxies in a fashion consistent with Bode's law for the solar system and the Tifft's measured quantization of red shifts.

9.11 Gravitational Bending of Starlight

Einstein's General Theory of Relativity predicts that the path of light is bent when it passes close to a massive body. Sir Arthur Eddington [43] claimed to verify this prediction when he observed the bending of starlight near the rim of the Sun during a solar eclipse in 1919. Thus the sun, quasars and other astronomical bodies should be able to serve as a kind of gravitational lenses.

As NASA began its space programs to investigate and confirm these crude findings by Eddington using a telescope as a star passed by the rim of the sun, a different picture emerged. The bending of the starlight observed by Eddington was caused by the thin plasma rim of the sun not General Relativity Theory. Furthermore at distances of 2, 3, 4 times the radius of the sun, which was beyond the plasma rim, no bending of starlight is observed. However, General Relativity Theory still predicts a measurable amount of bending at those distances. See Figure 9-27 below which shows what General Relativity Theory predicts. Figure 9-28 shows what is measured at various distances from the sun using the data of Lebach [44] as corrected by Dowdye. [45] Dowdye assumes conservation of energy will determine how an electromagnetic wave will propagate along a minimum path or a least time path in a plasma atmosphere under the influence of the gravitational gradient fields of the sun. Only the electrodynamics of the plasma rim of the Sun is needed to explain the data. There is no role left for General Relativity Theory!

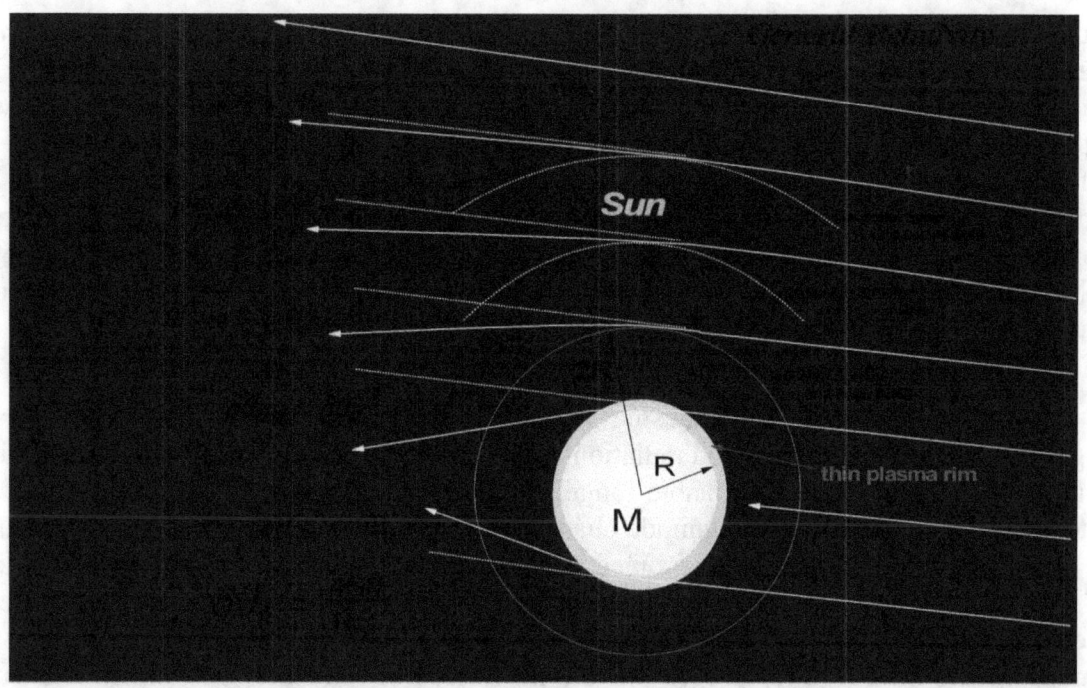

Figure 9-27 General Relativity Predicted Bending of Starlight by the Sun [44, 45]

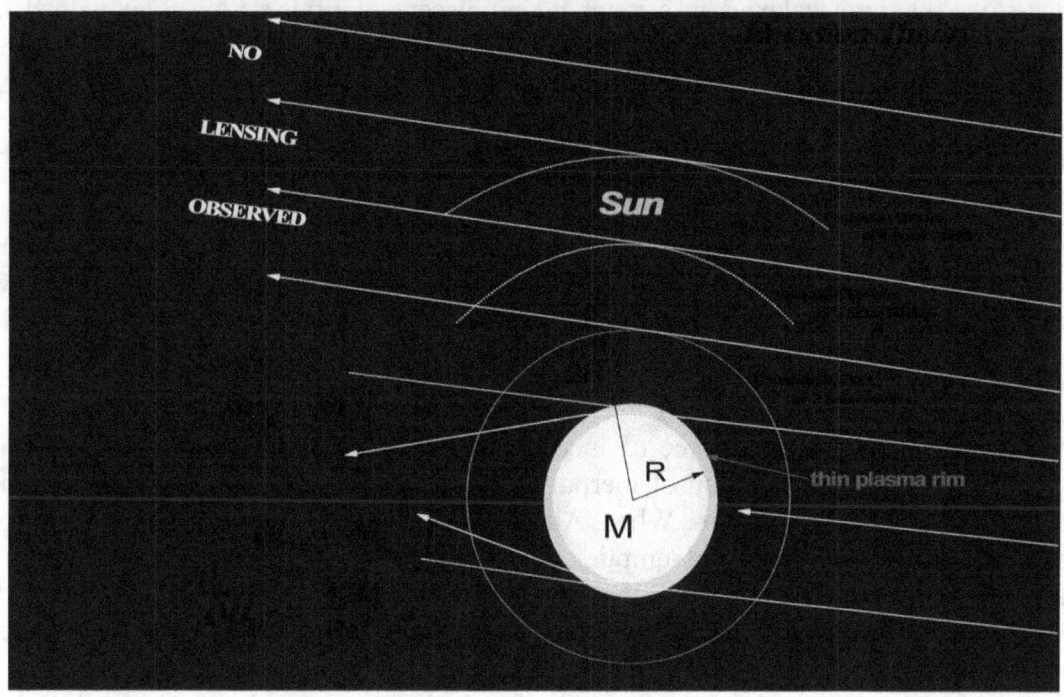

Figure 9-28 Observed Bending of Starlight by the Sun [44, 45]

9.12 Summary

From the derived universal electrodynamic contact force law for finite-size elastic particles of Chapter 4 equation (4-36), the force of gravity is identified as a small average

149

residual effect of the order $(v/c)^4$ due to vibration of atomic electrons with respect to the protons in the nucleus of neutral atoms. This electrodynamic force of gravity can also be derived from the action-at-a-distance covariant relativistic electrodynamic force law based on Maxwell's equations making it independent of version of electrodynamics. The derived gravitational force was found to have the customary radial term of Newton's Universal Law of Gravitation ($F = Gm_1m_2/R^2$) plus a new non-radial term. From the radial term the gravitational mass can be associated with certain electrodynamic parameters of vibrating neutral electric dipoles. The non-radial term gave rise to an $(\mathbf{R \cdot V}) \mathbf{R} \times (\mathbf{R} \times \mathbf{V})$ effect which causes the orbits of the planets about the sun to spiral around on a toroid about the sun giving the appearance of an elliptical orbit tilted with respect to the equatorial plane of the sun.

The vibrational mechanism causing the gravitational force was found to decay over time giving rise to the cosmic background radiation and Hubble's red shifts versus distance law due to gravitational red shifting. A reasonable range of vibrational amplitude for atomic electrons is able to explain many orders of magnitude change in the observed red shifts of light from distant stars. The vibrational mechanism combined with the $(\mathbf{R \cdot V}) \mathbf{R} \times (\mathbf{R} \times \mathbf{V})$ effect also explained Tifft's quantized red shifts as a type of Bode's Law indicating that there is a geometrical center to the universe. The vibrational mechanism by which gravity decays by radiation over time explained Tifft's measured rapid decay of the magnitude of red shifts over time. The Tulley-Fisher relationship for luminosity of spiral galaxies and Roscoe's observed quantization of the luminosity of 900 spiral galaxies in a manner reminiscent of Bode's Law is explained by the $(\mathbf{R \cdot V}) \mathbf{R} \times (\mathbf{R} \times \mathbf{V})$ term of the electrodynamic theory of gravity. The unexpectedly high velocity of the stars in the outer spiral arms of spiral galaxies is explained in Chapter 13 using Mach's Principle without resorting to the use of outlandish ideas such as dark matter and dark energy. The measured quantization of the luminosity and size of spiral galaxies is also explained by the new $(\mathbf{R \cdot V}) \mathbf{R} \times (\mathbf{R} \times \mathbf{V})$ term in the derived gravitational force.

This electrodynamic explanation of gravity appears to indicate that mass is not a fundamental quantity of nature! Thus the notions of mass that were associated with Newton's Universal Law of Gravitation and Einstein's General Relativity Theory appear to be false. This is further supported by the unexpectedly large number of diverse astronomical phenomena explained by this electrodynamic approach to gravity.

In chapter 10 where the force of inertia is derived from the universal electrodynamic force, the electrodynamic definition of inertial mass is found to be similar to the definition of gravitational mass of this chapter. When Albert Einstein developed his General Theory of Relativity, he started with the assumption that the correspondence between inertial and gravitational mass is not accidental and that no experiment will ever detect a difference between them. **In General Relativity Theory the effects of gravitation are ascribed to space-time curvature instead of a force. Thus in General Relativity Theory gravitation is not a force, and not subject to Newton's third law. So from the framework of General Relativity Theory the equality of inertial and gravitational mass is an unexplained mystery.**

9.13 Summary of the Electrodynamic Approach to Gravity

The neutral electric dipole to neutral electric dipole force is the first force term beyond the Coulomb force between individual charges to be explored in the hierarchy of electrodynamic interactions. Despite the fact that scientists in the field of fluid dynamics, plasma physics, and chemistry have been paying attention to electrical dipole effects, it appears that the Post-Modern philosophy of science has prevented this knowledge from being used directly in electrodynamics and the field of gravitation.

Some scientists object to the notion of an electrodynamic theory of gravity, because they claim that electrodynamic interactions can be shielded and gravity cannot. In his book **New Horizons in Electric, Magnetic and Gravitational Field Theory** [48], Hooper finds experimentally that there are three types of electric E fields, and they have different empirical properties. In Chapter 1 Hooper lists 14 different empirical properties of E fields. In Table 1 he gives what these properties are for electrostatic E fields, E fields dependent on dA/dt, and E fields dependent on motion v x B. In these tables Hooper notes in property 6 that motionally caused E fields cannot be shielded experimentally.

In the derivation of the improved version of the electrodynamic force law of this book, each of the types of electric fields are treated separately as distinctly different types of E fields. It is the term proportional to $\mathbf{R \times (R \times V)}$ in the universal force law that gives rise to the force of gravity. This field is equivalent to Hooper's motional E field.

Finally this approach to gravity, which is based on a derived universal electrodynamic force law, is more complete than previous approaches. First, it confirms the modern version of Bode's Law and the quantization of gravitation due to the $\mathbf{(R \cdot V)\, R \times (R \times V)}$ term which requires all physical systems involving motion to be quantized in order to have stability. The motion of the planets spiraling around the sun must return to the same starting point on the spiral or there is no stability or spatial coherence. Second, this approach is simpler, since it is based on a single universal force law. Third, this force is a local contact force based on the electromagnetic fields of a charge extending the range of the force instead of an action-at-a-distance concept like that used in Newton's Universal Force Law and Einstein's General Relativity Theory which employ unphysical point particles. Natural philosophers have known for thousands of years that there is no such thing as an action-at-a-distance force. Some mechanism is needed to transfer forces. Fourth, this approach explains more gravity-relevant data than all previous theories of gravity combined including Bode's law for the quantization of gravity, the tilts of the orbits of the planets about the sun, the expansion of the planets and moons of the solar system, the origin of the cosmic background radiation, Hubble's law for red shifts, the quantization of red shifts, the general decay of all red shifts, the quantized Tulley-Fisher relationship for luminosity and size of spiral galaxies, and the unexpectedly high velocity of the outer arms of spiral galaxies.

9.14 References

1. Isaac Newton, **The Principia, Mathematical Principles of Natural Philosophy: A New Translation**, translators I Bernard Cohen and Anne Whitman (University of California Press, Berkeley, 1999) Rules of Reasoning in Philosophy: : Rule 1.

2. Hubble Space Telescope image Eskimo Nebula NGC 2392
   ttp://www.stsci.edu/~inr/thisweek1/thisweek029.html

3. Hubble Space Telescope image Cat's Eye Nebula NGC 6543
   http://apod.nasa.gov/apod/ap080804.html

4. Lucas, Jr., Charles W., The Universal Electrodynamic Force, **Proceedings of the Natural Philosophy Alliance 18**, College Park, MD July 6-9, 2011.

5. J. D. Jackson, **Classical Electrodynamics-Second Edition** (John Wiley and Sons, Inc., New York, 1975).

6. J. D. Jackson, **Classical Electrodynamics-Third Edition** (John Wiley and Sons, Inc., New York, 1999).

7. Palmer, David. "Hydrogen in the Universe"
   http://imagine.gsfc.nasa.gov/docs/ask_astro/answers/971113i.html

8. **CRC Handbook of Chemistry and Physics 73rd edition** – edited by David R. Lide (CRC Press, Inc., Boca Raton, 1992).

9. Zumdahl, Steven S., **Chemistry – 4th Edition** (Houghton Mifflin Co., New York, 1997), p. 327.

10. COBE NASA Two Year Skymap (COBE2)
    http://www.astrosociety.org/education/publications/tnl/56/CMB.html

11. Lucas, Jr., Charles W., A Physical Model for Atoms and Nuclei Part 4 Blackbody Radiation and the Photoelectric Effect, **Foundations of Science Vol. 6 No. 3**, pp. 1-7 (2003).

12. US Office of Naval Research World Ocean Floor Map
    http://www.columbia.edu/cu/news/06/08/tharp.html

13. Sandwell-Smith NOAA satellite map of ocean floor
    http://gisremote.blogspot.com/2007/07/global-bathymetric-prediction-for-ocean.html

14. Movement of tectonic plates in mm/yr
    http://labspace.open.ac.uk/mod/resource/view.php?id=357255

15. Parallel magnetic ocean floor stripes
    http://eqseis.geosc.psu.edu/~cammon/HTML/Classes/IntroQuakes/Notes/plate_tect01.html

16. Expansion cracks in Jupiter's moon Ganymede http://www.pooai.com/ganymede/

17. Mares or seas of the moon
    http://www.amateurastronomy.org/blogs/index.php?blog=6&m=200712

18. Radar image of planet Venus surface showing expansion
    http://www.solarspace.co.uk/Venus/frmVenusPics.php

19. Spolter, Pari, **Gravitational Force of the Sun** (Orb Publishing Company, Granada Hills, CA, 1993) pp. 167-182.

20. Spiral orbits of Jupiter's moons about the orbit of Jupiter
    http://www.miqel.com/random_images/random-images-2.html

21. Pound, R. V. and Rebka, G. A., **Phys. Rev. Lett. 4**, 337 (1960).

22. Arp, Halton, **Olympia Conference on Fundamentals of Physics** – M. Barone and F. Selleri Editors (Plenum Press, 1994).

23. Edwin Hubble's Law for red shift vs. distance
    http://wapedia.mobi/en/Hubble's_law

24. Modern version of Bode's law
    http://www.astro.cornell.edu/academics/courses/astro2201/bodes_law.htm
    S.F. Dermott, "On the origin of commensurabilities in the solar system - II: the orbital period relation", **Monthly Notices of the Royal Astronomical Society, vol. 141**, pp. 363–376 (1968).
    S. F. Dermott, "On the origin of commensurabilities in the solar system - III: the resonant structure of the solar system", **Monthly Notices of the Royal Astronomical Society, vol. 142**, pp. 143–149 (1969).

25. Humphreys, D. Russell, "Our Galaxy is the Centre of the Universe, 'Quantized' Red Shifts Show", **The Journal, 16(2),** pp. 95-104 (2002).

26. Tifft, W. G., "Discrete States of Redshift and Galaxy Dynamics. I. Internal Motions in Single Galaxies". **Astrophysical Journal. 206**, pp. 38-56 (1976).

27. Tifft, W. G. and Cocke, W. J., "Global Redshift Quantization", **Astrophysical Journal, 287**, pp. 492-502 (1984).

28. Napier, W. M. and Guthrie, B. N. G., "Quantized Redshifts: A Status Report", **Journal of Astrophysics and Astronomy, 18**, pp. 455-463 (1997).

29. Tifft, W. G. "Evidence for Quantized and Variable Redshifts in the Cosmic Background Rest Frame", **Astrophysics and Space Science, 244**, pp. 29-56 (1996).

30. Tifft, W. G., "Redshift Quantization in the Cosmic Background Rest Frame", **Journal of Astrophysics and Astronomy, 18**, pp. 415-433 (1997).

31. Scott et al. in Allen's **Astrophysical Quantities, 4th Edition**, edited by A. N. Cox (Springer-Verlag, New York, 2000) pp. 658, 661. The sun is moving 370.6 ± 0.4 km/s with respect to the cosmic microwave background toward galactic longitude and latitude $(264.^{0}31 \pm 0.^{0}17, 48.^{0}05 \pm 0.^{0}10)$ or a right ascension and declination of about $(11^{h}, 9^{0}S)$. That direction is a little below the constellation Leo, in the lesser known constellation Sextans. From data in the reference one can calculate the following: the sun's velocity with respect to our galaxy's center is 240 km/s toward galactic coordinates $(88^{0}, 2^{0})$ and the velocity of the center of our galaxy with respect to the cosmic background is 556 km/s toward galactic coordinates $(266^{0},29^{0})$. The latter corresponds to right ascension and declination $(10^{h} 30^{m}, 24^{0}S)$ below the constellation Hydra. Note that the speeds above are much larger that the earth's average orbital velocity around the sun of 29.79 km/s.

32. Tifft as Anonymous, "Quantized Redshifts: What's Going on Here?" **Sky and Telescope Vol. 84**, p. 128 (1992).

33. McGaugh, S. S, "Seeing Through Dark Matter", **Science, Vol. 317**, pp. 607-608 (2007).

34. Sanders, R. H., "Modified Newtonian Dynamics: A Falsification of Cold Dark Matter", **Advances in Astronomy, Vol. 2009, Article ID 752439** NGC 2403 spiral galaxy with graph of rotational velocity vs. distance from the center. **http://www.hindawi.com/journals/aa/2009/752439/fig1/**

35. Milgrom, Mordehai, "A Modification of the Newtonian Dynamics as a Possible Alternative to the Hidden Mass Hypothesis", **Astrophysical Journal, Vol. 270**, pp. 365-370 (1983).

36. Tully, R. B. and Fisher, J. R., "A New Method of Determining Distances to Galaxies", **Astronomy and Astrophysics, Vol. 54**, pp. 661-673 (1977).

37. Roscoe, D. F., "An Analysis of 900 Optical Rotation Curves", **International Astronomical Union Symposium, Vol. 194**, pp. 379-383 (1999).

38. Lucas, Jr., Charles W., "The Electrodynamic Origin of the Force of Inertia ($F=m_i a$) Parts 1, and 2, and 3", **Foundations of Science Vol. 10, No. 4**, pp. 1-6 (2007), and **Vol. 11, No.1**, pp. 1-6 (2008), and **Vol. 11, No. 2**, pp. 1-6 (2008).

39. Assis, A. K. T. "Deriving Gravitation from Electromagnetism", **Canadian Journal of Physics, Vol. 70**, pp. 330-340 (1992).

40. Assis, A. K. T. "Gravitation as a Fourth Order Electromagnetic Effect" in **Advanced Electromagnetism - Foundations, Theory and Applications**, Singapore: World Scientific, pp. 314-331 (1995).

41. Poincaré, Henri, **Oeuvres, Vol. 9**, p. 497 (1954).

42. Einstein, Albert, "Kosmologische Betrachtungen zur allgemeinen Relativitätstheorie", **Sitzungsberichte der Preußischen Akademie der Wissenschaften:** p. 142 (1917).

43. Dyson, F.W.; Eddington, A.S., & Davidson, C.R. "A Determination of the Deflection of Light by the Sun's Gravitational Field, from Observations Made at the Solar Eclipse of May 29, 1919". **Phil. Trans. Roy. Soc. A 220** (571-581): 291–333 (1920).

44. D. E. Lebach, B. E. Corey, I. I. Shapiro, M. I. Ratner, J. C. Webber, A. E. E. Rogers, J. L. Davis, T. A. Herring, **Phys. Rev. Lett**. 75: 1439-1442 (1995). **http://www.cfa.harvard.edu/qpb/vlbi/lebach_prl1995.pdf**

45. Dowdye, Jr., Edward H., "Are the Conventional Concepts of Gravitational Lensing Adhering to the Observational Evidence and Mathematical Physics Fundamentals?", **Infinite Energy, Vol. 15 Issue 88**, p. 40 (2009) http://www.extinctionshift.com/SignificantFindings06.htm

46. http://www.rexresearch.com/hooper/horizon.htm

Chapter 10 Electrodynamic Origin of Inertial Forces

No more causes of natural things should be admitted than are both true, and sufficient to explain their phenomena. Isaac Newton [1]

10.1 Introduction

The inertial mass of a particle is defined by Newton's second law $F_i = m_i a$ [2] and by related mechanical quantities such as momentum $p = m_i v$. The mass m_i is termed inertial, since it is a measure of the persistence of the particle in its current state caused by its interactions with the rest of the universe.

The gravitational mass m_g is defined by Newton's Universal Gravitational Force Law $F_g = G m_{g1} m_{g2} / r_{12}^2$ and by related gravitational quantities such as an object's weight $W = m_g g$. The gravitational mass m_g is a measure of that physical property of a particle that gives rise to the gravitational force. In the previous chapter the origin of the gravitational force was found to be due to the $(v/c)^4$ terms in the electrodynamic force between vibrating neutral electric dipoles composed of electrons and protons in atoms.

$$\frac{F_{g1}}{F_{g2}} = \frac{m_{g1}g}{m_{g2}g} = \frac{Gm_{g1}m_E/R_E^2}{Gm_{g2}m_E/R_E^2} = \frac{m_{g1}}{m_{g2}} \qquad (10-1)$$

A particle's gravitational mass can be defined and measured by standard operational procedures. Suppose particles 1 and 2 with gravitational masses m_{g1} and m_{g2} are brought to the same location on the surface of the earth. Then the gravitational forces on each, F_{g1} and F_{g2}, are measured by weighing the two objects. Now by definition the ratio of the respective gravitational masses equals the ratio of the corresponding gravitational forces.

Suppose now that particles 1 and 2 are dropped at the same location and their accelerations a_1 and a_2 are measured. Both particles are found to have the same acceleration g. Thus the ratio of their inertial forces F_{i1} and F_{i2} is

$$\frac{F_{i1}}{F_{i2}} = \frac{m_{i1}a_1}{m_{i2}a_2} = \frac{m_{i1}g}{m_{i2}g} = \frac{m_{i1}}{m_{i2}} \qquad (10-2)$$

Now the inertial and gravitational forces are equal for particles 1 and 2 at rest on the surface of the earth so

$$\frac{F_{g1}}{F_{g2}} = \frac{m_{g1}}{m_{g2}} = \frac{F_{i1}}{F_{i2}} = \frac{m_{i1}}{m_{i2}} \qquad (10-3)$$

By experiment the ratio of the gravitational masses of two particles equals the ratio of the inertial masses by approximately 1 part in 10^{12}. [3]

In classical physics the equality of the inertial and gravitational masses is regarded as an extraordinary coincidence. However, in the general theory of relativity by Albert Einstein, this equality is taken as a basic assumption in the "principle of equivalence". The principle of equivalence relates to accelerated reference frames.

An observer in an accelerated reference frame can use Newton's second law to describe a particle's motion. To do so, such an accelerated observer must invoke, in addition to the "real" forces acting on the particle, a so-called "fictitious" inertial force, $F_i = -m_i a_i$. This inertial force depends on the particle's inertial mass m_i and on the acceleration a_i of the observer's reference frame relative to an inertial frame. The non-inertial observer writes Newton's second law as

$$\sum F_{Real} + F_i = m_i a \ with \ F_i = -m_i a_i \qquad (10-4)$$

where a, the particle's acceleration, is measured in the accelerated frame.

Now if a particle's gravitational mass m_g is precisely the same as its inertial mass m_i, then the inertial force can be written as

$$F_i = -m_i a_i = -m_g a_i \qquad (10-5)$$

Thus the "fictitious" inertial force is proportional to the observed particle's gravitational mass. This implies that

An inertial force arising in a non-inertial frame is altogether equivalent to, and indistinguishable from, a gravitational force as perceived by an observer at rest in this accelerated frame. [2]

This is Einstein's principle of equivalence in his General Theory of Relativity.

According to the principle of equivalence there is no experimental way of telling the difference between an inertial acceleration and a gravitational force. Actually this is an idealization based on point particles. In the real world of finite-size particles, there is a three dimensional gradient of the gravitational force over the volume of the particle, but a different gradient for an inertial acceleration. Thus for a non-spherical person in an elevator there is a difference between the gradient of the force of gravity and the acceleration due to the motion of the elevator! In the electrodynamic approach to gravity and inertia in this book the force of inertia and the force of gravity are both real forces on equal footing being due to separate terms in the electrodynamic force. The force of inertia is not a "fictitious" force, but is consistent with Mach's principle of **Chapter 13**.

10.2 Derivation of Force of Inertia from Universal Force Law

In this section the origin of the inertial forces will be shown to be due to the acceleration terms of the electrodynamic force. The coefficient of the acceleration term defines the inertial mass in the same way as the previous sections defined the gravitational mass. In this way the similarity of the gravitational and inertial mass will be demonstrated. In general the concept of mass will be shown to not be a fundamental quantity, but an approximately constant grouping of electrodynamic vibrating neutral dipole terms.

In the Gaussian system of units equation (5-7) for the universal electrodynamic force[4,5] was derived assuming that the electrodynamic potential was a regular well-behaved continuous function of r and v. Consider the possibility that the first acceleration term of equation (5-7) is the origin of Newton's second law F=m$_i$a. Expanding equation (5-7) in terms of spherical coordinates obtain equation (10-6).

$$\vec{F}(\vec{r},\vec{v},\vec{a}) = \frac{qq'}{\vec{r}^2}\left[\frac{\frac{2\vec{r}^2}{c^2}\vec{a}}{\left[\vec{r}^2 - \frac{\{\vec{r}\times(\vec{r}\times\vec{\beta})\}^2}{\vec{r}^2}\right]^{\frac{1}{2}}} - (1-\vec{\beta}^2)\frac{(\vec{r}\cdot\vec{r})\left\{\vec{r}\times\left(\vec{r}\times\frac{\vec{a}}{c^2}\right)\right\}}{\left[\vec{r}^2 - \frac{\{\vec{r}\times(\vec{r}\times\vec{\beta})\}^2}{\vec{r}^2}\right]^{3/2}}\right]$$

$$= \frac{qq'}{r}\frac{\frac{2\vec{a}}{c^2}}{(1-\beta^2 sin^2\theta)^{1/2}} - \frac{qq'}{r}\frac{(1-\vec{\beta}^2)\left\{\hat{r}\times\left(\hat{r}\times\frac{\vec{a}}{c^2}\right)\right\}}{(1-\beta^2 sin^2\theta)^{3/2}}$$

$$= \frac{qq'}{r}\frac{2\vec{a}}{c^2}\left[1 + \frac{\beta^2 sin^2\theta}{2}\right]$$

$$- \frac{qq'}{r}\left\{\hat{r}\times\left(\hat{r}\times\frac{\vec{a}}{c^2}\right)\right\}(1-\vec{\beta}^2)\left[1 + \frac{3}{2}\beta^2 sin^2\theta\right.$$

$$\left. + \frac{\left(\frac{3}{2}\right)\left(\frac{5}{2}\right)}{2}\beta^4 sin^4\theta\right]$$

$$= \frac{qq'}{r}\frac{2\vec{a}}{c^2}\left[1 + \frac{\beta^2}{2} - \frac{\beta^2 cos^2\theta}{2}\right]$$

$$- \frac{qq'}{r}\left\{\hat{r}\times\left(\hat{r}\times\frac{\vec{a}}{c^2}\right)\right\}\left[1 + \frac{\beta^2}{2} - \frac{3}{2}\beta^2 cos^2\theta + \frac{3}{8}\beta^4 - \frac{9}{4}\beta^4 cos^2\theta\right.$$

$$\left. + \frac{15}{8}\beta^4 cos^4\theta\right] \qquad\qquad (10-6)$$

where sin$^2\theta$ = 1 − cos$^2\theta$. Assume that the force of Newton's second law is between a charge and some vibrating neutral electric dipoles consisting of positive protons and negative electrons where the dipole is defined by

$$\vec{r}_{2+,1+} = \vec{r}_{2+} - \vec{r}_{1+} \quad and \quad A_1 f_1 = v_1$$

$$\vec{r}_{2+,1-} = \vec{r}_{2+} - \vec{r}_{1-} - \vec{A}_1 \cos(\omega_1 t + \varphi_1)$$

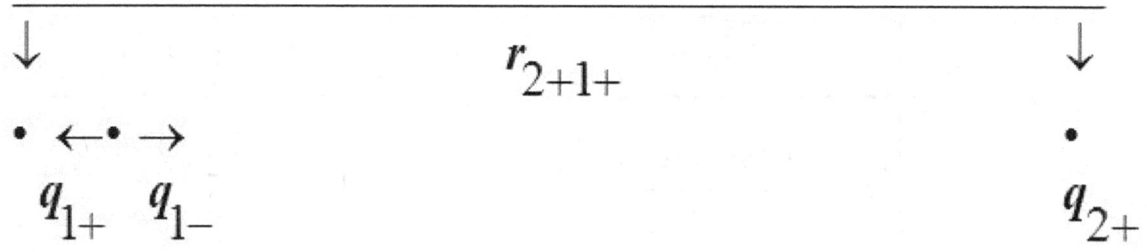

$\downarrow$ r_{2+1+} $\downarrow$

$\bullet \;\; \leftarrow\bullet\rightarrow$ $\bullet$

q_{1+} q_{1-} q_{2+}

Figure 10-1 Oscillations of Electron in Vibrating Neutral Electric Dipole

The amplitude of the oscillation A_1 is of the order of the size of the atom, i.e. 10^{-12} cm. The frequency of oscillation ω_1 is in the microwave range of 10^{10} per second. At time t = 0 the amplitude of the vibration is maximum as determined by $\cos(\omega t + \varphi)$.

In order to simplify the calculations assume that the positively charged proton is much more massive than the negatively charged electron such that the vibratory motion of the dipole can be considered as due primarily to the motion of the electron. Since the mass of the proton is 1836 times the mass of the electron, this is a reasonable approximation.

In order to calculate a quantity comparable to Newton's second law $F = m_i a$ [6], it will be necessary to perform a number of averages. Each oscillating dipole has a different phase that must be averaged over. Each oscillating dipole may have a different spatial orientation that needs to be averaged over for all the dipoles in the material body. In order to obtain a time independent value, it will be necessary to perform a time average on the oscillating dipoles. Thus the force to be compared with Newton's second law is

$$\vec{F} = \frac{1}{\tau_1} \int_0^{\tau_1} dt \frac{1}{\pi} \int_0^{\pi} sin\theta d\theta \frac{1}{2\pi} \int_0^{2\pi} d\varphi \vec{F}(r, \theta, \varphi, A_1, \omega_1, t) \qquad (10-7)$$

For simplicity assume that the collection of dipoles has spherical symmetry. In this case the integral over φ becomes just 2π giving

$$\vec{F} = \frac{1}{\tau_1} \int_0^{\tau_1} dt \frac{1}{\pi} \int_0^{\pi} sin\theta d\theta \vec{F}(r, \theta, \varphi, A_1, \omega_1, t) \qquad (10-8)$$

Note that there are two fundamental type terms in the force given in equation (10-6). The first term is proportional to the acceleration a. The second term is perpendicular to the acceleration, a, and r. This term will lead to a new (previously unknown?) inertial force which causes corkscrew spiraling motion perpendicular to the acceleration.

10.3 Derivation of Newton's 2nd Law from 1st Acceleration Term

The force terms for the first acceleration term to order β^2 in equation (10-6) are given below

$$\vec{F}_{2+,1+} = \frac{e^2 2\vec{a}}{|\vec{r}_2 - \vec{r}_1|c^2} [1]$$

$$\vec{F}_{2+,1-} = \frac{-e^2 2\vec{a}}{|\vec{r}_2 - \vec{r}_1|c^2}\left[1 - \left\{\frac{A_1\omega_1}{c}sin(\omega_1 t_1 + \varphi_1)\right\}^2\left(\frac{1 - cos^2\theta}{2}\right)\right] \quad (10-9)$$

One can see that the sum of the first terms in the [ ] of the two forces is just 0. Thus the total force is

$$\vec{F}(r,\theta,\varphi,A_1,\omega_1,t) = \vec{F}_{2+,1+} + \vec{F}_{2+,1-}$$

$$= \frac{e^2 2\vec{a}}{|\vec{r}_2 - \vec{r}_1|c^2}\left\{\frac{A_1\omega_1}{c}sin(\omega_1 t_1 + \varphi_1)\right\}^2\left(\frac{1 - cos^2\theta}{2}\right) \quad (10-10)$$

Now the integrals of equation (10-8) can be evaluated using equation (10-11)

$$\vec{F} = \frac{1}{\tau_1}\int_0^{\tau_1} dt\frac{1}{2\pi}\int_0^{2\pi} d\varphi_1\frac{1}{\pi}\int_0^{\pi} sin\theta d\theta \vec{F}(r,\theta,\varphi,A_1,\omega_1,t) \quad (10-11)$$

$$= \frac{1}{\tau_1}\int_0^{\tau_1} dt\frac{1}{2\pi}\int_0^{2\pi} d\varphi_1\frac{1}{\pi}\int_0^{\pi} sin\theta d\theta \frac{e^2 2\vec{a}}{|\vec{r}_2 - \vec{r}_1|c^2}\left\{\frac{A_1\omega_1}{c}sin(\omega_1 t_1 \right.$$

$$\left. + \varphi_1)\right\}^2\left(\frac{1 - cos^2\theta}{2}\right)$$

$$= \frac{1}{\tau_1}\int_0^{\tau_1} dt\frac{1}{2\pi}\int_0^{2\pi} d\varphi_1\frac{e^2 2\vec{a}}{|\vec{r}_2 - \vec{r}_1|c^2}\left\{\frac{A_1\omega_1}{c}sin(\omega_1 t_1 + \varphi_1)\right\}^2\left(\frac{2}{3\pi}\right)$$

$$= \frac{\omega_1}{2\pi}\int_0^{\frac{2\pi}{\omega_1}} dt\frac{e^2 2\vec{a}}{|\vec{r}_2 - \vec{r}_1|c^2}\left\{\frac{A_1\omega_1}{c}\right\}^2\left(\frac{1}{2}\right)\left(\frac{2}{3\pi}\right) = \left(\frac{2e^2}{3\pi}\right)\left(\frac{A_1^2\omega_1^2}{|\vec{r}_2 - \vec{r}_1|c^2}\right)\vec{a} = m_{i1}\vec{a}$$

where $\tau_1 = 2\pi/\omega_1$ and

$$\frac{1}{2\pi}\int_0^{2\pi} sin(\omega t + \varphi)d\varphi$$

$$= \frac{1}{2\pi} \int_0^{2\pi} (sin^2\omega t cos^2\varphi + 2cos\omega t sin\omega t sin\varphi cos\varphi + cos^2\omega t sin^2\varphi)d\varphi$$

$$= \frac{1}{2\pi}(\pi cos^2\omega t + 0 + \pi sin^2\omega t) = \frac{1}{2}(cos^2\omega t + sin^2\omega t = \frac{1}{2} \quad (10-12)$$

Thus in equation (10-11), we have derived an electrodynamic formula for the inertial mass m_{i1}. Note that it has a dependence on r. In Chapter 13 we will see from Mach's Principle that this r is the average distance between the vibrating neutral electric dipole and all the rest of the charges in the universe. For a universe with spherical symmetry this average distance is the distance to the center of the universe R_{UC} from the position of the vibrating neutral electric dipole. Using this definition we can evaluate equation (10-2) at a point in the universe such as a point on the surface of the earth, i.e.

$$\frac{F_{i1}}{F_{i2}} = \frac{m_{i1}a}{m_{i2}a} = \frac{N_1 Z_1 A_1{}^2 \omega_1{}^2/R_{UC}}{N_2 Z_2 A_2{}^2 \omega_2{}^2/R_{UC}} \quad (10-13)$$

From the derivation of the force of gravity given in equation (9-16) we have

$$F_G = -\frac{2}{5\pi}\frac{e^2}{|\vec{r}_2 - \vec{r}_1|^2}\frac{N_1{}^2 Z_1{}^2 A_1{}^2 \omega_1{}^2}{c^2}\frac{N_2{}^2 Z_2{}^2 A_2{}^2 \omega_2{}^2}{c^2}$$

$$= -G\frac{m_{g1}m_{g2}}{|\vec{r}_2 - \vec{r}_1|^2} \quad (10-14)$$

Substituting equation (10-14) into equation (10-1) one obtains the ratios of gravitational and inertial masses for the same two bodies at the same point on the surface of the earth to be equal.

$$\frac{F_{g1}}{F_{g2}} = \frac{m_{g1}g}{m_{g2}g} = \frac{Gm_{g1}m_E/R_E{}^2}{Gm_{g2}m_E/R_E{}^2} = \frac{m_{g1}}{m_{g2}} = \frac{N_1{}^2 Z_1{}^2 A_1{}^2 \omega_1{}^2}{N_2{}^2 Z_2{}^2 A_2{}^2 \omega_2{}^2} = \frac{m_{i1}}{m_{i2}} \quad (10-15)$$

Thus we have shown that the gravitational and inertial masses at any point in the universe are equal to within a constant k of one another and a radial factor R_{UC}, i.e.

$$m_g = kR_{UC}m_i \quad (10-16)$$

Using the definition of gravitational mass in equation (10-16) and the force of gravity in equation (10-14), one might expect the value of Newton's universal gravitation constant G could be determined from electrodynamics to be

$$G = \frac{9\pi c^4 R_{UC}{}^2}{10e^2} \quad (10-17)$$

However, this is over simplified. If there is an asymmetry in the near universe, it can dominate over the rest of the universe. The 1/R factor for the Earth may be larger than the average of the spherically symmetric contribution from the rest of the universe. In volume 2 of this series on an electrodynamic model of elementary particles, the proton and neutron contain + and – charge elements that can act as vibrating neutral electric dipoles. Also in volume 3 on an electrodynamic model of the atom and the nucleus, there are additional forms of rotation and vibration of the charged particles in the atomic and the nuclear shells. All of these must be taken into account to predict a proper value for G.

In summary we have shown that Newton's second law can be derived from electrodynamics. We defined the inertial mass m_i for the vibration of neutral electric dipoles consisting of atomic electrons and protons in the atomic nucleus. We showed that the ratio of two inertial masses is exactly the same as the ratio of two gravitational masses for the same vibration. Then we showed in principle that we can derive an electrodynamic value for Newton's universal gravitation constant G. Thus we have a classical electrodynamic explanation for "why the value of the ratios of gravitational and inertial masses are identical". They are both due to electrodynamic forces involving vibrating neutral electric dipoles. **This is a strong indication that the electrodynamic force law is the "universal" force law.**

10.4 Additions to Newton's 2nd Law from 2nd Acceleration Term

Equation (10-6) above has a second acceleration term. This term is unknown in classical mechanics, although it has been seen in some gyroscope experiments such as those of Eric Laithwaite. [6] This term will lead to a corkscrew type of motion just like the electrodynamic case of particle beams and the gravitational case of planetary orbits. It will be one more piece of evidence that the electrodynamic force law is indeed the "universal" force law.

The force terms for the second acceleration term to order $(v/c)^4$ in equation (10-6) are given below. For the velocity terms in the [ ] of the expressions for the force above, consider the case $\beta_2 = \beta_1$. In this case only the $A_1\omega_1$ terms are left giving

$$\vec{F}_{2+,1+} = \frac{e^2 \hat{r} \times \left(\hat{r} \times \dfrac{\vec{a}}{c^2}\right)}{|\vec{r}_2 - \vec{r}_1|c^2}\,[1]$$

$$\vec{F}_{2+,1-} = \frac{-e^2 \hat{r} \times \left(\hat{r} \times \dfrac{\vec{a}}{c^2}\right)}{|\vec{r}_2 - \vec{r}_1|c^2}\left[1 - \left\{\frac{A_1\omega_1}{c}\,sin(\omega_1 t_1 + \varphi_1)\right\}^2\left(\frac{1 - 3cos^2\theta}{2}\right)\right.$$
$$\left. - \left\{\frac{A_1\omega_1}{c}\,sin(\omega_1 t_1 + \varphi_1)\right\}^4\left(\frac{3}{8} - \frac{9}{4}cos^2\theta - \frac{15}{8}cos^4\theta\right)\right]\,(10-18)$$

One can see that the sum of the first terms in the [ ] of the two forces is just 0. Thus the total force is

$$\vec{F} = \vec{F}_{2+,1+} + \vec{F}_{2+,1-}$$

$$= \frac{e^2 \hat{r} \times \left(\hat{r} \times \frac{\vec{a}}{c^2}\right)}{|\vec{r}_2 - \vec{r}_1|c^2} \left[\left\{\frac{A_1\omega_1}{c} sin(\omega_1 t_1 + \varphi_1)\right\}^2 \left(\frac{1 - 3cos^2\theta}{2}\right) \right.$$

$$\left. + \left\{\frac{A_1\omega_1}{c} sin(\omega_1 t_1 + \varphi_1)\right\}^4 \left(\frac{3}{8} - \frac{9}{4}cos^2\theta - \frac{15}{8}cos^4\theta\right)\right] \quad (10-19)$$

Now the integrals of equation (10-19) need to be evaluated. Note that the integral over θ for the $(1 - 3cos^2\theta)/2$ term averages to zero. That is why we had to include terms up to order β^4 for this term.

$$\frac{1}{\pi}\int_0^\pi sin\theta d\theta \frac{(1 - 3cos^2\theta)}{2}$$

$$= \frac{-1}{\pi}\int_0^\pi dcos\theta \frac{(1 - 3cos^2\theta)}{2} = \frac{-1}{2\pi}\left(cos\theta - \frac{3cos^3\theta}{3}\right)\Bigg|_0^\pi$$

$$= \frac{-1}{2\pi}(-1 - 1 - (-1) + 1) = 0 \quad (10-20)$$

Thus the total force due to the second acceleration term in equation (10-6) is

$$\vec{F} = \frac{1}{\tau_1}\int_0^{\tau_1} dt \frac{1}{2\pi}\int_0^{2\pi} d\varphi_1 \frac{1}{\pi}\int_0^\pi sin\theta d\theta \vec{F}(r, \theta, \varphi, A_1, \omega_1, t) \quad (10-21)$$

$$= \frac{1}{\tau_1}\int_0^{\tau_1} dt \frac{1}{2\pi}\int_0^{2\pi} d\varphi_1 \frac{1}{\pi}\int_0^\pi sin\theta d\theta \frac{-e^2 \hat{r} \times \left(\hat{r} \times \frac{\vec{a}}{c^2}\right)}{|\vec{r}_2 - \vec{r}_1|c^2}\left\{\frac{A_1\omega_1}{c} sin(\omega_1 t_1 + \varphi_1)\right\}^4$$

$$\left(\frac{3}{8} - \frac{9}{4}cos^2\theta - \frac{15}{8}cos^4\theta\right)$$

$$= \frac{1}{\tau_1}\int_0^{\tau_1} dt \frac{1}{2\pi}\int_0^{2\pi} d\varphi_1 \frac{-e^2 \hat{r} \times \left(\hat{r} \times \frac{\vec{a}}{c^2}\right)}{|\vec{r}_2 - \vec{r}_1|c^2}\left\{\frac{A_1\omega_1}{c} sin(\omega_1 t_1 + \varphi_1)\right\}^4 \frac{3}{2\pi}$$

$$= \frac{\omega_1}{2\pi}\int_0^{\frac{2\pi}{\omega_1}} dt \frac{e^2 \hat{r} \times \left(\hat{r} \times \frac{\vec{a}}{c^2}\right)}{|\vec{r}_2 - \vec{r}_1|c^2}\left(\frac{A_1{}^4\omega_1{}^4}{c^4}\right)\left(\frac{3}{8}\right)\left(\frac{3}{2\pi}\right)$$

$$= \frac{\omega_1}{2\pi} \frac{2\pi}{\omega_1} \frac{e^2 \hat{r} \times \left(\hat{r} \times \frac{\vec{a}}{c^2}\right)}{|\vec{r}_2 - \vec{r}_1|c^2} \left(\frac{A_1^4 \omega_1^4}{c^4}\right) \left(\frac{3}{8}\right) \left(\frac{3}{2\pi}\right)$$

$$= \frac{9e^2}{16\pi |\vec{r}_2 - \vec{r}_1|c^2} \left(\frac{A_1^4 \omega_1^4}{c^4}\right) \hat{r} \times \left(\hat{r} \times \frac{\vec{a}}{c^2}\right)$$

$$= \frac{27}{32} \left(\frac{A_1^4 \omega_1^4}{c^4}\right) m_{i1} \hat{r} \times \left(\hat{r} \times \frac{\vec{a}}{c^2}\right)$$

where

$$\frac{1}{\pi} \int_0^\pi sin\theta d\theta \left(\frac{3}{8} - \frac{9}{4} cos^2\theta - \frac{15}{8} cos^4\theta\right)$$

$$= -\frac{1}{\pi} \int_{-1}^1 dcos\theta \left(\frac{3}{8} - \frac{9}{4} cos^2\theta - \frac{15}{8} cos^4\theta\right)$$

$$= -\frac{1}{\pi} \left(\frac{3}{8} - \frac{9}{4} cos^2\theta - \frac{15}{8} cos^4\theta\right) \Big|_{-1}^1 = \frac{3}{2\pi} \qquad (10-22)$$

and

$$\frac{1}{2\pi} \int_0^{2\pi} sin^4(\omega t + \phi) d\varphi = \frac{1}{2\pi} \int_0^{2\pi} [sin\omega t cos\varphi + cos\omega t sin\varphi]^4 d\varphi$$

$$= \frac{1}{2\pi} \int_0^{2\pi} [sin^2\omega t cos^2\varphi + 2sin\omega t cos\varphi cos\omega t sin\varphi + cos^2\omega t sin^2\varphi]^2 d\varphi$$

$$= \frac{1}{2\pi} \int_0^{2\pi} [sin^4\omega t cos^4\varphi + 4sin^3\omega t cos^3\varphi cos\omega t sin\varphi$$

$$+ 6sin^2\omega t cos^2\varphi cos^2\omega t sin^2\varphi + 4sin\omega t cos\varphi cos^3\omega t sin^3\varphi$$
$$+ cos^4\omega t sin^4\varphi] d\varphi$$

$$= \frac{1}{2\pi} \left[\frac{6\pi}{8} sin^4\omega t + \frac{6\pi}{4} sin^2\omega t cos^2\omega t + \frac{6\pi}{8} cos^4\omega t\right]$$

$$= \frac{3}{8} [sin^2\omega t + cos^2\omega t]^2 = \frac{3}{8} \qquad (10-23)$$

Note the odd powers of sin φ and cos φ in equation (10-23) integrate to zero. Thus the inertial force law is given by

$$\vec{F}_I = m_i \vec{a} + \frac{27}{32} \frac{A^2 \omega^2}{c^2} m_i \hat{r} \times \left(\hat{r} \times \frac{\vec{a}}{c^2} \right) \qquad (10-24)$$

where

$$m_i = \left(\frac{2e^2}{3\pi R c^2} \right) \left(\frac{A^2 \omega^2}{c^2} \right) \qquad (10-25)$$

Note that in equation (10-25) the inertial mass m_i depends on $1/R$ where R is the distance to the center of the masses of the universe if it has a center. Thus the definition of inertial mass satisfies Mach's Principle in that the value of the local mass depends on the structure and matter distribution of the universe. If the universe is homogenous and isotropic as assumed by Einstein's Special and General Relativity theories, this electrodynamic definition of the force of inertia and inertial mass is invalid. Fortunately the Microwave Cosmic Background Radiation distribution coupled with the atomic red and blue Doppler shifts identify the center of the universe. And the atomic red shifts about that defined center obey the modern version of Bode's Law formulated by Stanley Dermott confirming a spherical distribution of galaxies about that center.

10.5 Summary

In summary, the first acceleration term of the electrodynamic force to order β^2 is Newton's second law F = ma for non-relativistic velocities. The second term, of order β^4, is a new term that gives rise to a corkscrew type of spiraling inertial motion. The strength of the second term is much less than the first due to the $A^2\omega^2/c^2$ factor. The first term is the inertial term we are normally accustomed to see. The second term can be large enough to observe for fast spinning large size gyroscopes where $A^2\omega^2/c^2$ gets large. Eric Laithwaite [6] has performed public demonstrations of high speed large size gyroscopes that appear to defy Newton's second law. It is the second term of the inertial force causing the unexpected results.

The electric charge to neutral electric dipole force is the second force term beyond the Coulomb force between individual charges to be explored in the hierarchy of electrodynamic interactions. Despite the fact that scientists in the field of fluid dynamics, plasma physics, and chemistry have been paying attention to electrical dipole effects, it appears that it has been overlooked in the field of electrodynamics and the fields of gravitation and inertia.

The inertial force law derived from the electrodynamic force law is more complete than Newton's 2nd law [2], because it describes the non-radial inertial forces as observed in gyroscope experiments. Also it can be derived from the electrodynamic force law. The theory is based on finite-size particles instead of fictitious point particles. It satisfies Mach's Principle (see Chapter 13). The theory is based on a local contact force instead of a fictitious action-at-a-distance type of force. Also it defines inertial mass and shows that it is not a fundamental quantity of nature. Finally it explains why the ratios of inertial and gravitational masses are equal for the same two bodies which no previous theory has been able to do.

10.6 References

1. Isaac Newton, **The Principia, Mathematical Principles of Natural Philosophy: A New Translation**, translators I Bernard Cohen and Anne Whitman (University of California Press, Berkeley, 1999) Rules of Reasoning in Philosophy: : Rule 1
2. Newton, Isaac, **Principia** (1687).
3. Weidner, Richard T., **Physics** (Allyn and Bacon, Inc., Boston, 1985) pp. 330-331.
4. Lucas, Charles W. Jr. and Joseph C. Lucas, "Weber's Force Law for Realistic Finite-Size Elastic Particles", **J. of New Energy Vol. 5, No. 3**, pp. 70-89 (2001).
5. Lucas, Charles W. Jr. and Joseph C. Lucas, "Weber's Force Law for Finite-Size Elastic Particles", **Galilean Electrodynamics Vol. 14,** pp. 3-10 (2003).
6. Eric Laithwaite's gyroscope experiments at http://gyroscopes.org/1974lecture.asp and http://www.youtube.com/watch?v=MHlAJ7vySC8

Chapter 11 Evidence for Strong and Weak Forces in Nuclei

No more causes of natural things should be admitted than are both true, and sufficient to explain their phenomena. Isaac Newton [1]

11.1 Does Nuclear Data Require the Existence of Strong or Weak Force?

Modern physics, as currently taught in the scientific community, is based upon four fundamental forces, i.e. the electrodynamic force holding atoms, molecules and crystals together, the force of gravity holding the solar system and galaxies together, the strong force holding protons and neutrons together in the nucleus plus quarks together in heavy hadronic elementary particles, and the weak force governing nuclear and elementary particle decays.

In previous chapters an improved version of the electrodynamic force was derived by perfecting the union of the axiomatic and empirical scientific methods. This resulted in a version of electrodynamics that appears to be superior to the current relativistic Maxwellian version of electrodynamics in that it includes acceleration a and radiation reaction da/dt terms which are missing from previous versions of electrodynamics. From this version of electrodynamics was derived the force of gravity and the force of inertia by taking into account the hierarchy of electromagnetic interactions between multipoles. It appears that these versions of the gravitational and inertial forces are superior to all previous versions of these forces. The only other claimed forces in nature are the strong and the weak interaction forces. These remaining forces have a very short range on the order of the size of the proton and neutron. They also incorporate the point particle idealization. Thus these forces could be due to finite-size effects.

If the atomic and nuclear data can be completely explained by the electrodynamic force and conservation of energy, then the need for a strong interaction force and weak interaction force would be removed clearing the way for the declaration of the improved electrodynamic force as the universal force. The purpose of this chapter is to show that the precise authorative NIST atomic and nuclear data can indeed be completely explained by the electrodynamic force with no need for a strong interaction force or a weak interaction force. **It is important to note that the highly precise NIST atomic and nuclear data did not exist at the time that the neutron was "discovered" and the strong and weak interaction forces were "invented".**

11.2 An Analysis of Isoelectronic Data for One Electron Atoms

In atomic spectroscopy there are several types of sequences of elements that are useful because they reveal regularities in the progressive values of parameters relating to the internal structure and forces in atoms. One of these sequences is known as the isoelectronic sequence. An isoelectronic sequence consists of a neutral atom and those atoms of other elements having the same number of electrons as the neutral atom.

Consider the isoelectronic sequence for the hydrogen atom with one electron. The other atoms in the sequence consist of the helium atom ionized to have only one remaining electron, the lithium atom ionized to have only one remaining electron, etc.

The National Institute of Science and Technology (NIST) has given grants to various laboratories in the world to measure to high precision various properties of isoelectronic sequences. One of those properties is the ionization energy. For the hydrogen one electron isoelectronic sequence the ionization energy is the energy to remove the remaining electron in each member of the sequence. These measured ionization energies are given on the NIST website. [2] By searching on each element of the hydrogen isoelectronic sequence on this website, one may construct a table of the ionization energy for each member of the isoelectronic sequence. This data may be entered into a Microsoft Excel spreadsheet and analyzed graphically as shown in Figure 11-1 below.

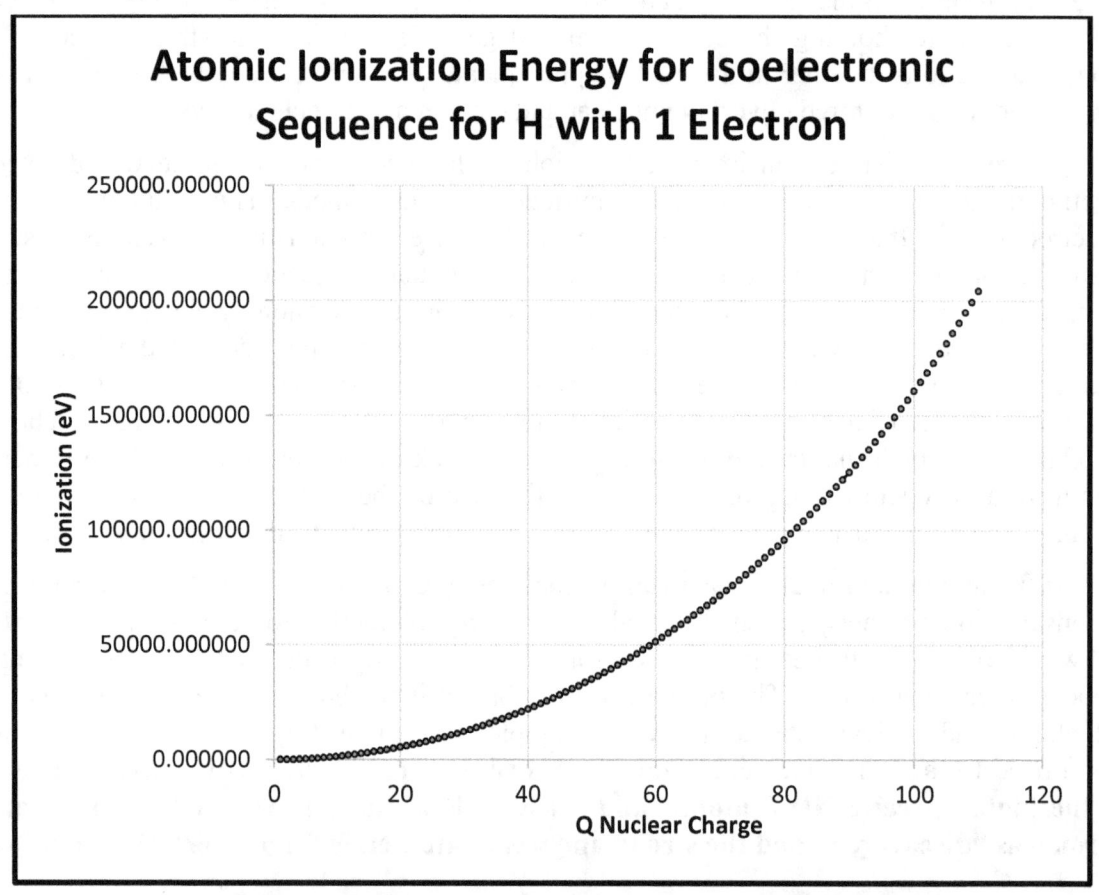

Figure 11-1 Ionization Energies for H Isoelectronic Sequence

Now the ionization energy of a one electron atom such as hydrogen is the electromagnetic energy that must be given to the electron to remove it from the atom so that it is no longer bound to that atom's nucleus. Coulomb's Law says that the electrical attraction between two oppositely charged particles, such as the electron and the nucleus, is directly proportional to the charge of the electron and the charge of the nucleus divided by the square of the distance between them. For the Bohr model of the atom the simplest atom is the hydrogen atom with one negative electron orbiting a nucleus with one positive charge. The energy of attraction between the opposite charges is measured to be 13.598434 electron-volts. For the second member of the isoelectronic sequence, i.e. the helium ion, the nucleus contains two positive charges with one electron orbiting it. According to Coulomb's Law the attraction between the electron and the nucleus should be twice as great. However, when this is measured it turns out to be 4.00177 times as great as that of hydrogen (54.417763 electron-volts). For the third member of the isoelectronic sequence, i.e. the twice-ionized Lithium atom, there is a single electron orbiting a nucleus with three charges in it. Is the attraction three times as much as the hydrogen atom as predicted by Coulomb's Law? No! It is experimentally 9.00503 times as great as that of hydrogen (122.454350 electron-volts). The graph in Figure 11-1 shows that the ionization energy of the hydrogen isoelectronic sequence is a function of the charge of the nucleus squared for all members of the isoelectronic sequence for nuclear charge 1 – 110. The curve defined by the data is a parabola and is due entirely to the electrodynamic $1/R^2$ force.

11.3 Explanation of Z^2 Dependence of Atomic Ionization Energy

Coulomb's Law says that the ionization energy to remove the electron from the hydrogen isoelectronic sequence of atoms is proportional to the charge of the electron times the charge of the nucleus divided by the relative distance between them. What is observed in Figure 1 is that the ionization energy is proportional to the charge of the nucleus squared. How can this be explained?

Consider an atom consisting of a nucleus of charge +Ze and mass M and a single electron of charge −e and mass m. Assume that the electron is in a circular orbit about the nucleus. Since the mass of the electron is very small compared to the mass of the nucleus, assume that the nucleus remains fixed in space as the electron orbits. Thus the condition for mechanical stability for the electron orbit is that the Coulomb electrical force is equal to the centripetal force.

$$\frac{Ze^2}{r^2} = \frac{mv^2}{r} \quad (11-1)$$

where v is the velocity of the electron in its orbit and r is the radius of the orbit.

Now the orbital angular momentum of the electron must be a constant, because the force acting on the electron is entirely in the radial direction. The angular momentum L is

$$L = mvr \quad (11-2)$$

Applying Bohr's quantum postulate to the angular momentum obtain

$$L = mvr = n\hbar \quad n = 1, 2, 3, \dots \quad (11-3)$$

Solving equation (11-1) for Ze^2 and using equation (11-3) obtain

$$Ze^2 = \frac{mv^2r^2}{r} = \frac{L^2}{mr} = \frac{n^2\hbar^2}{mr} \quad n = 1, 2, 3 \dots \quad (11-4)$$

Solving equation (11-4) for r obtain

169

$$r = \frac{n^2 \hbar^2}{mZe^2} \quad n = 1, 2, 3, \ldots \quad (11-5)$$

Consider the total energy of an atomic electron moving in one of its allowed orbits. If we define the potential energy of the electron to be zero when the electron is infinitely distant from the nucleus, then the potential energy V at any finite distance r can be obtained by integrating the energy imparted to the electron by the Coulomb force acting from infinity to r:

$$V = \int_{\infty}^{r} \frac{Ze^2}{r^2} \, dr = -\frac{Ze^2}{r} \quad (11-6)$$

From equation (11-1) the kinetic energy T of the electron is

$$T = \frac{1}{2}mv^2 = \frac{Ze^2}{2r} \quad (11-7)$$

The total energy E of the electron is then from equations (11-6) and (11-7)

$$E = T + V = \frac{Ze^2}{2r} - \frac{Ze^2}{r} = -\frac{Ze^2}{2r} \quad (11-8)$$

Substituting for r from equation (11-5) into equation (11-8) obtain

$$E = -\frac{Ze^2}{2r} = -\frac{Ze^2}{2\frac{n^2\hbar^2}{mZe^2}} = -\frac{mZ^2e^4}{2n^2\hbar^2} \quad n = 1, 2, 3, \ldots \quad (11-9)$$

Thus the ionization potential energy which is equal to the binding energy for the n=1 state is proportional to the charge Z^2. This result holds for the Bohr model of the atom and the toroidal ring model of the electron atom. The method of explaining these results for the non-relativistic Schrödinger wave equation model of the atom and the relativistic Dirac wave equation model of the atom is given in reference [3].

11.4 Analysis of Nuclear Isotopic Masses

In the past nuclear isotopes have always been examined as parts of a family of same-element isotopes. The approach of this paper will be to analyze them instead as members of families of the same atomic weight isotopes. This approach will reveal very strong empirical evidence that neutrons do not actually exist as neutrons within atomic nuclei. Neutrons appear to exist in free space, but they do not exist within the atomic nuclei. The nuclear isotopic mass data is only consistent with neutrons existing as separate protons and electrons within the nucleus.

Nuclear physicists agree that the heaviest nuclei are generally all unstable, because they contain too many neutrons. Such a statement seems to imply that a common method of nuclear decay should be the spontaneous emission of a neutron in order to enable a nucleus to become more stable. However it has long been known that none of the heaviest nuclei decay by emitting a neutron. Only three relatively common isotopes decay by emitting a neutron, i.e. $^{88}Br_{35}$, $^{87}Br_{35}$, and $^5He_2$. The first two of these isotopes do not naturally occur and are only produced as fission products of $^{235}U_{92}$. If discrete neutrons actually existed within nuclei, it seems that at least some nuclei would decay naturally by releasing some of them. Instead, virtually the only time that neutrons are released from any nuclei is as the result of an external disturbance by an external thermal neutron from nuclear fission or incoming radiation.

11.5 1st Proof that Neutrons and Weak Force Do Not Exist in Nuclei

Consider the natural decay of Tritium $^3H_1$ with a half-life of 12.33 years into $^3He_2$ and an escaping electron which then is captured as an orbiting electron of the Helium nucleus. This situation is clearly one where the exact same amount and number of objects are involved, i.e. three protons and three electrons, but where some of the protons and electrons are allegedly bound together as neutrons in the Tritium nucleus. The laws of conservation of mass and energy should apply, so a strict energy accounting for this decay should show exactly the same total energy and mass before and after the decay. Initially there is one neutron in the Tritium nucleus which no longer exists in the final Helium-3 nucleus. The difference in the total mass and energy of these two nuclear atomic masses should therefore include the 0.782 MeV of binding energy of the one neutron inside the Tritium nucleus which is no longer a neutron in the final Helium-3 nucleus. However, using the accepted NIST data [4] for the atomic masses, the difference in atomic masses is (3.016 049 2777 AMU − 3.016 029 3191 AMU = 0.000019958 AMU = 0.018589 MeV). This is the total amount of energy available to get released in the decay conserving energy. Experimentally the escaping electron carries away 0.01859 MeV of kinetic energy. Thus there is no energy produced that even suggests that there had been an initial neutron binding energy of 0.782 MeV. The energy accounting for this decay exactly accounts for the kinetic energy of the escaping electron. No possible neutron binding energy is involved.

In the 1930s when the neutron and neutrino were originally discovered, physicists tried to explain how a neutron having spin of ½ could decay into a proton having spin ½ and an electron also having spin ½. For some reason these early physicists assumed that spin angular momentum was a scalar quantity and not the vector quantity which it actually is! The fact that neutrons apparently do not actually exist in atomic nuclei seems to suggest that there could be no source for the multitude of neutrinos which most scientists assume fill the universe. Conservation of energy and mass requires that there should have been 0.782 MeV of neutron binding energy released in the Tritium decay plus whatever energy would have been necessary to create the anti-neutrino plus whatever kinetic energy that anti-neutrino would carry away. Even if the neutrino has zero rest mass, some amount of energy must be provided to give it motion energy. For example photons do not have rest mass, but they still have some measureable radiation energy. In the analysis above we found that less than 10 eV was available from Tritium decay for the neutrino. This amount of energy is too small to allow a neutrino to be emitted from Tritium beta decay with any energy related to motion.

There is also the matter of the Weak Nuclear Force which is universally assumed to exist within atomic nuclei and controlling the occurrence of beta decays and electron captures by the nucleus. The Weak Nuclear Force was assumed to be required by physicists in the 1930s in order to have conservation of energy and mass. However, the analysis above shows that there is no energy or mass which is not fully accounted for when neutrons do not exist within atomic nuclei. These results based on the highly precise NIST nuclear data [4] suggest that the Weak Nuclear Force also does not exist within atomic nuclei.

11.6 2nd Proof That Neutrons and Weak Force Do Not Exist in Nuclei

There is a second nuclear process supposedly governed by the Weak Nuclear Force where an orbiting electron is captured into the nucleus of an atom. This is known simply as Electron Capture or EC. A careful examination of the energy accounting in such processes shows that no neutron self-binding energy exists in such nuclei.

Consider $^4Be_7$ which decays with a half-life of 54 days by Electron Capture. When the nucleus is considered to contain electrons and protons and no neutrons, no difference of mass occurs in the process. The result of the Electron Capture is $^3Li_7$. It is experimentally determined that radiation of 0.8618 MeV energy is released in this decay. An examination of the NIST isotopic masses reveals that the 7.0169292 AMU mass of $^4Be_7$ is converted into the 7.0160040 AMU mass of $^3Li_7$. This is a difference of 0.0009252 AMU mass which disappears. Converting this mass to MeV gives 0.8618 which provides the energy of the radiation given off by the decay.

If a newly formed neutron within the $^3Li_7$ nucleus had required 0.782 MeV to bind that electron to a proton to form the new neutron, then most of the 0.8618 MeV would have been used up. But this is not the case. The energy accounting shows that the emitted radiation is exactly the amount of the mass difference of the initial and final nuclei. This situation is also true for many EC decays such as $^6C_{11}$ releasing 1.982 MeV of radiation which exactly accounts for the mass difference. EC for $^{19}K_{40}$ (1.5048 MeV), $^{20}Ca_{41}$ (0.4213 MeV), $^{23}V_{49}$ (0.6018 MeV), $^{24}Cr_{51}$ (0.7527 MeV), $^{25}Mn_{53}$ (0.597 MeV), $^{26}Fe_{55}$ (0.2314 MeV), and $^{32}Ge_{68}$ (0.106 MeV) are additional examples. Thus the only reasonable interpretation of these Electron Capture processes is that there are no neutrons in nuclei and their decay is not caused by the weak interaction force.

11.7 Proof that the Strong Interaction Does Not Exist Within Nuclei

Consider the two isotopes $^{75}Re_{181}$ and $^{76}Os_{181}$. Both have an atomic weight of 181. Physicists assume that they each contain 181 protons and neutrons inside the nucleus. The first atom contains 75 protons and 106 neutrons in the nucleus with 75 electrons orbiting the nucleus. The second atom contains 76 protons and 105 neutrons in the nucleus with 76 electrons orbiting the nucleus. As a result they are very similar. The first atom has an additional neutron inside its nucleus. Now the neutron is unstable decaying into a proton and an electron with a half-life of a few minutes. Thus it could be said that these two atoms have the same constituent parts as long as that one extra neutron is identified as a proton plus an electron plus binding energy and a neutrino.

These two atoms have the same nominal atomic weight, but their precise weight is different. According to the NIST [4] data, the atomic weight of $^{75}Re_{181}$ is 180.950068 AMU and that of $^{76}Os_{181}$ is 180.95324. Since the actual components (considered as protons and electrons) are in the same exact numbers, this difference must be completely due to differences in (1) the binding energy holding each of the atomic nuclei together, (2) the binding energy holding each of the neutrons together, (3) the energy due to the neutrinos associated with individual neutrons inside the nucleus, and (4) the energy equivalent of whatever pions are present inside the nucleus.

According to standard nuclear physics, there are a lot of different binding energies that must be taken into account. The primary one is called the Strong Nuclear Force. It was invented by physicists as a way to overcome the incredibly powerful electrostatic repulsion between the positively charged protons in the nucleus and make it stable. Many isotopes are observed to be stable. Since the electrostatic repulsion is extremely strong and has an inverse-square dependence on distance, the Strong Nuclear Force was postulated to have an inverse-cube or higher distance dependence so as to be extremely powerful at short distances between protons but not to have any measureable effect beyond the nucleus.

In addition to the Strong Nuclear Force binding energy, there must be a binding energy that holds internal neutrons together. From the study of the decay of free neutrons that binding energy is known to be 0.78235 MeV or 0.000841 AMU. There is also a very small factor due to there being one less electron orbiting the $^{75}Re_{181}$ nucleus. The difference in ionization-related binding energy due to that electron is rarely higher than 0.0001 MeV or 0.0000001 AMU. Conventional nuclear physics also postulates that there are also many other binding energies inside the nucleus resulting from the presence of various pions, neutrinos and other objects as well as the energy equivalents of the rest mass of these particles.

Thus the measured difference of NIST atomic weights must be due to a combination of these many contributions, but primarily due to the Strong Nuclear Force. This suggests that if we graph all the NIST atomic weights of isotopes [4] of any one atomic weight, such as the 13 known isotopes of atomic weight 181, we should get a very complex graph. Figure 11-2 gives a graph of the NIST isotopic masses of all the known isotopes of nominal odd atomic weight 181. Notice that the graph of the data gives an amazingly pure parabolic shape just like that obtained for the one electron isoelectronic sequence of ionization energies. Why is it that we obtain such a well-defined parabolic shape dependent on the square of the charge in the nucleus?

A simple curve fit (See Figure 11-3) shows that indeed the curve is a parabola with a fit of 0.9992. This graph is representative of all same-atomic weight isotope families. Such graphs have been generated for every atomic weight where more than two isotopes are known. The group for 181 was chosen for this chapter, because it contains many known isotope members. Also this group provides enough data points to get a statistically reliable curve shape. Note that the isotopes at the bottom of the parabola are always stable nuclei.

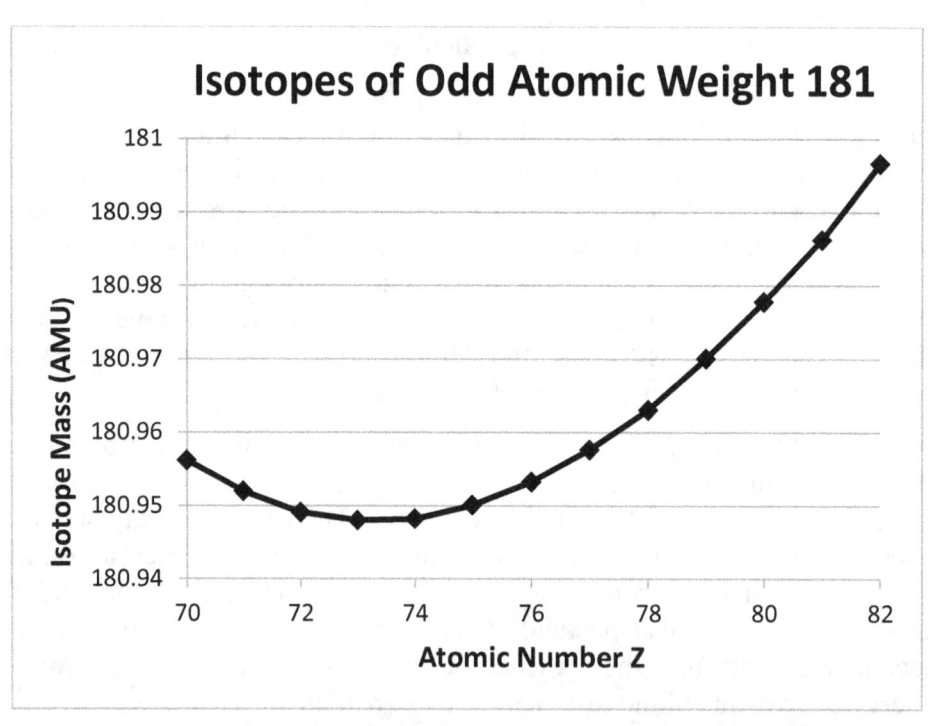

Figure 11-2 Nuclear Isotopic Masses for Odd Atomic Weight 181

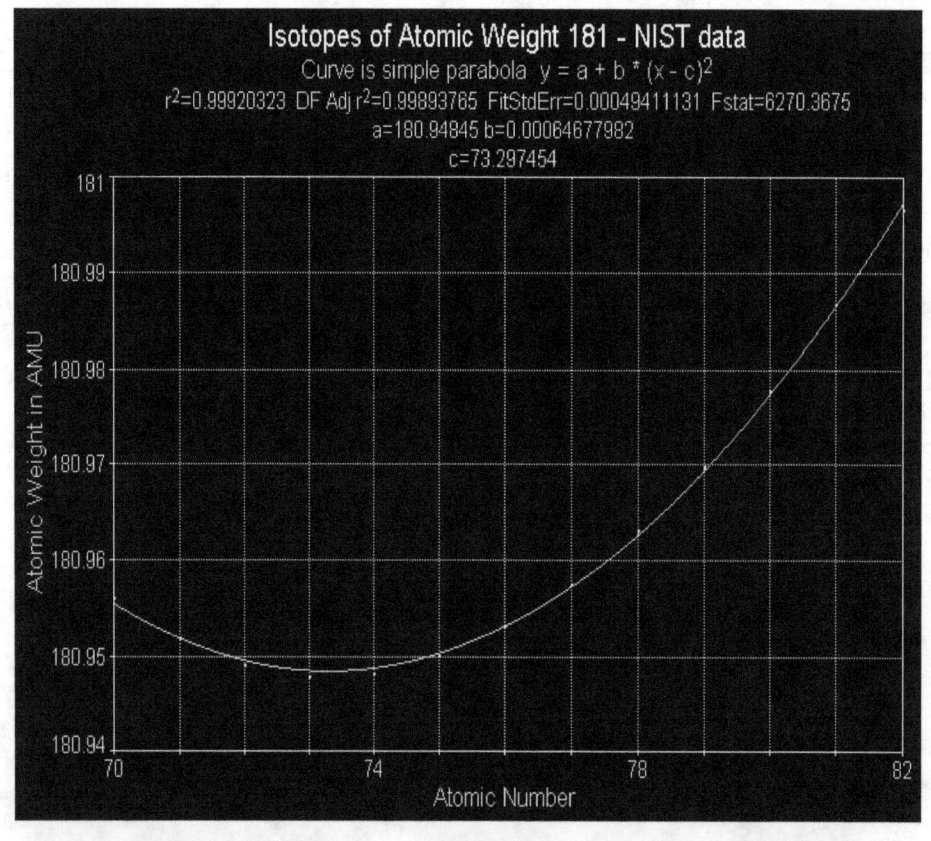

Figure 11-3 Isotopic Masses of Isotopes of Odd Atomic Weight 181

If the constituent neutrons are considered to be a combination of a proton and electron, then each of the 13 isotopes represented have the same exact quantity of electrons and protons. Therefore the amount of actual mass attributable to actual protons and electrons is identical in each case. Thus the different actual atomic weights are then purely due to differences in the binding energies, energy equivalents of pions, neutrinos, etc.

The parabolic shape, that fits the data very precisely, strongly implies a second power $1/r^2$ dependence of the electrodynamic force and definitely not a third power or higher dependence attributed to the Strong Nuclear Force. Thus it seems reasonable to interpret the parabola as due to the electrodynamic force involving $1/r^2$ dependence just as in the case for atomic ionization potentials. An examination of over 200 such graphs of the NIST data for each atomic weight always shows this extremely prominent parabolic shape. The lack of any curve distortions due to any other very strong binding force source seems to deny the possibility of the Strong Force existing in nuclei.

The data shown in Figures 11-2 and 11-3 was for odd atomic weight. For even atomic weights such as 104 as displayed in Figure 11-4 there appear to be two sets of data.

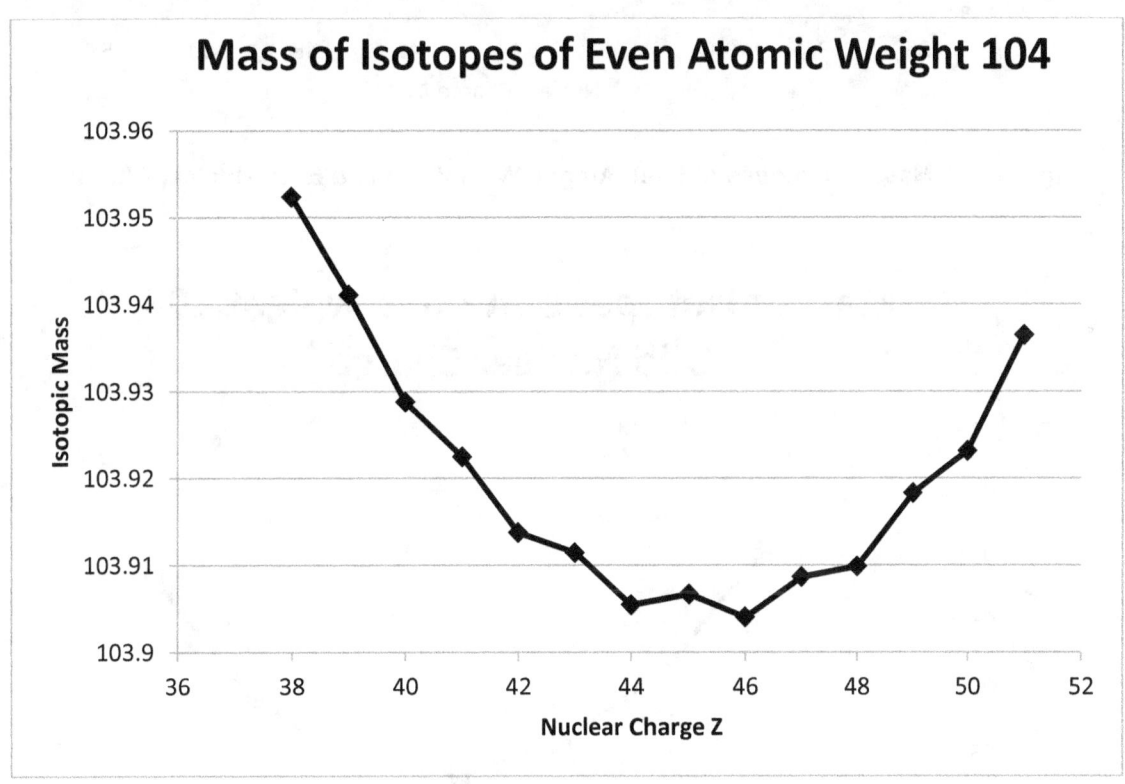

Figure 11-4 Isotopic Weights of Isotopes of Odd Atomic Weight 104

One is for even nuclear charge and one is for odd nuclear charge. When plotted separately they both have a precise parabolic shape as shown in Figures 11-5 and 11-6. This effect is due to the magnetic moment of the protons causing a different binding for even or odd numbers of protons. In Figures 11-2 and 11-3 there is always an odd proton or neutron in the nucleus. If the neutron is really an electron and proton, then there is always an odd proton in the nucleus so there is only one parabola describing the data.

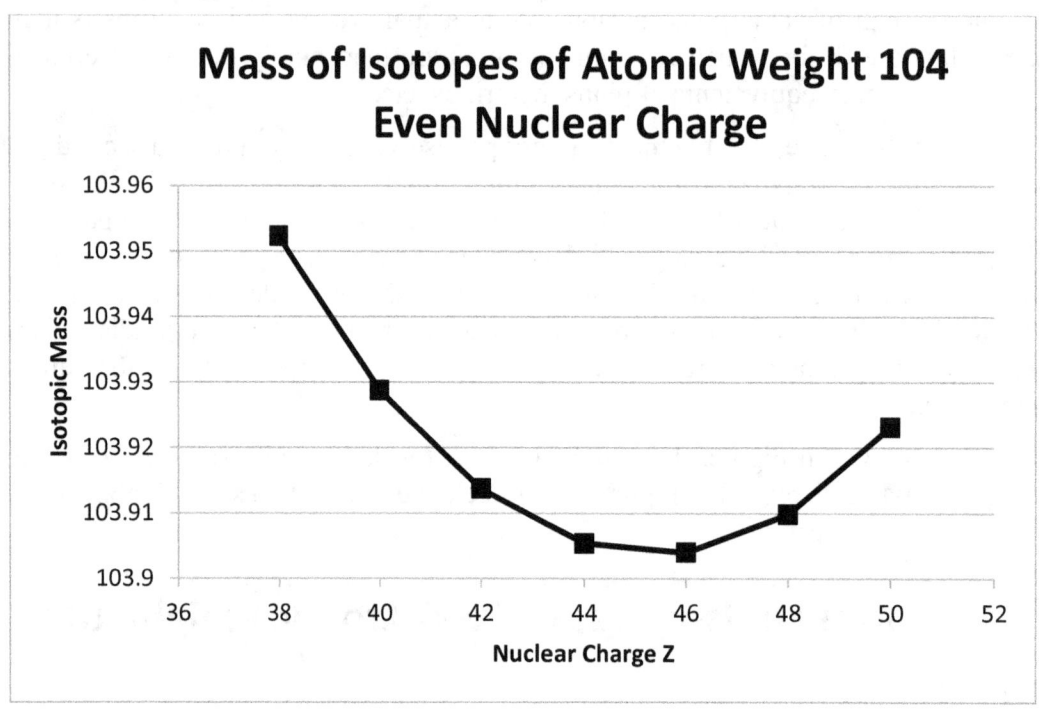

Figure 11-5 Mass of Isotopes of Even Atomic Weight 104 and Even Nuclear Charge

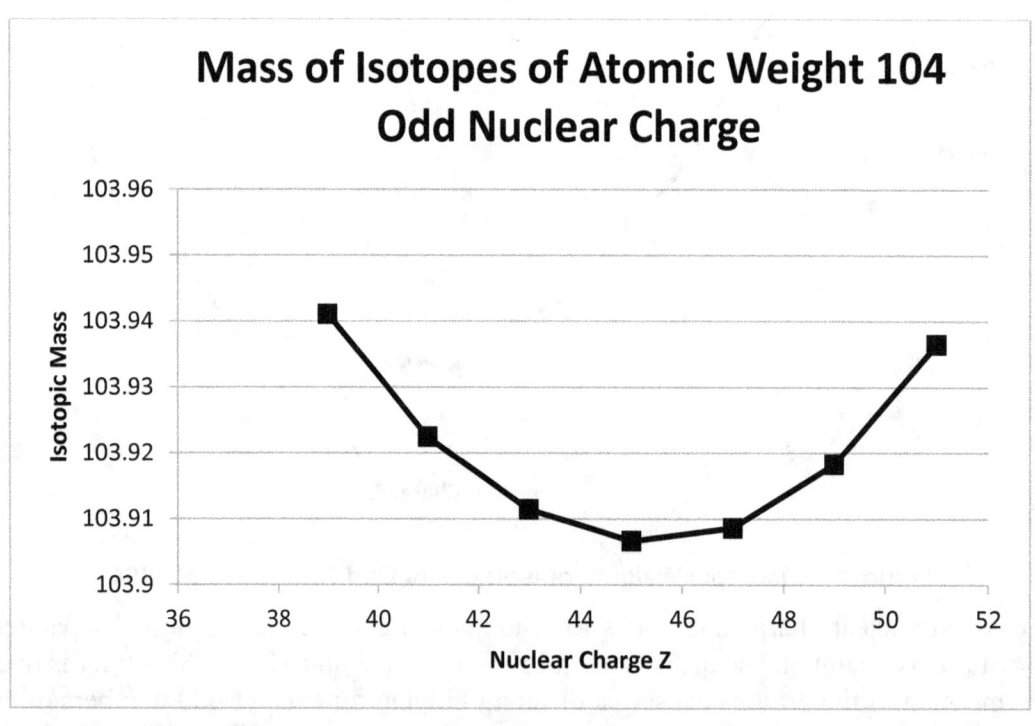

Figure 11-6 Mass of Isotopes of Even Atomic Weight 104 and Odd Nuclear Charge

11.8 Conclusions

Modern science is ultimately based upon the forces of nature and the models for elementary particles, atoms, nuclei and molecules. In this paper an analysis of the NIST ionization energies of the hydrogen one electron isoelectronic sequence has identified a precise Z^2 nuclear charge dependence in the electrodynamic ionization energy. A second analysis of the NIST isotopic masses of various nuclei has discovered that there is no data in the entire table of isotopic masses to justify (1) the existence of neutrons inside nuclei, (2) the existence of the Strong Force inside nuclei to keep the protons bound together in the nucleus, and (3) the existence of the Weak Force controlling the decay of various nuclei. Only evidence for the electrodynamic force is found.

These conclusions are compatible with the work on an improved electrodynamic force derived in this book. In this book the forces of gravity and inertia plus the centripetal force were obtained directly from the improved electrodynamic force law. In declaring the improved electrodynamic force to be the universal force, it was assumed that the very short range strong and weak interaction forces were due to finite-size effects, since those forces were based on the point particle idealization and had very short range. The analysis of this chapter challenges the very existence or necessity for the strong and weak interaction forces in nuclei.

Also this work further confirms the author's purely electrodynamic model of the nucleus which explains more nuclear data than any previous model. [5-12] **Modern science based on modern physics is due for a major reformation!**

11.9 References

1. Isaac Newton, **The Principia, Mathematical Principles of Natural Philosophy: A New Translation**, translators I Bernard Cohen and Anne Whitman (University of California Press, Berkeley, 1999) Rules of Reasoning in Philosophy: : Rule 1

2. **http://physics.nist.gov/PhysRefData/ASD/ionEnergy.html**

3. **http://www.hindawi.com/journals/tswj/2013/157412/**

4. **http://physics.nist.gov/cgi-bin/Compositions/stand_alone.pl?ele=&ascii=html&isotype=all**

5. C. W. Lucas, Jr., "A Physical Model for Atoms and Nuclei", **Galilean Electrodynamics Vol. 7**, pp. 3-12 (1996).

6. C. W. Lucas, Jr., "A Physical Model for Atoms and Nuclei", **Proceedings of the Physics Workshop**, held August 24-31, 1997 in Cologne, Germany

7. Charles W. Lucas, Jr. and Joseph Lucas, "A Physical Model for Atoms and Nuclei Part 1, 2, 3, 4" **Foundations of Science Vol. 5, No. 1**, pp. 1-7 (2002), **Vol. 5. No. 2**, pp. 1-8 (2002), **Vol. 6, No. 1**, pp. 1-10 (2003), **Vol. 6, No. 3**, pp. 1-8 (2003).

8. Lucas Jr., Charles W., "Derivation of the Classical Universal Electrodynamic Force Law", "The Electrodynamic Origin of the Force of Inertia", "The Electrodynamic Origin of the Force of Gravity", "A Classical Electrodynamic Theory of the Atom", "A Classical Electrodynamic Theory of the Nucleus", "A Classical Electrodynamic String Theory of Elementary Particles", "The Electrodynamic Origin of Life in

Organic Molecules Such as DNA and Proteins" **13th Annual Conference of the Natural Philosophy Alliance (NPA) "Science for the Next Generation"**, University of Tulsa, at Tulsa, OK April 3-7, 2006.

9. Charles W. Lucas, Jr., Roger A. Rydin, "Letter to the Editor" **Nuclear Science and Engineering Vol. 161**, pp. 1-2 (2009).

10. Charles W. Lucas, Jr., Roger A. Rydin, "Electrodynamic Model of the Nucleus", **Nuclear Science and Engineering Vol. 161**, pp. 255-256 (2009).

11. Roger A. Rydin, "A New Approach to Finding Magic Numbers for Heavy and Super-Heavy Elements" **Annals of Nuclear Energy, Vol. 38**, pp. 238-242 (2011).

12. Charles W. Lucas, Jr., Eric C. Baxter, Edward A. Boudreaux, and Roger A. Rydin, "A Classical Electro-Dynamic Theory of the Nucleus" **Physics Essays Vol. 26, No. 3**, pp. 392-400 (2013).

Chapter 12 Structure and Symmetry of the Universe

According to Plato of the ancient Greeks, mathematics is the mystical tool for the disclosure of the ultimate symmetry and structure of the universe.

To the man who pursues his studies in the proper way, all geometric constructions, all systems of numbers, all duly constituted melodic progressions, the single ordered scheme of all celestial revolutions, should disclose themselves, and disclose themselves they will, if, as I say, a man pursues his studies aright with his mind's eye fixed on their single end. As such a man reflects, he will receive the revelation of a single bond of natural interconnection between all these problems. Plato [1]

12.1 Importance of the Structure and Symmetry of the Universe

One might ask, "Why is the structure and symmetry of the universe important?" The answer appears to be that the current pillars of modern science, i.e. Maxwell's electrodynamics, the Copenhagen version of quantum mechanics, and Einstein's special and general relativity theories appear to make key assumptions regarding the structure and symmetry of the universe in strong disagreement with reality.

According to Maxwell's electrodynamics the universe consists of structureless point particles. Also the universe on a larger scale is structureless, or homogeneous and isotropic such that only linear field effects need to be taken into account. However, there are a number of electrodynamic phenomena such as lasers which appear to be based on non-linear field effects associated with structure.

The Copenhagen version of quantum mechanics also assumes that the universe consists of structureless point particles. In quantum mechanics the emission and absorption of radiation on particles is considered to be different from the emission and absorption of radiation on radio antennas which have a strong structural dependence. Philosophically quantum mechanics claims that "**reality is in the observations not in the Structure**". Thus quantum mechanics claims that there is no **Law of Cause and Effect** in the universe.

Einstein's special and general relativity theories assume that all particles are point-like on the small scale allowing unique reference frames to be defined. On the larger scale relativity theory claims the universe is homogeneous and isotropic with no structure. This does not appear to be in agreement with the distribution of matter in our solar system or the Milky Way galaxy. Also mass is assumed to be a fundamental quantity of nature independent of structure. This assumption does not agree with the concept of atoms, molecules and crystals which appear to have observable and measureable structure.

These three pillars of modern science were developed during the reign of the existential philosophy in science and freely incorporated many idealizations into science, since a majority of the scientists of the time did not believe that reality was knowable. The purpose of this chapter is to show that the electrodynamic force, when properly derived, predicts the symmetry of the universe on all size scales that is a combination of spherical and chiral symmetry. This symmetry is found in the structure of all elementary particles, atoms, nuclei, and molecules on the microscopic scale. On the macro scale of crystals, plant flowers, plant leaves, plant seed pods, animal body structures, orbits of the planets about the sun, orbits of the moons about the planets, the structure of the Milky way galaxy, and the structure of the universe as a whole about its center there is a combination of spherical and chiral symmetry in the structures. Thus it appears that the three pillars of modern science based on existential philosophy have missed the grand picture of the universe. They have missed its symmetry. They are out of step with reality.

Newton was different from his contemporaries who practiced science by making hypotheses from which specific predictions were made which were then tested by performing experiments. An examination of the mathematical theories of Galileo and Huygens shows that the propositions that they were pursuing were ones that made a distinctive empirical prediction that provided an answer to some practical question, or explained some known phenomena. **This is the modern scientific method of the Post-Modern philosophy of science in our time.** Newton in the **Principia** took a more general approach. He was not so interested in conjecturing hypotheses and then testing the implications of those hypotheses, but rather in using experimental observations and mathematics to discover the basic force axioms from which science can attain by deductive logic more general force laws from which the most general fundamental force law can be obtained by inductive generalization. Newton argued that this was the best way to make progress in science towards its ultimate goal of understanding the universe.

12.2 Structure Is From Symmetry of Universal Force

Newton acknowledged the risk of inductive generalization in his famous methodological passage in the **Opticks**, in the discussion of the methods of "analysis and synthesis" in the next to last paragraph of the final Query, which was added in 1706:

> **The Analysis consists in making Experiments and Observations, and in drawing general Conclusions from them by Induction, and admitting of no Objections against the Conclusions, but such as are taken from Experiments, or other certain Truths. Hypotheses are not to be regarded in experimental Philosophy. And although the arguing from Experiments and Observations by Induction be no Demonstration of general Conclusions; yet it is the best way of arguing which the Nature of Things admits of, and may be looked upon as so much the stronger, by how much the Induction is more general. And if no Exceptions occur from Phenomena, the Conclusion may be pronounced generally. But if at any time afterwards any Exception shall occur from Experiments, it may then begin to be pronounced with such Exceptions as occur. By this way of Analysis we may proceed from Compounds to Ingredients and from Motions to the Forces producing them;**

and in general, from Effects to their Causes, and from particular Causes to the more general ones, till the Argument end in the most general. [2]

Newton's arguments for a universal force of gravity and a universal force of inertia illustrated his new empirical approach to natural philosophy and physics in general. This new approach was based on a generic mathematical approach, the roles of deduced theory and empirical axioms in ongoing research, and the insistence on pushing theory far beyond its original experimental domain. Newton did not allow the incorporation of unproved hypotheses or idealizations in science.

In that same spirit the improved version of the electrodynamic force derived in this work by means of a more perfect union of the axiomatic and empirical scientific methods is conjectured to be the universal force. This conjecture is based on the ability of this newly derived force to explain the electrodynamic force and the forces of gravity and inertia more completely than any previous theory. It is based on the ability of this newly derived force to reveal the local contact nature of these forces based on the electromagnetic fields of a charge remaining attached and extending the range of the force. Also it is based on its inherent ability to explain what gravitational and inertial mass are for the first time in history. Finally it is based on the force's support for the Law of Cause and Effect, conservation of energy and momentum, and Mach's Principle.

Some scientists might object to this conjecture, because the strong interaction force within elementary particles and the weak nuclear force responsible for beta decay have not yet been explained. However, both of these forces are very short-ranged with a range on the order of the measured size of elementary particles. These forces were invented to compensate for incorporating the point particle approximation in electrodynamics, relativity theory and quantum mechanics. **The Universal Force Volume 2 - An Electrodynamic Model of Elementary Particles** and **The Universal Force Volume 3 - An Electrodynamic Model of the Atom and Nucleus** will explain the experimental data more completely and without resort to these imaginary forces.

Due to the generic mathematical approach that Newton introduced into science, there are now some new possibilities to identify and confirm a universal force that were not possible in science before his time. This can be seen by examination of the mathematical expressions for the electrodynamic potential and the electrodynamic force displayed in Equations (11-1) and (11-2) below.

$$U(r,v) = \frac{qq'}{r}\frac{(1-\beta^2)}{(1-\beta^2 sin^2\theta)^{1/2}} = \frac{qq'(1-\vec{\beta}^2)}{\left[\vec{r}^2 - \frac{\{\vec{r}\times(\vec{r}\times\vec{\beta})\}^2}{\vec{r}^2}\right]^{\frac{1}{2}}} \quad (12-1)$$

where $\beta = v/c$ and

$$\frac{dU(\vec{r},\vec{v})}{dt} = -\vec{v}\cdot\vec{F}(\vec{r},\vec{v},\vec{a})$$

$$\vec{F}(\vec{r}, \vec{v}, \vec{a}) = \frac{qq'}{\vec{r}^2} \left[\frac{(1 - \vec{\beta}^2)\vec{r} + \frac{2\vec{r}^2}{c^2}\vec{a}}{\left[\vec{r}^2 - \frac{\{\vec{r} \times (\vec{r} \times \vec{\beta})\}^2}{\vec{r}^2} \right]^{\frac{1}{2}}} \right.$$

$$\left. - (1 - \vec{\beta}^2) \frac{(\vec{\beta} \cdot \vec{r})\vec{r} \times (\vec{r} \times \vec{\beta}) + (\vec{r} \cdot \vec{r})\vec{r} \times \left(\vec{r} \times \frac{\vec{a}}{c^2}\right)}{\left[\vec{r}^2 - \frac{\{\vec{r} \times (\vec{r} \times \vec{\beta})\}^2}{\vec{r}^2} \right]^{3/2}} \right] \qquad (12-2)$$

Notice the symmetry of the terms. Those terms proportional to **r** and **a** have spherical symmetry. Those terms proportional **r x (r x β)** and **r x (r x a)** have chiral symmetry.

The word "chiral" comes from the Greek for "hand". The most common example of chiral symmetry is the mirror symmetry shown by your left and right hand. Chirality is already an important concept in quantum field theory, nuclear physics, chemistry of molecules, and biology. In chemistry chiral symmetry produces a left- and right-handedness and a spiraling of fibers or polymers in organic molecules. In quantum field theory, chiral symmetry is one of many possible symmetries of the Lagrangian. [3]

The combination of right and left handed mirror symmetry, quantization, and spiraling produce some tell-tale signs of chiral symmetry. Structures tend to consist of a prime number of identical sub-structures. Often these substructures are arranged with spherical symmetry.

If the derived electrodynamic force law of this work is indeed the universal force, one would expect to see a combination of spherical and chiral symmetry in structures of the universe on all size scales. This appears to be the case. Evidence will now be presented for the structure and symmetry of the universe from the structure of elementary particles, atoms, nuclei, molecules, crystals, plant leaves, plant flowers, plant seed heads, animal body structures, orbits in the solar system, structure of the Milky Way galaxy, and the structure of the universe as a whole.

12.3 Symmetry of Structure of Elementary Particles

The universal electrodynamic force leads to a new model of finite size elementary particles that consist of closed charge loops. The various elementary particles are composed of 1, 3, 5, 7, etc. primary toroidal structures which may be complex and consist of some secondary closed charge loop structures which may also be complex and consist of some tertiary closed charge loop structures. Below is a diagram of a single closed charge loop structure that is the building block of all elementary particles. Note that the toroidal ring is just there to guide the eye to see that the charge fiber is spiraling. For the electron there are three of these secondary loops equally spaced on the surface of a larger toroid to form one primary fiber. These secondary fibers are bound together in a stable configuration to form the more complex primary fiber. See the book **The Universal Force Volume 2 – An Electrodynamic Model of Elementary Particles** for more details.

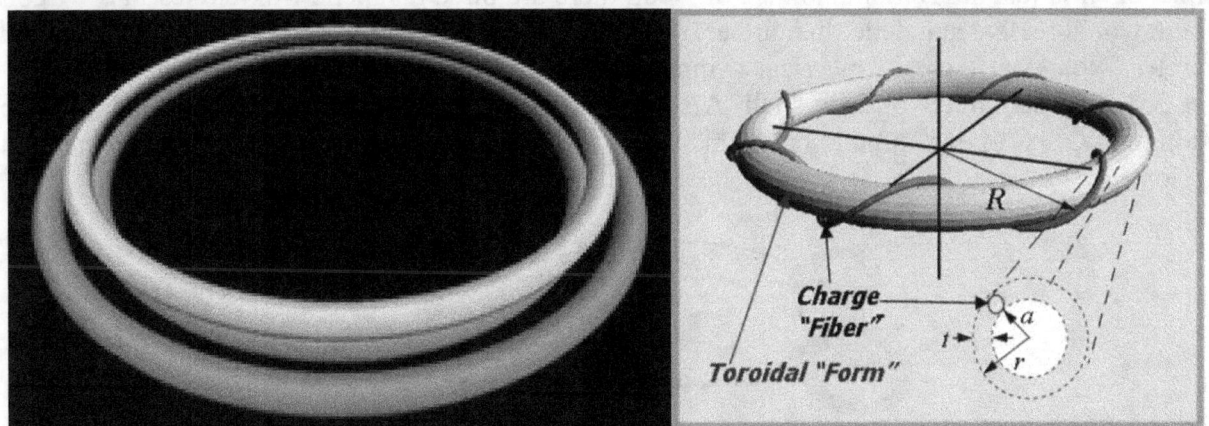

Figure 12-1 Spherical and Chiral Symmetry of the Electron and Toroidal Rings

12.4 Symmetry of Structure of Nuclei

The universal electrodynamic force leads to a new model of the nucleus that exhibits spherical and chiral symmetry in its shell structure. Below is a diagram of the oxygen-16 nucleus. Note the polarizing of the neutrons to form something like an electron-proton pair that aligns with another proton to form proton-electron-proton triplets. The overall symmetry of the nucleon shells is spherical and they are arranged in triplet sets for chiral symmetry. The structures of the other nuclei are similar. See **The Universal Force Volume 3 – An Electrodynamic Model of Atoms and Nuclei**.

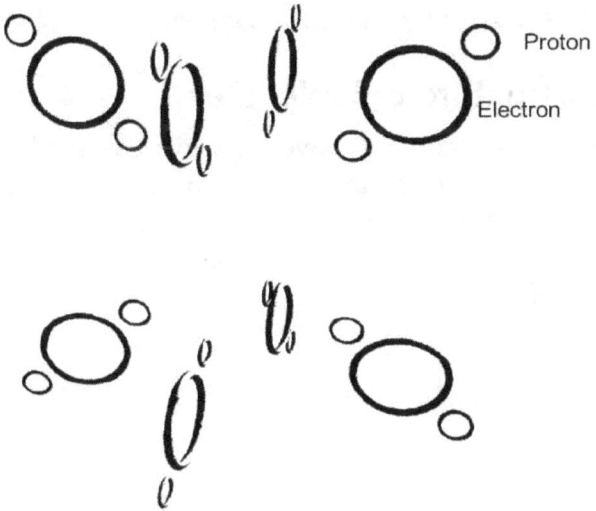

Figure 12-2 Spherical and Chiral Symmetry of the Oxygen-16 Nucleus

12.5 Symmetry of Structure of Atoms

The universal electrodynamic force leads to a new model of the atom that exhibits spherical and chiral symmetry. See **The Universal Force Volume 3 – An Electrodynamic Model of the Atom and the Nucleus.** Below is a diagram of the Neon atom. The innermost shell of electrons has two ring electrons with a magnetic flux circle through their centers. This inner shell is just the Helium atom. For Neon there are three magnetic flux circles. For Argon there would be five magnetic flux circles. Thus there is chiral symmetry in atomic magnetic flux circles. Note that there is spherical symmetry in the electron distributions which are equally spaced on great circles of each shell. Also note that this is identical to Parson's model of the atom that Gilbert Lewis used to create his very successful dot diagram showing how atoms bond to form molecules.

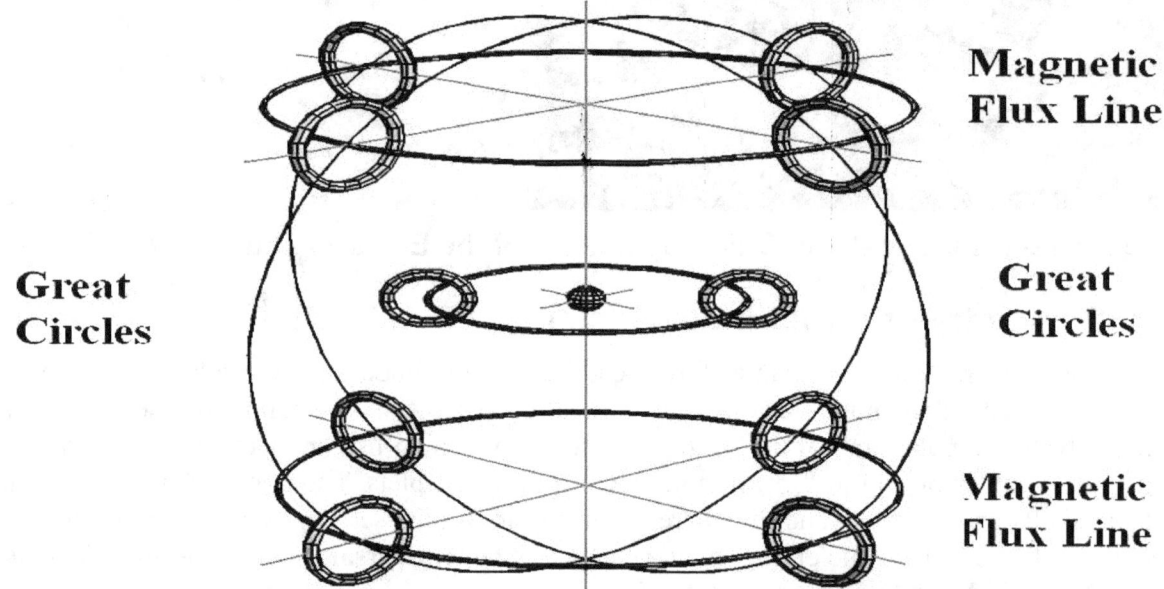

Figure 12-3 Spherical and Chiral Symmetry of Neon-20 Atom

12.6 Symmetry of Structure of Molecules

The universal electrodynamic force above leads to a new model of molecule formation in which the atoms of the molecule are bound together by magnetic bonding of electron shells. Samples of simple molecules below show the role of the cubic or 3 chiral symmetry in these simple molecules. See the book The **Universal force Volume 4 – An Electrodynamic Model of Molecules and the Origin of Life**.

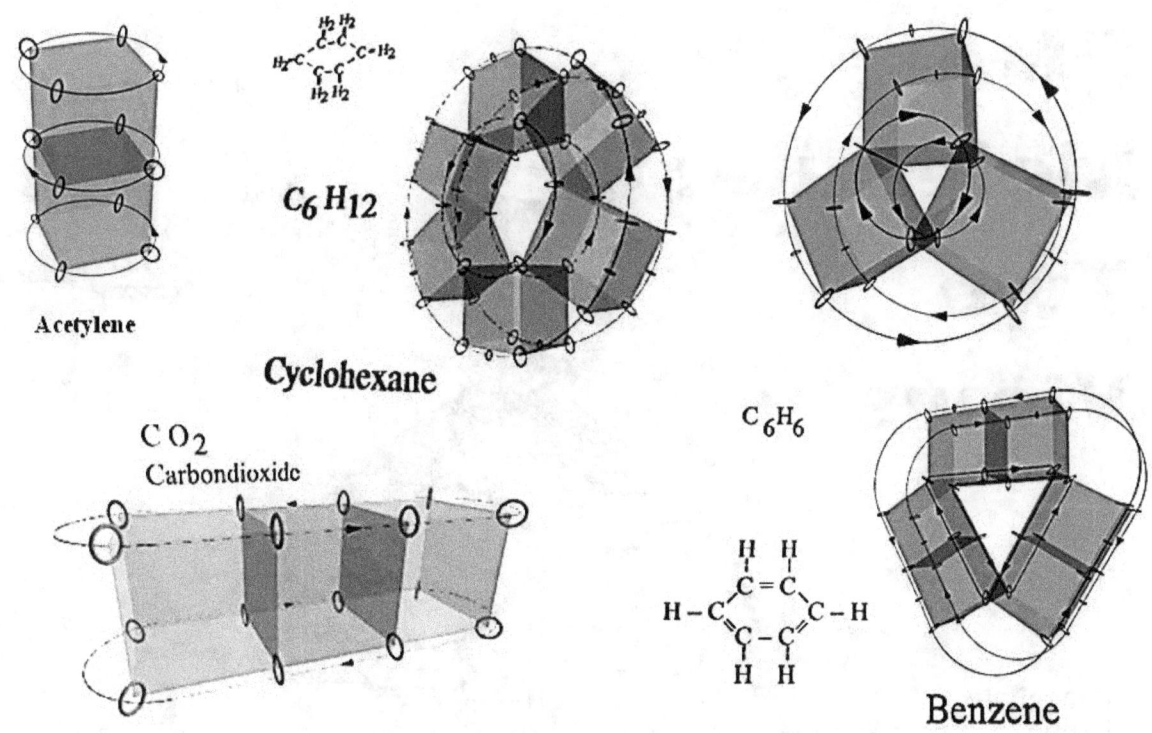

Figure 12-4 Chiral Symmetry of Simple Molecules

For more complex organic molecules such as starches, proteins, and DNA we see 1, 3, 5 spiraling fibers in the structure.

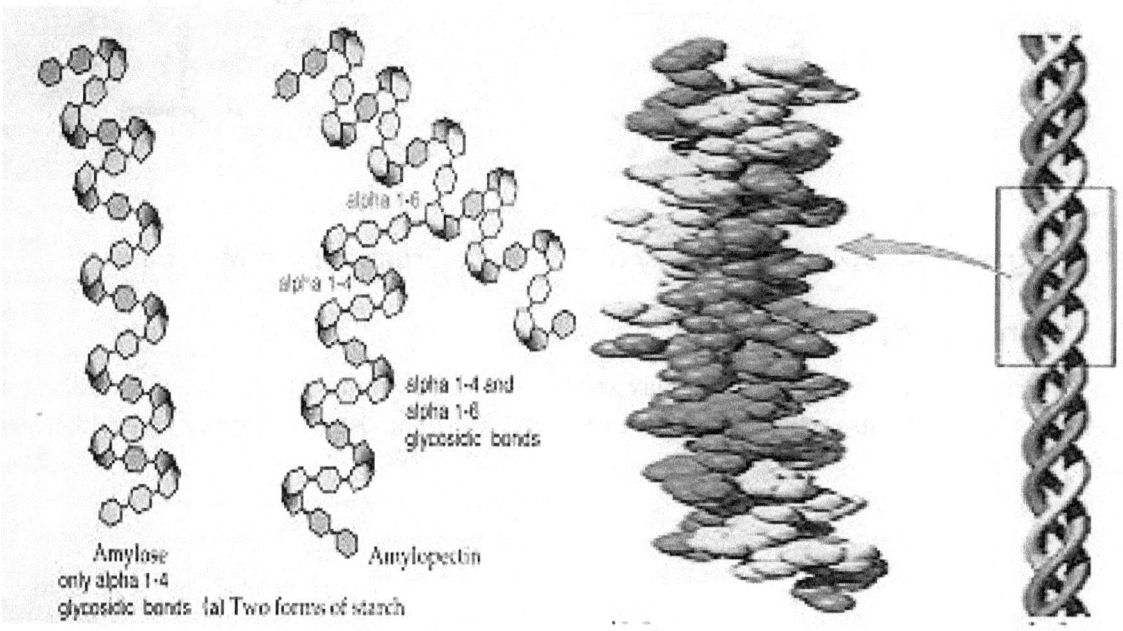

Figure 12-5 Chiral Symmetry of Complex Organic Starch and Protein Molecules

DNA Molecule: Two Views

Sugar —

Bases —

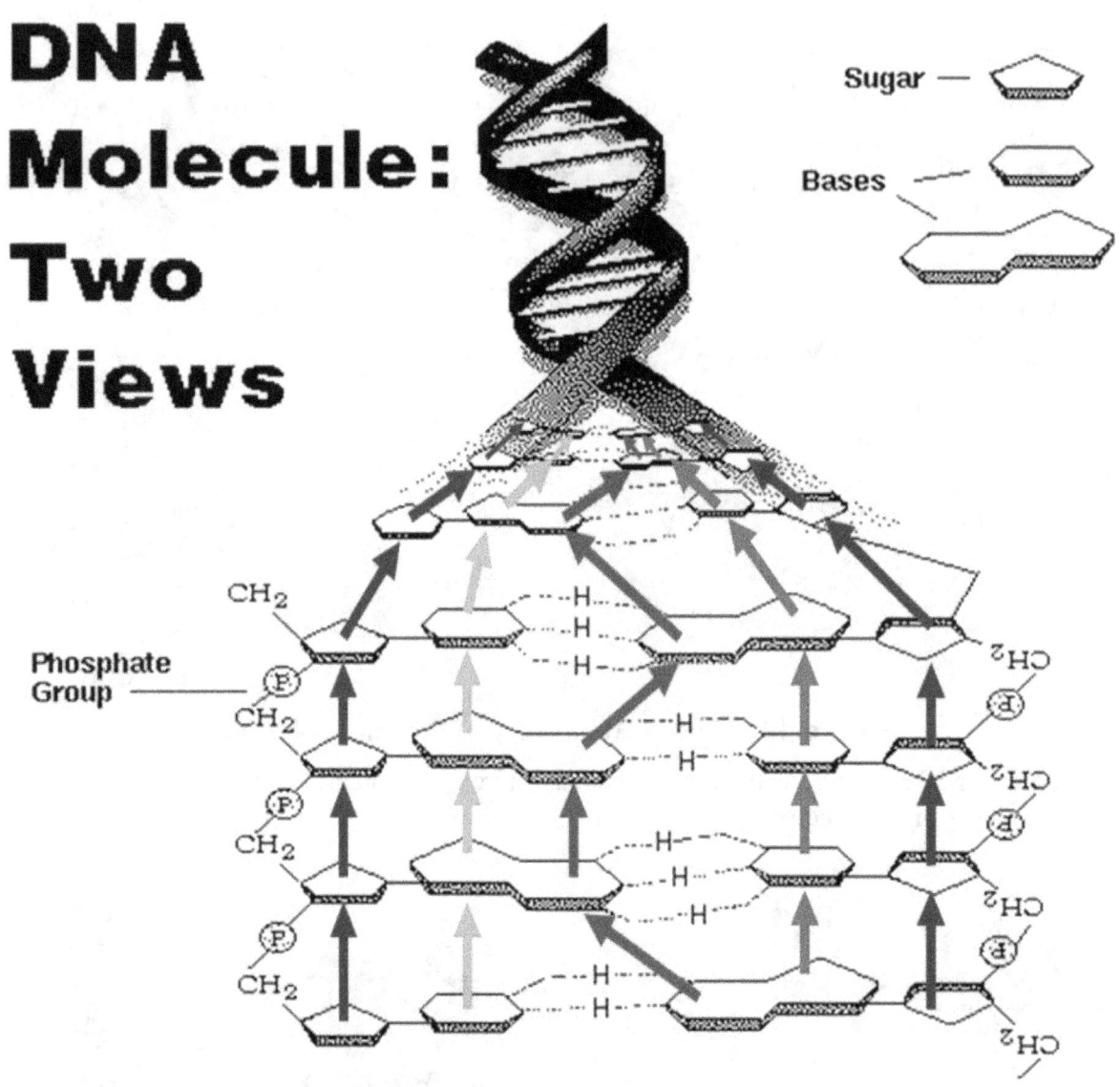

Phosphate Group —

Figure 12-6 Chiral Symmetry of Complex Organic DNA Molecule

12.7 Symmetry of Structure of Crystals

Crystals of various types also display spherical and chiral symmetry in a plane as can be seen in these pictures of snow flake crystals with 1 and 3 principal axes of symmetry and 1, 3, 5, and 7 secondary axes of symmetry.

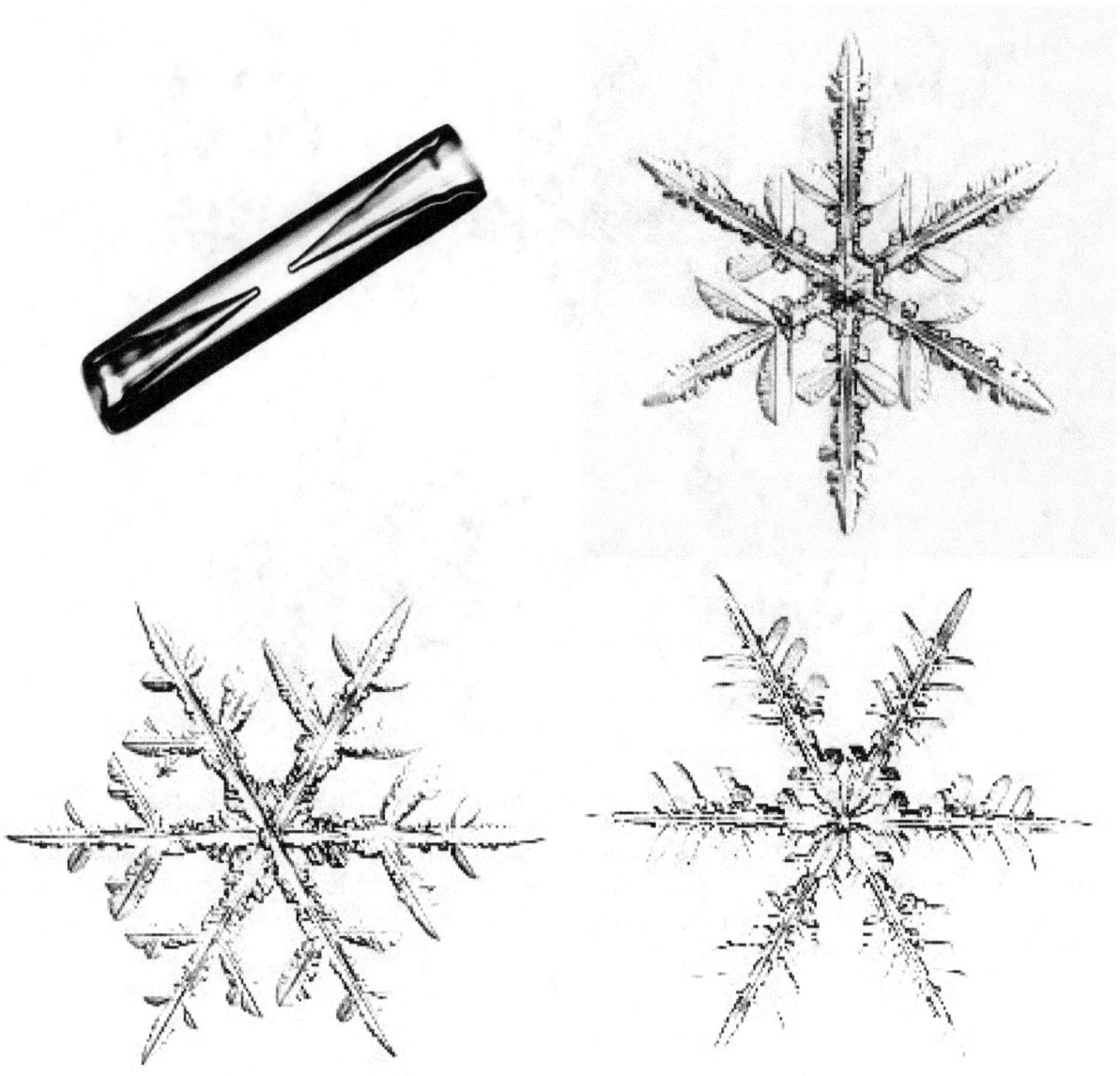

Figure 12-7 Spherical and Chiral Symmetry of Snowflakes

Even though no two snowflakes are identical, they all seem to have a prime number of principal axes, a prime number of secondary axes and spherical symmetry.

12.8 Symmetry of Structure of Plant Leaf Patterns

The pictures below show the spiraling nature of leaf patterns on the plant and the chiral symmetry of 1, 3, 5, and 7 leaves in a cluster. Note the top left picture shows the spiraling pattern from the top down. The picture to the right of it shows the chiral symmetry of 3 axes in the leaf pattern from the top down.

Figure 12-8 Spherical and Chiral Symmetry of Plant Leaf Patterns

The chiral symmetry of leaves is not confined to the pattern on the stem or branch. The internal structure of the leaf also reveals chiral symmetry. In the leaves depicted below note the number of lobes in the structure of the leaves is 1, 5, and 7.

Figure 12-9 Chiral Symmetry of Leaf Shapes

12.9 Symmetry of Structure of Plant Flower Petal Patterns

The flowers of plants also exhibit spherical and chiral symmetry in the petals of the flowers in a plane. Note that the flowers have 3, 5, and 7 petals in a plane with spherical symmetry. Some flowers, such as the rose, have multiple symmetry patterns with 1, 3, 5 petals in plane layers of the same flower.

Figure 12-10 Spherical and Chiral Symmetry of Flower Petal Patterns

12.10 Symmetry of Structure of Plant Seed Head Patterns

Not only do plant leaf structures and flower structures have chiral symmetry, but also the seed head patterns have chiral symmetry. This can be seen in the sunflower seed head and the pine cone seed head below. Note the spiraling pattern.

Figure 12-11 Spherical and Chiral Symmetry of Seed Head Patterns

12.11 Symmetry of Man and Animals

An examination of man and other animals shows that they have chiral symmetry. From Figure 12-12 one sees that man has a left and right eye, ear, arms, hands, breasts, legs, feet, etc.

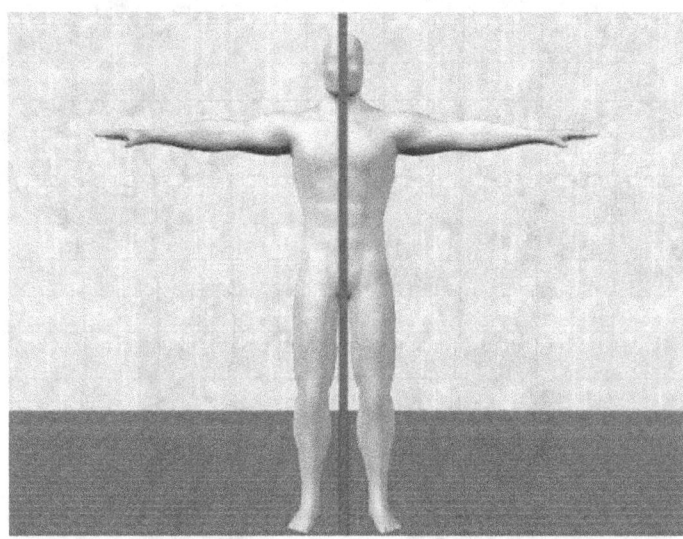

Figure 12-12 Chiral Symmetry of Human Body

12.12 Symmetry of Structure of Orbits in Solar System

The force of gravity as derived from the universal electrodynamic force law has a spherical first term and a chiral second term that is missing from Newton's Universal Law of Gravitation and Einstein's General Relativity Theory. Those theories do not predict spiraling quantized orbits for planets and moons about the planets. The first diagram below shows the spiraling of a planet on the surface of a toroidal ring in which the planet goes around the cross section of the toroid once in one revolution of the sun producing what appears to be an ellipse tilted with respect to the equatorial plane of the sun.

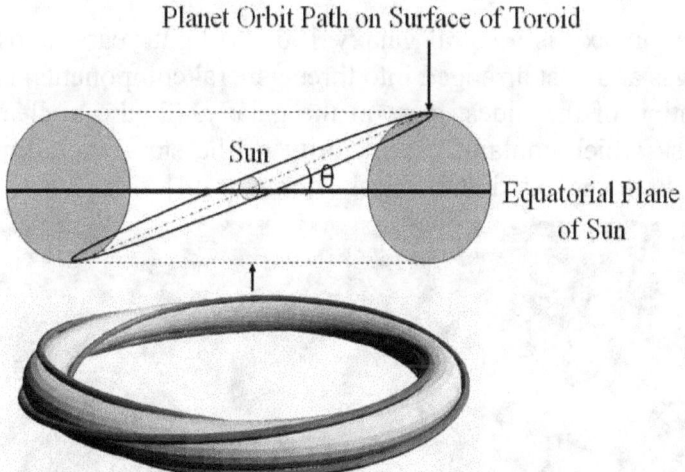

Figure 12-13 Spherical and Chiral Symmetry of Planetary Orbits

Figure 12-14 shows the orbits of four of Jupiter's moons about the planet's orbit. Here the spiraling is very noticeable. Note that the periods and radii of the orbits of different moons are integer multiples of each other. This is known as the modern version of Bode's Law. Also note that there are 5 bodies spiraling about each other.

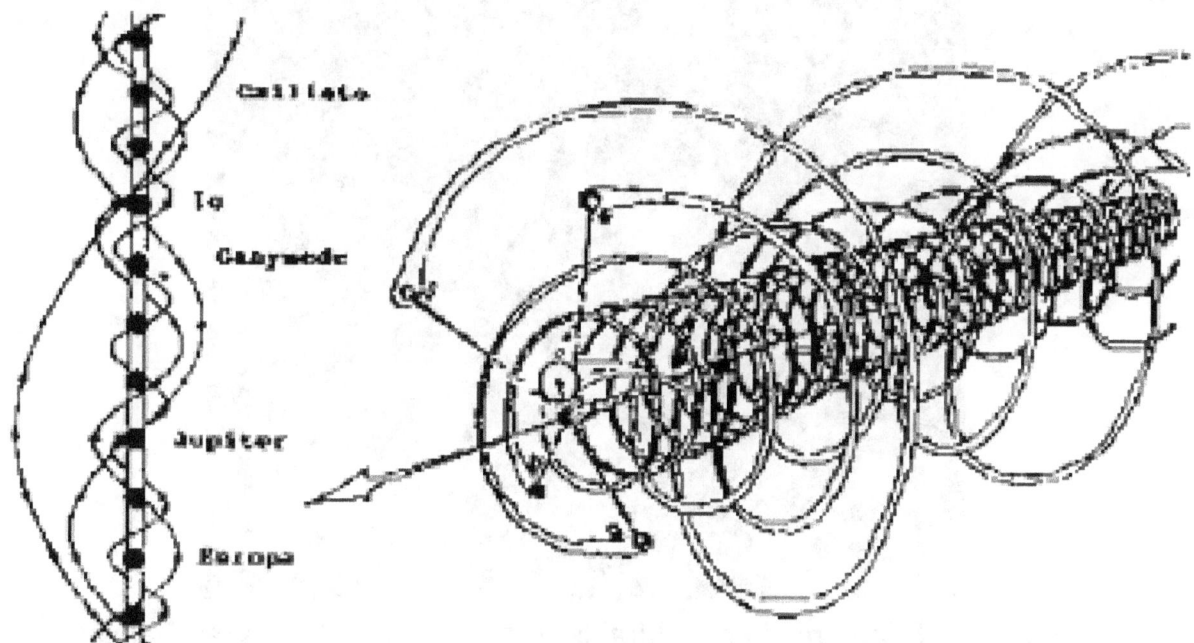

Figure 12-14 Spherical and Chiral Symmetry of the Orbits of Jupiter and Its Moons

Also note that there are 9 planet orbits in the solar system, if one assumes that the fifth planet came apart to produce the asteroid belt.

12.13 Symmetry of Structure of Milky Way

The Milky Way galaxy is a spiral galaxy 100,000 light years across consisting of over 400 billion stars plus gas and dust arranged into three general components, i.e. (1) the halo which is a spherical distribution of the oldest stars in the galaxy, (2) the nuclear bulge and galactic center, and (3) the disk which contains the majority of the stars, including the sun. Note that there are 7 spiral arms in the galaxy due to chiral symmetry. [4]

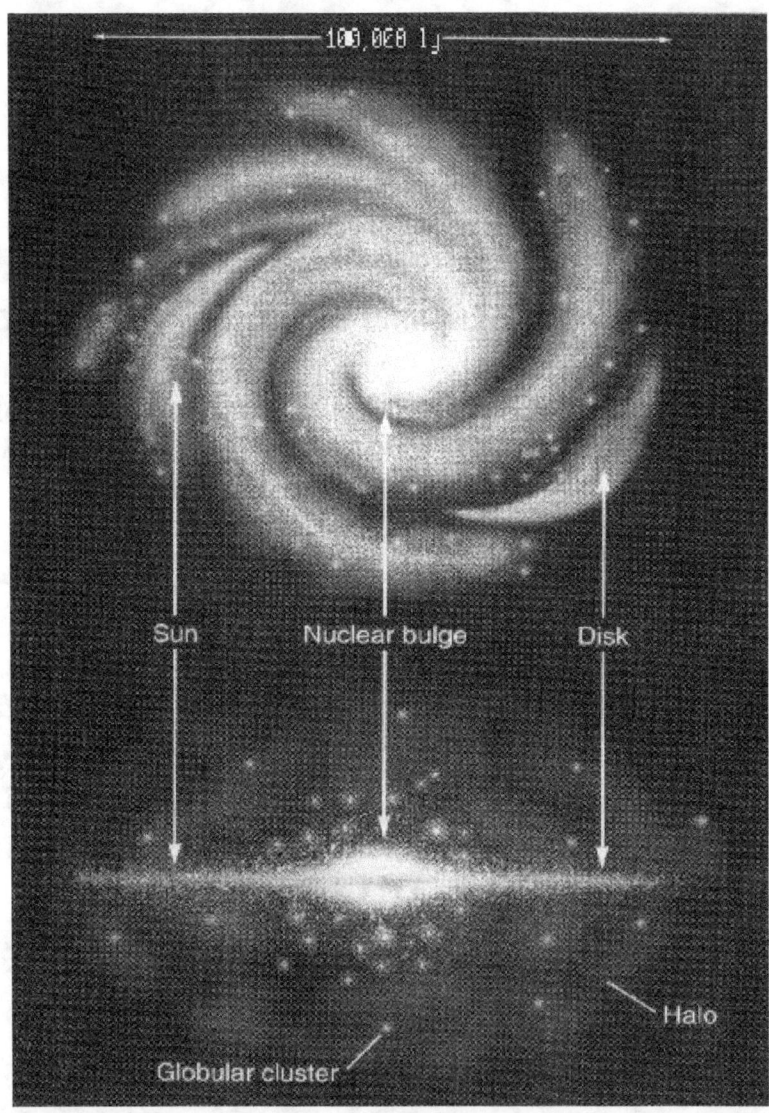

Figure 12-15 Spherical and Chiral Symmetry of the Milky Way Galaxy

Galaxies also come in the shape of a ring. Here the ring is composed of 3 fibers composed of billions of stars orbiting around one another just like the moons of Jupiter. Note that there are 3 orbits just like in the diagram to the right.

192

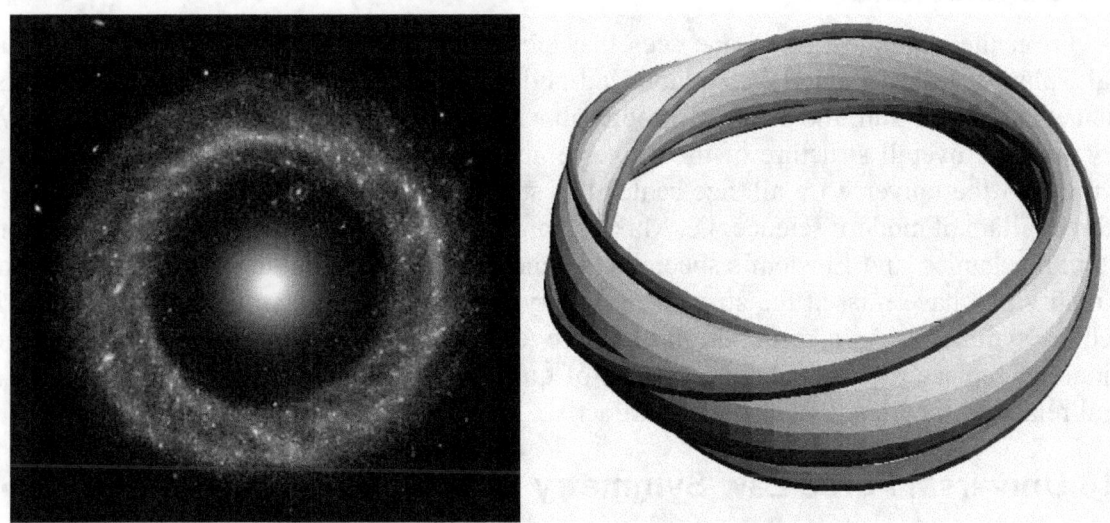

Figure 12-16 Spherical and Chiral Symmetry of Hoag's Ring Galaxy

12.14 Symmetry of Structure of Whole Universe

Chiral symmetry gives rise to quantization in the structure of elementary particles, the nucleus, the atom, the solar system, and galaxies.   Thus we expect to see some sort of quantization of the motions of galaxies about the center of the universe. In Chapter 9 section 8 and 9 on the origin of Hubble's Law due to gravitational red shifting, the argument is given that the universe has a center due to the quantization of red shifts about a point in the universe showing that the galaxies move in quantized orbits obeying the modern version of Bode's Law. [5] The number of these quantized spherical orbits about the center of the universe could also be a prime number according to the data shown below. [5]

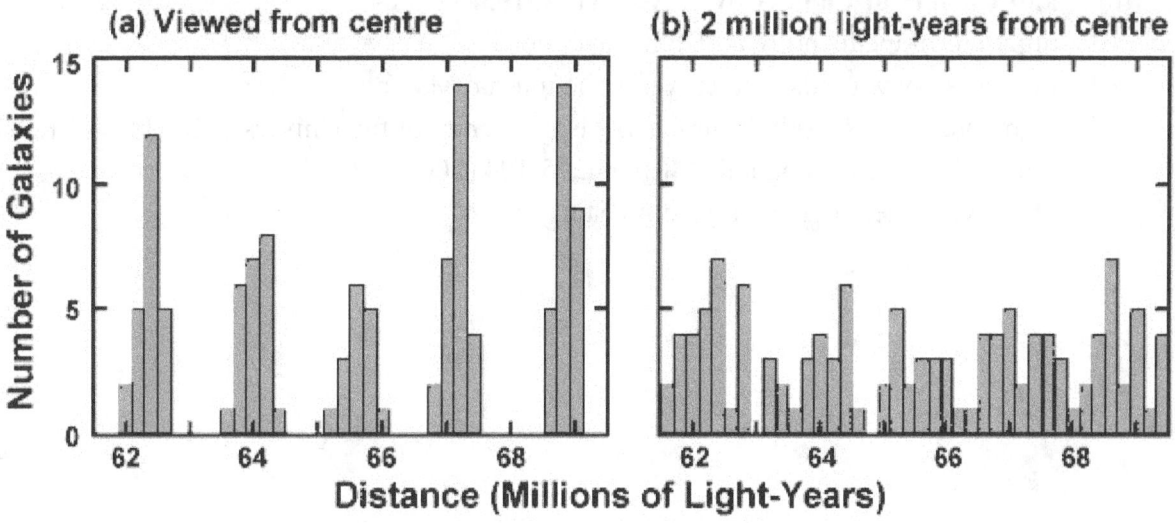

Figure 12-17 Spherical and Chiral Symmetry of the Universe from Red Shift Data

12.15 Conclusions

From the examples given one sees that elementary particles, atoms, nuclei, molecules, crystals, plant flowers, plant leaves, plant seed pods, animal body structures, man, the orbits of the planets about the sun, the orbits of moons about the planets, the structure of the Milky Way galaxy, and the overall structure of the universe about the center of the universe indicates that everything in the universe on all size scales has structure with spherical and chiral symmetry. The three pillars of modern science, i.e. Maxwellian electrodynamics, the Copenhagen version of quantum mechanics, and Einstein's special and general theory of relativity, are not in agreement with reality and have missed the structure of everything in the universe including the spherical and chiral symmetry. As a result the new improved universal electrodynamic force is a good candidate to replace the Copenhagen version of quantum mechanics and Einstein's special and general relativity theories. It is also more attractive, because it is a simpler theory.

12.16 Universal Force Law Symmetry Is Source of All Beauty

Beauty is that quality that gives pleasure to the mind or senses and is associated with harmony or symmetry of form or color. [6]

From the sections above one can see that the combination of spherical and chiral symmetry is unique and is the property that creates the pleasing harmony of form and color throughout the universe. Thus one may say that the symmetry in nature resulting from the universal electrodynamic force law is the source of all natural beauty in nature.

12.17 References

1. Plato, **Epinomis**, translated by A. E. Taylor (Nelson and Sons, 1956) 991e-992a.
2. Isaac Newton, **Opticks: or A Treatise of the Reflections, Refractions, Inflections and Colours of Light** (Dover, New York, 1952) p. 404.
3. http://en.wikipedia.org/wiki/Chiral_symmetry
4. http://casswww.ucsd.edu/archive/public/tutorial/MW.html
5. Humphreys, D. Russell, "Our Galaxy is the Centre of the Universe, 'Quantized' Red Shifts Show", **The Journal, 16(2),** pp. 95-104 (2002).
6. http://www.thefreedictionary.com/beauty

Chapter 13 Mach's Principle and the Concept of Mass

Inertia originates in a kind of interaction between bodies.

Albert Einstein [1]

In theoretical physics, especially in inertial and gravitational theories, Mach's Principle is the name given by Einstein to a general principle credited to the physicist and philosopher Ernst Mach. Mach's Principle is based on the observed forces of gravity and electrodynamics having a $1/R^2$ type force with infinite range. It assumes that the motion of any mass must take into account the interactions with all the other masses in the universe.

Typically when scientists calculate the forces of gravity and inertia, they only take into account the local anisotropic masses nearby. Thus for the motion of a satellite around the Earth a scientist would take into account the effect of the Earth, the moon, the sun, and perhaps the planet Jupiter with the effect of each of these bodies getting smaller in order of significance. The effect of the rest of the universe is ignored as being too small to measure.

Mach's Principle is based on the notion that there is a very large number of masses very far away and if the universe is not isotropic and homogeneous but has a symmetry such as a spherical symmetry with all masses rotating about the center of the universe like the planets rotate about the sun and the stars rotate about the center of the Milky Way galaxy, these masses will make a measurable contribution. Thus for the forces of inertia and gravity, there is a second term to take into account the masses in the rest of the universe. For stability in the universe the forces of inertia must be in equilibrium with the forces of gravity and electrodynamics.

In 1883 Mach wrote:

For me only relative motions exist … When a body moves relatively to the fixed stars, centrifugal forces are produced; when it moves relatively to some different body, and not relatively to the fixed stars, no centrifugal forces are produced. I have no objection to calling the first rotation "absolute" rotation, if it be remembered that nothing is meant by such a designation except relative rotation with respect to the fixed stars. [Ernst Mach, The Science of Mechanics 1883]

A very general statement of Mach's Principle is

Local physical laws are determined by the large scale structure of the universe. [2]

This basic idea also appeared before Mach's time in the writings of George Berkeley. [3] The book **Absolute or Relative Motion?** (1896) by Benedict Friedländer and his brother Immanuel contained ideas similar to Mach's Principle.

In Mach's own words, the principle was expressed as follows:

[The] investigator must feel the need of … knowledge of the immediate connections, say, of the masses of the universe. There will hover before him

as an ideal insight into the principles of the whole matter, from which accelerated and inertial motions will result in the same way. [4]

Einstein seemed to view Mach's Principle as something along the lines of

Inertia originates in a kind of interaction between bodies (masses) ... [1]

In some sense Mach's Principle is related to philosophical holism. It appears to require that inertial and gravitational theories should be relational theories depending on relative coordinates. In his book **The Science of Mechanics [4]** Mach criticized Newton's idea of absolute space based on his bucket argument.

In his book **Philosophiae Naturalis Principia Mathematica** [5] Latin for "Mathematical Principles of Natural Philosophy" Newton tried to demonstrate that one can always decide if one is rotating with respect to the absolute space by measuring the apparent forces that arise only when an absolute rotation is performed. If a bucket is filled with water, and made to rotate, initially the water remains still, but then, gradually, the walls of the bucket communicated their motion to the water making it curve and climb up the sides of the bucket, because of the centrifugal forces produced by the rotation. Newton says that this experiment demonstrates that the centrifugal forces arise only when the water is in rotation with respect to the absolute space (represented here by the Earth's reference frame or the distant stars). Alternatively, when the bucket was rotating with respect to the still water no centrifugal forces were produced, thus indicating that the water was still with respect to the absolute space.

Mach says in his book that the bucket experiment only demonstrates that when the water is in rotation with respect to the bucket no centrifugal forces are produced, and that we cannot know how the water would behave if in the experiment the bucket's walls were greatly increased in depth and width. In Mach's approach the concept of absolute motion should be substituted with a total relativism in which every motion, uniform or accelerated, has sense only in reference to other bodies. Thus one cannot simply say that the water is rotating, but must specify if it is rotating with respect to the vessel or to the Earth or something more massive. Furthermore one should take into account the particular asymmetry that exists in our local reference frame between the small bodies (like buckets) and the bodies like the earth and distant stars that are overwhelmingly bigger and more massive.

Mach's Principle was never developed into a quantitative physical theory that could explain a mechanism by which the stars can have such an effect. Although Einstein was intrigued and inspired by Mach's Principle, his formulation of the principle is not a fundamental assumption of General Relativity Theory. However, before completing his development of the General Theory of Relativity, Einstein found an effect which he interpreted as being evidence of Mach's Principle. In a thought experiment Einstein considered a fixed background of the stars for conceptual simplicity and constructed a large spherical shell of mass, and set it spinning in that background. According to Mach's Principle the reference frames in the interior of the mass shell will precess with respect to the fixed background. This effect has been measured and is known as the Lense-Thirring effect. Einstein was so satisfied with this manifestation of Mach's Principle that he wrote a letter to Mach saying,

It ... turns out that inertia originates in a kind of interaction between bodies, quite in the sense of your considerations on Newton's pail experiment... If one rotates [a heavy shell of matter] relative to the fixed stars about an axis

going through its center, a Coriolis force arises in the interior of the shell; that is, the plane of the Foucault pendulum is dragged around (with a practically unmeasurably small angular velocity).[1]

Another form of the Lense-Thirring effect is the Schiff precession or spin-spin precession known as the Lense-Thirring precession of an orbiting spinning gyroscope. It is caused by the **R x (R x A)** effect of the second term of the electrodynamic force of inertia of Equation (9-24) on a gyroscope orbiting a spinning body. See Figure 13-1. Two effects—**R x (R x A)** (frame-dragging in relativity theory) and the local variations in mass with R (geodetic effect in relativity theory)—were expected to cause a precession (at ninety degree angles with respect to one another) of the gyroscopes aboard the Gravity Probe B satellite. Predicted precessions, given in milliarcseconds (mas), are compared to those measured by Gravity Probe B. The Gravity Probe B was designed by NASA and Stanford University to measure two key predictions of Einstein's General Theory of Relativity by monitoring the orientations of ultra-sensitive gyroscopes relative to a distant guide star.

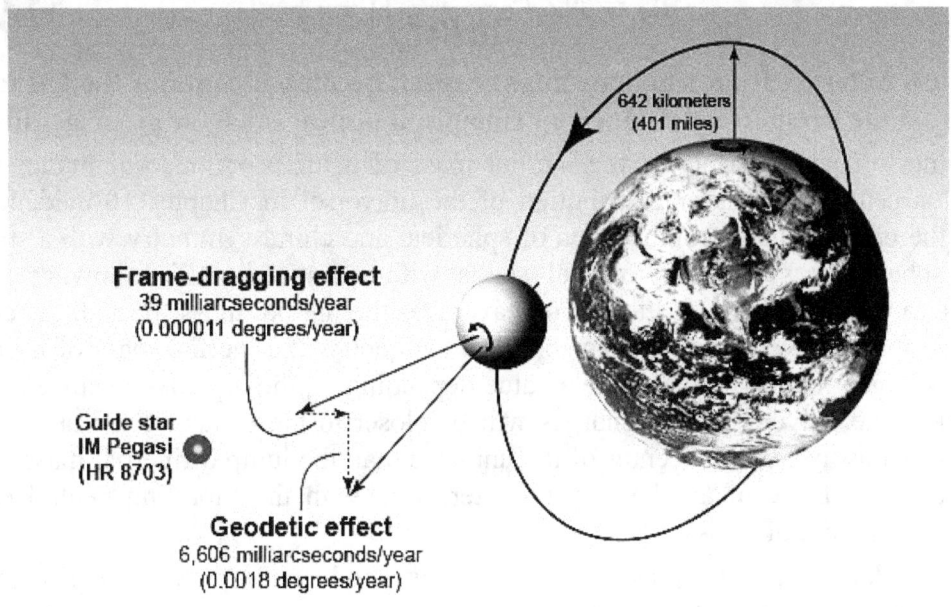

	Measured	Predicted
Geodetic precession (mas)	6602 ± 18	6606
Frame-dragging (mas)	37.2 ± 7.2	39.2

Figure 13-1 Schiff Spin-Spin Precession and Gravity Probe B [6]

In this chapter the concepts of inertial and gravitational mass, as defined by the universal electrodynamic force law, will be developed for the first time into a quantitative theory to support or prove Mach's Principle.

13.1 Inertial Mass

In Equation (9-11) the inertial mass m_i of a single vibrating neutral electric dipole with amplitude A and frequency ω consisting of an atomic electron and a nuclear proton was derived to be

$$m_i = \left(\frac{2e^2}{3\pi Rc^2}\right)\left(\frac{A^2\omega^2}{c^2}\right) \qquad (13-1)$$

Or in terms of a lump of some element containing N atoms each having Z protons and electrons the more general result is

$$m_i = NZ\left(\frac{2e^2}{3\pi Rc^2}\right)\left(\frac{A^2\omega^2}{c^2}\right) \qquad (13-2)$$

Now this result is different from what one might expect, because it contains the 1/R term in it. However, from the perspective of Mach's Principle, it appears to have great significance!

According to Mach's Principle the inertial mass of a lump or piece of matter should depend on the structure and matter distribution of the universe. In Chapter 10 evidence was presented that the universe has a combination of spherical and chiral symmetry with a structure consisting of spherically symmetric toroidal shells with quantized radii following Stanley Dermott's modern version of Bode's Law. If one averages the inertial mass from all the charges in all the atoms of the universe, which has spherical symmetry, the inertial mass of a lump of matter on a very large grand scale will have a value depending on the effective average radius R from the center of the universe. Some charges will be closer to the center of the universe and some will be farther away from the center of the universe than the lump of inertial mass. This R will be the average distance of all the charges interacting with the vibrating neutral electric dipoles in the inertial lump of mass.

On a more local scale the presence of a massive body nearby can give rise to local asymmetric effects in addition to the grand scale effects. One of these type effects is the Lense-Thirring effect or spin-spin effect. This effect was confirmed by the NASA Gravity Probe B data.

Thus the interaction force that Einstein referred to above that gives rise to the force of inertia is the electric charge to vibrating neutral electric dipole force. In the next section the interaction force that gives rise to the force of gravity will be considered.

13.2 Gravitational Mass

In Equation (9-16) the electrodynamic force giving rise to the radial gravitational force was derived as the average force between neutral vibrating electric dipoles. This force can be factored such that the terms corresponding to the mass are the same as the expression above in Equation (12-1) for the inertial mass to obtain

$$\vec{F}_G = -\left(\frac{2}{5\pi}\right)\frac{e^2}{R^2}\left(\frac{A_1^2\omega_1^2}{c^2}\right)\left(\frac{A_2^2\omega_2^2}{c^2}\right)\hat{R} = G\frac{m_{g1}m_{g2}}{R^2}\hat{R}$$

$$= -\frac{9\pi c^4}{10e^2}\left\{\frac{2e^2}{3\pi Rc^2}\frac{A_1^2\omega_1^2}{c^2}\right\}\left\{\frac{2e^2}{3\pi Rc^2}\frac{A_2^2\omega_2^2}{c^2}\right\}\hat{R}$$

$$= -\frac{9\pi c^4}{10e^2}m_{i1}m_{i2}\hat{R} \qquad (13-3)$$

Note that the force of gravity has been defined in the past as a $1/R^2$ attractive force using a different definition of mass than the inertial mass. However, if one writes the equation for the force of gravity in terms of the inertial mass as shown in the { } of the second line of Equation (12-3), one sees that the force of gravity on the grand large scale is a constant attractive force throughout the universe. On the other hand on the local scale where asymmetry exists, the force of gravity empirically appears to be a $1/R^2$ attractive force between two bodies.

In a large spiral galaxy the asymmetric force of gravity dominates near the center of the galaxy, but far out from the center it gets very weak such that the constant force of gravity on the grand scale predominates. Since that force does not diminish with distance, but remains constant except for decaying over time allowing expansion, the velocity v_s of the rotation of the outer spiral arms remains constant beyond the distance that the asymmetric gravitational force dominates.

Consider the forces of inertia and gravity for a lump of matter m in the outer spiral arm of a spiral galaxy of mass M. There are two terms in the force of inertia. The first term represents the force of inertia due to motion a_s with respect to the center of the spiral galaxy of mass M in direction r. The second terms represents the force of inertia due to motion a_0 with respect to the center of the universe in direction R. See equation (12-4)

$$\vec{F}_I = m\vec{a} = ma_s\hat{r} + ma_0\hat{R} \qquad (13-4)$$

where the magnitude of the observed acceleration a when $a_s < a_0$ is

$$a = \sqrt{\vec{a}_s^2 + \vec{a}_0^2} = a_0\sqrt{1 + \frac{a_s^2}{a_0^2}} = a_0\left(1 + \frac{1}{2}\frac{a_s^2}{a_0^2} + \cdots\right) = a_0 + \frac{1}{2}\frac{a_s^2}{a_0} \quad (13-5)$$

There are two terms for the force of gravity. The first term represents the force of gravity with respect to the center of the spiral galaxy of mass M_S. The second term represents the force of gravity with respect to the mass of the rest of the universe M_U. See equation (12-6).

$$\vec{F}_G = -\frac{GmM_S}{r^2}\hat{r} - \frac{GmM_U}{R^2}\hat{R} \quad (13-6)$$

For stability the forces of inertia must be in equilibrium with the forces of gravity.

$$\vec{F}_I = m\left(a_0 + \frac{1}{2}\frac{a_s^2}{a_0} + \cdots\right) = -\vec{F}_G = \frac{GmM_U}{R^2}\hat{R} + \frac{GmM_S}{r^2}\hat{r} \quad (13-7)$$

199

Now in equation (12-7) the first term for the force of inertia is approximately equal to the first term for the force of gravity. Subtracting an equal term from each side of the equation causes it to reduce to

$$m\frac{1}{2}\frac{a_s^2}{a_0} = \frac{GmM_S}{r^2} \qquad (13-8)$$

Solving for the acceleration of the lump of mass about the center of the spiral galaxy obtain

$$a_s = \frac{\sqrt{2GM_Sa_0}}{r} \qquad (13-9)$$

Now using the relationship for the acceleration in terms of the velocity for circular orbits gives

$$a_s = \frac{v_s^2}{r} = \frac{\sqrt{2GM_Sa_0}}{r} \qquad (13-10)$$

Solving for v_s obtain

$$v_s = \sqrt[4]{2GM_Sa_0} \qquad (13-11)$$

or

$$M_S \propto v_s^4 \qquad (13-12)$$

Note that equation (12-11) gives a constant value for the velocity when a_s becomes smaller than a_0 as shown in Figure 13-2. Equation (12-11) also allows one to calculate a_0 from the observed v_s. Milgrom [8, 9] calculated a value of $a_0 = 6.0 \times 10^{-11}$ m/sec^2 that fits the spiral rotational velocity as shown in Figure 13-2. (Note that the background picture of the spiral galaxy was not centered at the origin of the graph.)

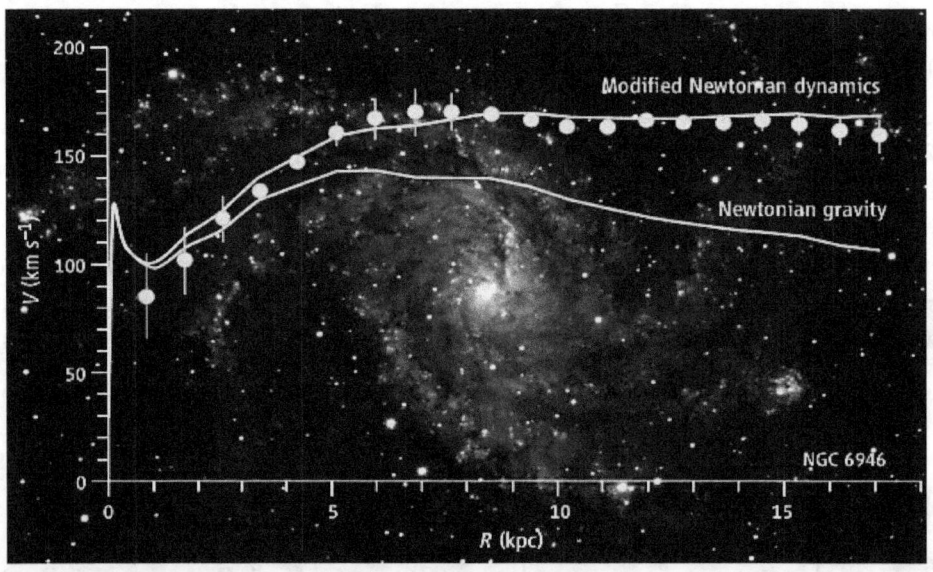

Figure 13-2 NGC 6946 Spiral Galaxy Graph of Rotational Velocity vs. Distance R from Center [7]

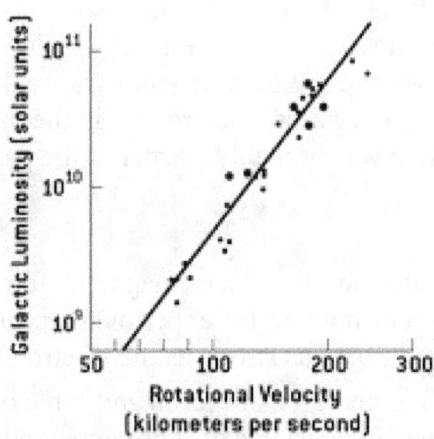

Figure 13-3 Relationship of Rotational Velocity to Galactic Luminosity [10]

Equation (12-12) predicts the Tully-Fisher relationship between the rotation velocity in spiral galaxies and the luminosity as shown in Figure 13-3. [10] The luminosity is proportional to the spiral galaxy mass in agreement with $M_S \propto V_S^4$ as derived.

In the Pioneer 10 and 11 missions an anomalous acceleration was measured as shown in Figure 13-4. [11] It was found to have a constant value of $(8.74 \pm 1.33) \times 10^{-10}$ m/s^2. This is of the same order of magnitude as the acceleration observed in the spiral galaxies above. Also the acceleration increases with distance from the center of our solar system until it reaches a maximum in the same manner as the difference between the MOND and Newtonian velocities of Figure 13-2. Thus Mach's Principle along with the electrodynamic definition of the inertial and gravitational mass appears to be able to explain both phenomena.

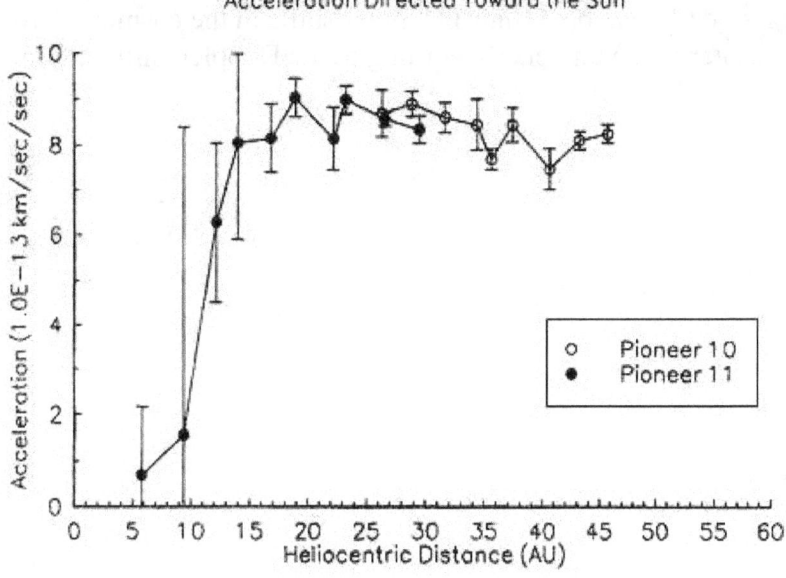

Figure 13-4 Anomalous Acceleration on Pioneer 10 and 11 [11]

Since the vibrating neutral electric dipoles included in the mass definitions of Equation (12-3) are decaying, the constant gravitational force is slowly decaying. This causes the universe to expand on the grand scale. Even on the local asymmetric scale the force of gravity between the moon and the earth is decaying over time. This results in the moon moving further away from the earth in its orbit. The decay rate was probably greater in the past than the present decay rate.

13.3 Conclusions

Mach's Principle originates in the electromagnetic force involving vibrating neutral electric dipoles. This force was not noticed by experimenters in the past, because the dipole-dipole force is a $(v/c)^4$ effect that is 10^{-39} as strong as the electrostatic Coulomb force.

When Newton was deriving his force of inertia and force of gravitation, the mass in those equations appeared to be fundamental constants of those apparently different fundamental forces. When Newton was questioned about what mass is, he said toward the end of his life that he did not know. Also Newton said that he did not know what caused the force of inertia or the force of gravity. However, Newton did specify a process to combine the axiomatic scientific method with the empirical scientific method that would decrease the number of independent fundamental forces in the course of scientific investigation and hopefully allow the logic of induction to discover the one universal force and proper set of terms. That has happened in our time as explained in this book.

Thus from the universal electrodynamic force derivations of the force of inertia and the force of gravity, mass is not a fundamental entity. What was called mass in the past represents a grouping of nearly constant electrodynamic parameters associated with vibrating neutral electric dipoles. The energy of these vibrating neutral electric dipoles is decaying by giving off radiation. The resulting radiation has been identified with the cosmic microwave background radiation. This radiation has been measured and found to be distributed throughout the universe in a pattern corresponding to the matter distribution as shown in Figure 13-5. This figure shows the distribution of the red (dark) and blue (light) Doppler shifts in the cosmic microwave background radiation about the center of the universe according to the Doppler shift pattern.

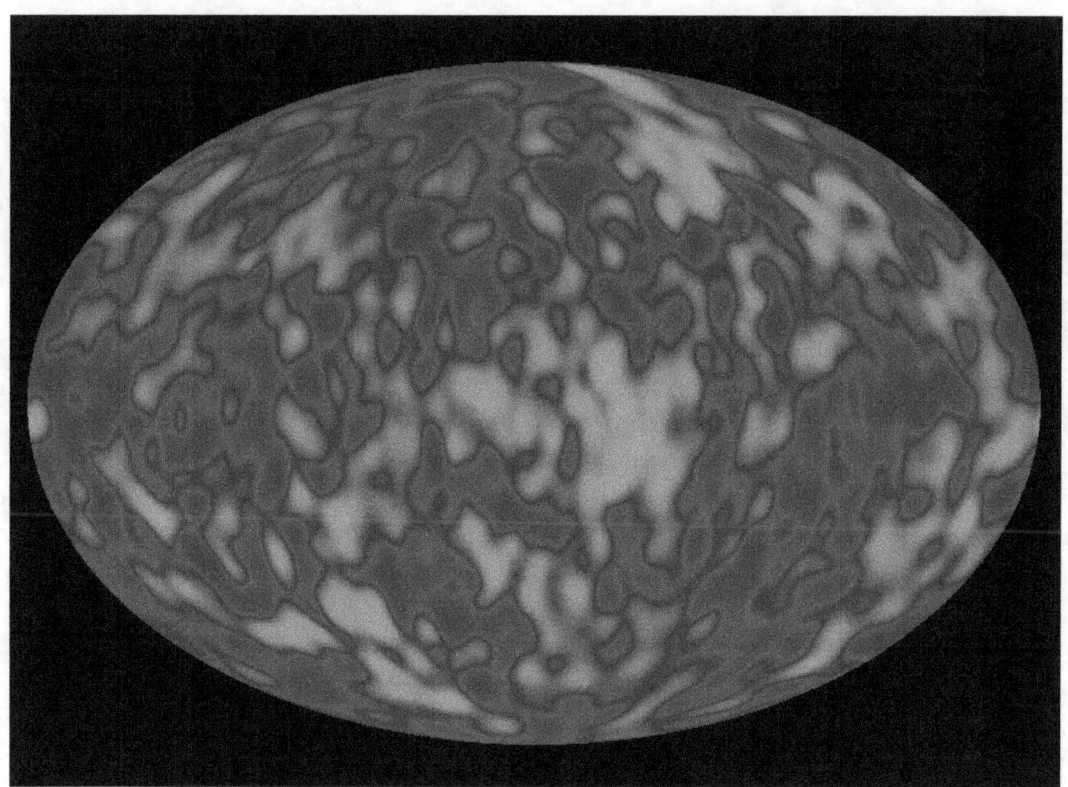

Figure 13-5 COBE Cosmic Microwave Background Radiation

Red & Blue Doppler Shift Data [12]

It appears that what was called mass in the past is not really constant. It is decaying over time. Also the inertial mass changes with distance from the center of the universe. Local asymmetries also cause mass to change with distance from the center of spiral galaxies and with distance from a star to a planet or a space probe like Pioneer 10 and 11. Unlike the Standard Model of Cosmology based on General Relativity Theory and quantum mechanics, the universal electrodynamic force approach does not need to invent dark matter and dark energy to explain the higher than expected constant velocities of the outer spiral arms of spiral galaxies, the anomalous acceleration of Pioneer 10 and 11, or the expansion of the universe. These phenomena, for which a halo of dark matter and energy were invented, are explained directly from the universal electrodynamic force law and Mach's Principle.

General Relativity Theory assumes that the universe is homogeneous and isotropic without any shape. The Microwave Cosmic Background Radiation plus the associated Doppler red and blue shifts of Atomic Spectra reveals that the universe does have a structure. Mach's Principle reveals that Einstein's **Principle of Equivalence** in General Relativity Theory is invalid by showing how the structure of the universe enters into the physical laws for gravity and inertia. Gravitational and inertial mass are not the same and do not have the same values. General Relativity Theory's imperfections are further revealed in the necessity to have over 95% of the universe consisting of dark matter and dark energy in order to explain the phenomena that Mach's Principle explains so simply. Also dark matter and energy, unlike regular mass and energy, are not observable in the laboratory.

13.4 References

1. A. Einstein, letter to Ernst Mach, Zurich, 25 June 1923, in Misner, Charles; Thorne, Kip S.; and Wheeler, John Archibald (1973). **Gravitation**. San Francisco: W. H. Freeman. **ISBN** **0-7167-0344-0.**

2. Stephen W. Hawking & George Francis Rayner Ellis, **The Large Scale Structure of Space–Time**, Cambridge University Press. p. 1 (1973). **ISBN 0-521-09906-4**.

3. G. Berkeley, **The Principles of Human Knowledge**. See paragraphs 111–117, 1710 (1736).

4. Mach, Ernst, **The Science of Mechanics; a Critical and Historical Account of its Development**. (Open Court Pub. Co., LaSalle, IL, 1960). LCCN 60010179. This is a reprint of the English translation by Thomas H. MCormack (first published in 1906) with a new introduction by Karl Menger.

5. Isaac Newton's **Philosophiae Naturalis Principia Mathematica: the Third edition (1726) with variant readings**, assembled and ed. by Alexandre Koyré and I Bernard Cohen with the assistance of Anne Whitman (Cambridge, MA, 1972, Harvard UP). Note the book is in Latin.

6. http://physics.aps.org/assets/7ae6c34b01556a35

7. **http://francisthemulenews.wordpress.com/2008/01/**

8. Milgrom, M. "A Modification of the Newtonian Dynamics as a Possible Alternative to the Hidden Mass Hypothesis", **Astrophysical Journal vol. 270**, pp. 365–370 (1983).

9. Milgrom, M. "A Modification of the Newtonian Dynamics - Implications for Galaxies". **Astrophysical Journal Vol. 270**, pp. 371–389 (1983).

10. Tully, R. B. and Fisher, J. R., "A New Method of Determining Distances to Galaxies", **Astronomy and Astrophysics, Vol. 54**, pp. 661-673 (1977).

11. http://arxiv.org/pdf/gr-qc/0104064v5.pdf

12. http://static.bbc.co.uk/universe/img/ic/640/sights/cosmic_microwave_ background_radiation/cosmic_microwave_background_radiation_large.jpg

Chapter 14 Conclusions

A theory is the more impressive the greater the simplicity of its premises, the more different kinds of things it relates, and the more extended is its area of applicability. Albert Einstein [1]

The philosopher Poincaré argued from the logic of meta-theory (theory of theories) that any two fundamental theories that employ the same fundamental constants or the same mathematical form are not both fundamental. It is interesting to note that this argument prevented Einstein from receiving the Nobel Prize for relativity theory. Einstein did receive the Nobel Prize for his wife Mileva's work on the photoelectric effect, and he gave her all the money when they divorced.

In this work the axiomatic method for obtaining "Maxwell's Equations of Electrodynamics" by using only the empirical electric and magnetic laws of Gauss, Ampere, and Faraday as the axioms or postulates has been improved by

1. **Removing the point particle idealization**
2. **Adding Lenz's Law for induction of moving charges satisfying conservation of energy, Newton's 3rd Law for magnetic phenomena, and for non-linear field effects**
3. **Recognizing that induced fields are non-linear velocity dependent and so must be treated separately from linear static fields**
4. **Allowing only contact forces - based on experiments by Hooper showing that the fields of a charged particle remain attached to the particle when it moves and have tensile strength**
5. **Using Galilean invariance for true relativity**
6. **Satisfying Mach's Principle via Lenz's Law**
7. **Adding conservation of kinetic and radiation energy via Lenz's Law**
8. **Adding conservation of momentum including radiation reaction via Lenz's Law**

From chapter 4 equations (4-39 and 4-43) we can write the fully complete definition of Lenz's Law for the first time as

$$\vec{E}_i(\vec{r}, \vec{v}) \propto -\vec{E}_0(\vec{r})$$

$$= -\lambda(\vec{v})\vec{E}_0(\vec{r}) = \frac{(1 - \beta^2)\vec{E}_0(\vec{r})}{(1 - \beta^2 sin^2\theta)^{\frac{3}{2}}} - \vec{E}_0(\vec{r}) \quad (14 - 1)$$

The induced electric field goes to 0 as β goes to 0 as it should. Note that the static field $\mathbf{E_0(r)}$ depends only on **r** while the induced field $\mathbf{E_i(r, v)}$ depends on both **r** and **v**.

From the derivation in Chapter 4 one sees the significance of handling the non-linear induced fields separately from the linear static fields and the contributions from the nonlinear effects in producing the series of terms that were summed using the binomial theorem to produce the so-called "relativistic effects". Thus the improved version of the electrodynamic force resulting from a more perfect union of Euclid's axiomatic method and Newton's empirical scientific method is able to explain the so-called "relativistic effects" in electrodynamics through the use of the complete set of the empirical laws of electrodynamics and the removal of all the idealizations, especially the point particle idealization, such that the relativistic effects are due to electrical feedback effects on finite size electrodynamic particles. This work confirms Poincaré's arguments from meta-theory that relativistic effects must be of electrodynamic origin due to the use of the fundamental constant c in the theory.

Quantum mechanics also uses c and should be of electrodynamic origin according to Poincaré's meta-theory. In Chapters 2 and 3 is outlined the history of the finite-size structure of the electron from Ampere's magnetic molecule down to the present day toroidal ring. The structure of the atom consisting of shells of finite-size toroidal electrons was first put forward by Parson and used by Lewis in chemistry in his famous dot diagrams to explain in a simple way the geometrical bonding of atoms in molecules. The emission spectra of the atom composed of finite-size toroidal electrons in non-orbiting shells about the nucleus was calculated for hydrogen. Due to the continuity of the toroidal ring and tensile forces on the charge in the toroidal ring n=1/2, 1/3, 1/4, etc. emission spectral lines were predicted in addition to the n=1, 2, 3, etc. spectral lines. NASA missions found up to 7 orders for all the 12 emission spectral lines predicted for hydrogen in the ultra violet spectral range measured in NASA rocket experiments. Since the Dirac quantum electrodynamic model of the atom, which incorporates point particles orbiting the nucleus, cannot support or predict the existence of any emission spectral lines for n = 1/2, 1/3, 1/4, etc. in hydrogen, the electrodynamics of finite-size closed-ring charged particles is the proper basis for quantum effects in the atom. Also the problem that the Dirac model of the atom has with point electrons orbiting the nucleus without continually give off radiation due to circular acceleration as observed in all circular accelerators is removed. Thus this work further confirms Poincaré's argument from meta-theory that quantum effects must be of electrodynamic origin also due to the use of c in the theory.

Planck's description of blackbody radiation was based on the notion that point-charges undergoing simple harmonic motion in the blackbody were absorbing and emitting radiation. This picture led to oscillations of point-electron charges that were too big to remain in the atom or the lattice of the solid. Also the empirical laws of electrodynamics were violated by Planck's theory. Both Ampere's Law and Faraday's Law require continuous emission and absorption of radiation for simple harmonic motion of point-electron charges. Furthermore the quantum theory of blackbody radiation was not compatible with optical reflection, refraction, and diffraction phenomena due to its emission of radiation that is discontinuous in time. In Appendix C is documented that the most satisfying explanation of Planck's blackbody radiation formula comes from electrons composed of finite-size toroidal charge rings which do not suffer from any of these inconsistencies. Thus this work further confirms Poincaré's argument from meta-theory that quantum effects must be of electrodynamic origin due to the use of c in the theory.

Poincaré's observation that the mathematical form of the electric force and the gravitational force is the same $1/R^2$ suggesting a common force was confirmed by deriving the force of gravity from the electrodynamic force between vibrating neutral electric dipoles in Chapter 9. The force of gravity obtained was more complete than Newton's Universal Force of

Gravitation and Einstein's General Theory of Relativity in that it also explained the decay of gravity and the quantization of gravity. Also the derived electrodynamic theory of gravity eliminated the need for dark matter and dark energy, because it explained Hubble's Law for Red Shifts and the higher than expected velocity of stars in the outer arms of spiral galaxies when Mach's Principle was taken into account. Thus this work further confirms Poincaré's argument from meta-theory that gravity must be of electrodynamic origin due to the use of the same mathematical form $1/R^2$ of the force.

In Chapter 10 the force of inertia was derived from the electrodynamic force law. From this force law it was possible to define mass as an electrodynamic quantity associated with vibrating neutral electric dipoles. Then one could show from the derived electrodynamic gravitational force, that the ratios of gravitational and inertial masses are identical for the same two particles. No previous theories have been able to define mass and show that the ratios of the gravitational and inertial masses for the same two bodies are equal. The next volume in this series (**The Universal Force Volume 2 – An Electrodynamic Model of Elementary Particles**) will define and predict the masses of all the elementary particles from electrodynamics and combinatorial geometry.

The success in deriving the force of gravity and the force of inertia from the new electrodynamic force law gives credibility to the conjecture that the newly derived electrodynamic force is the universal force in agreement with the predictions of metatheory. This is further supported by the fact that the two remaining fundamental forces, the strong interaction force between elementary particles and the weak nuclear force responsible for beta decay, can be explained in terms of finite-size elementary particles, since the range of these forces is approximately the experimentally measured size of elementary particles such as the proton and neutron.

Following Newton's rule that no more causes of natural things are to be allowed than such as are both true and sufficient to explain the experimental data, we are forced to reject Einstein's Theories of Special and General Relativity and the Copenhagen version of Quantum Mechanics. Also we are to reject them, because they are not in agreement with the fundamental experiments on which they were originally based, such as the modified Fizeau experiment by Michelson and Morley, the photoelectric experiment for amorphous metals, the velocity of the stars in the outer arms of spiral galaxies, and the expansion of the universe as measured by Hubble's Law for Red Shifts.

Furthermore, this improved version of electrodynamics is able to describe radiation and radiation reaction directly from the electrodynamic force law which the covariant version of electrodynamics fails to do properly, because of the inherent limitation of constant velocity in Einstein's Special Relativity Theory and Maxwell's electrodynamics and the point-particle idealization. The newly derived electrodynamic force law is able to explain future states of charged particles, such as the electron, due to having acceleration a and da/dt terms in it to describe acceleration, radiation emission and absorption, and radiation reaction. This improved version of electrodynamics supports the continuity equation for charge with local conservation of charge and energy. The relativistic covariant version of Maxwellian electrodynamics and the Copenhagen version of quantum mechanics do not support the continuity equation for charge with local conservation of charge and energy.

The conjecture that the improved version of electrodynamics is the universal force law was further bolstered by arguments from symmetry in Chapter 10. The new electrodynamic force law has a unique combination of spherical and chiral symmetry. Confirmation of this symmetry on all size scales in the universe is a necessary but not completely sufficient argument to prove the improved electrodynamic force law is the correct universal force law. Data was presented to show that this unique symmetry is found in structures on all size scales in the universe. The structures included elementary particles, nuclei, atoms, molecules, crystals, plant leaf patterns, flower petal patterns, plant seed head patterns, animal body structures, solar system orbits, Milky Way structure, and the structure of the whole universe about its center.

In Chapter 13 Mach's Principle was shown to originate in the electromagnetic force involving vibrating neutral electric dipoles. This force was not noticed by experimenters in the past, because the dipole-dipole force is a $(v/c)^4$ effect that is typically 10^{-39} as strong as the electrostatic Coulomb force.

When Newton was deriving his force of inertia and force of gravitation, the masses in those equations appeared to be fundamental constants of different fundamental forces. When Newton was questioned about what mass is, he said up to the end of his life that he did not know. Also Newton said that he did not know what caused the force of inertia or the force of gravity.

According to the universal electrodynamic force derivation of the force of inertia and the force of gravity, mass is not a fundamental entity. What was called mass in the past represents a grouping of nearly constant electrodynamic parameters associated with vibrating neutral electric dipoles. The energy of these vibrating neutral electric dipoles is decaying by giving off radiation. The resulting radiation has been identified with the cosmic microwave background radiation. This radiation has been measured and found to be distributed throughout the universe in a pattern corresponding to the matter distribution.

What was called mass in the past is not really constant. It is decaying over time. The value of the mass changes universally with distance from the center of the universe. Local asymmetries cause mass to change with distance from the center of spiral galaxies, with distance from a star to a planet, and with distance from a planet to a moon. Unlike the Standard Model of Cosmology based on General Relativity Theory and quantum mechanics, the universal electrodynamic force approach does not need to invent dark matter and dark energy to explain the higher than expected constant velocities of the outer spiral arms of spiral galaxies and the expansion of the universe. These phenomena, for which dark matter and energy were invented to explain, are explained directly by the universal electrodynamic force law.

The kind of interaction between bodies to which Einstein attributed the force of inertia has been found. It is the electrodynamic force involving vibrating neutral electric dipoles. The concept of mass upon which Einstein founded his General Theory of Relativity is no longer valid.

Albert Einstein in his Nobel Prize award address said

A theory is the more impressive the greater the simplicity of its premises, the more different kinds of things it relates, and the more extended is its area of applicability. [1]

Thus Einstein would be the first to discard his theories of Special and General Relativity in favor of an improved electrodynamics. **Note that he never did like Heisenberg's quantum mechanics anyway!**

14.1 References

1. Albert Einstein, Nobel Prize in Physics Award Address, 1921.

Chapter 15 Epilogue

It appears to me, that the study of electromagnetism in all its aspects has now become of the first importance as a means of promoting the progress of science. James Clerk Maxwell [1]

Before Newton the small residual discrepancies between theory and the real world were dismissed as being of no practical importance. After Newton every systematic deviation from current theory automatically has the status of a pressing unsolved problem. How does one proceed to advance science and bring it closer to completeness? The answer is to continue to perform experiments and discover new phenomena that cannot be explained by the current set of empirical laws. Then one or more of the current empirical laws must be updated by induction to explain all the observed data or a new empirical law added to the others. After that is done the axiomatic method of deducing the general force law from the complete set of empirical laws must be revised to take into account the revised empirical laws.

There are some experiments, promoted in alternative science journals, which are claimed to not be compatible with Maxwell's equations. These experiments need to be examined to see if they are described by the extended and improved version of electrodynamics presented in this book. Some of these experiments are as follows:

15.1 Hooper-Monstein Experiment

This experiment was originally conducted by Hooper and Monstein in 1992 and claims to show that an electromagnetic induction voltage can be produced in a wire which is centered between two moving parallel bar magnets arranged with their axes in opposing directions and moved inwards towards the central wire. See Figure 15-1 below.

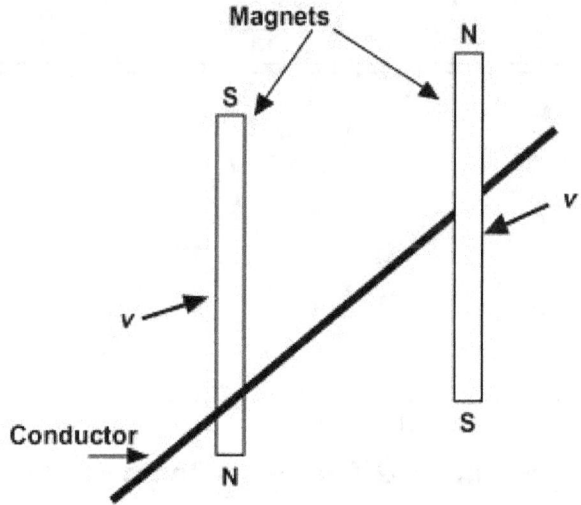

Figure 15-1 Hooper-Monstein Experiment

The experimenters claim that Faraday's induction law (relating the voltage induced in a closed circuit to the time rate of change of the magnetic flux through the circuit) can not apply to this experiment. Presumably this is because the experimenter's attention is focused on the wire towards which the magnets are travelling. However, the experiment must involve a closed circuit in order to measure the voltage on a meter. Although the net magnetic field is zero at the position of the wire of interest, elsewhere it is not zero, and the field lines do pass through the closed loop. As the magnets move the total flux through the loop will change. A detailed analysis of the total flux change will reveal that Faraday's Law is obeyed. Also Lenz's Law gives the velocity terms for induction and these depend on v^2/c^2 and do not cancel out between the two moving magnets.

15.2 Aharonov-Bohm Effect

The Aharonov-Bohm effect is claimed to be a quantum mechanical phenomenon in which an electrically charged particle is affected by an electromagnetic field (E, B) despite being confined to a region in which both the magnetic field B and electric field E are zero. The underlying mechanism is assumed to be the coupling of the electromagnetic potential vector A with the complex phase of the charged particle's wave function. Thus the Aharonov-Bohm effect is illustrated by interference experiments.

In the most commonly described case, the wave function of a charged particle passes around a long solenoid and experiences a phase shift as a result of the enclosed magnetic field, despite the strength of the magnetic field being negligible in the region through which the particle passes and the particle's wave function being negligible inside the solenoid. The resulting interference pattern caused by this phase shift has been observed experimentally. [2] See Figure 15-2 below for the double slit interference experiment.

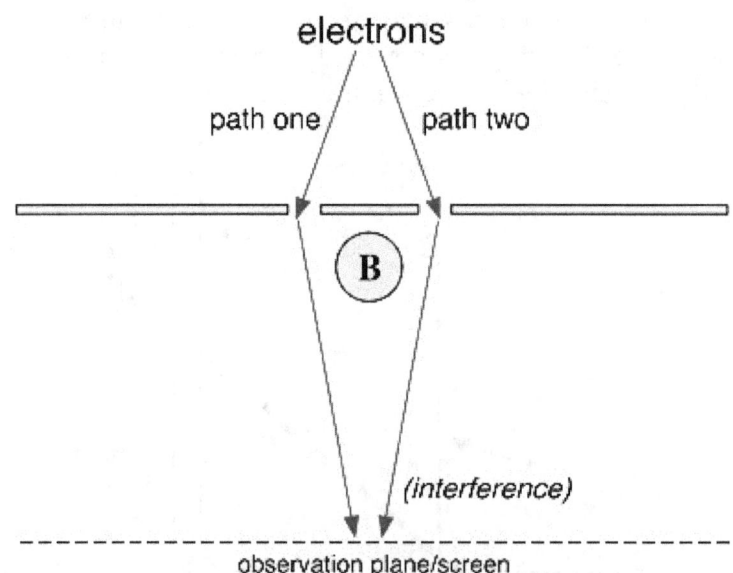

Figure 15-2 Aharonov-Bohm Effect

This experiment touches upon a number of significant issues that need to be addressed. In the 18th and 19th centuries, physics was dominated by Newton's empirical scientific method with its emphasis on forces. As a result electromagnetic phenomena were investigated by a series of experiments involving the measurement of forces between charges, currents and magnets in various configurations. Eventually, a description arose according to which charges, currents and magnets acted as local sources of force fields, which then acted on other charges and currents locally through the Lorentz force law.

One of the observed properties of the electric field was that it was irrotational, $\nabla \times E = 0$. One of the observed properties of the magnetic field was that it was divergenceless, i.e. $\nabla \cdot B = 0$. From vector calculus it is possible to express a vector such as the electrostatic field E as the gradient of a scalar potential, i.e. $E = -\nabla \varphi$, where φ is analogous to Coulomb's electrostatic potential and also the classical gravitational potential. **The notion of the Coulomb potential and the gravitational potential is based on conservation of energy.** From vector calculus one can also express a stationary magnetic field as the curl of a vector potential A, i.e. $B = \nabla \times A$. The vector potential is a new concept not based on conservation of energy. In fact Maxwell's equations do not conserve magnetic energy, because they do not include Lenz's Law which handles conservation of magnetic energy and Newton's 3rd Law.

The interpretation of the Aharonov-Bohm effect is important conceptually, because it is used to justify the following:

1. **Interconnection of Special Relativity Theory and quantum mechanics with electrodynamics via the relativistic four-vector potential**
2. **To recast Maxwell's classical electromagnetic theory as a gauge theory suitable for the quantum electrodynamic theory of elementary particles**
3. **To determine that action principles based on energy are more fundamental than local forces**
4. **To abandon the principle of locality by claiming that the electromagnetic scalar and vector potential offers a more complete description of electromagnetism than the electric and magnetic field approach.**

15.2.1 Potentials vs. Fields

The use of the vector and scalar potentials of vector calculus is different than the use of physical fields. For instance all physical effects are describable in terms of the physical fields. However using the potentials of vector calculus, physical effects are described by the derivatives of the scalar and vector potentials. Thus potentials are not uniquely determined by physical effects. Potentials can only be defined to within an arbitrary additive constant electrostatic potential and an irrotational stationary magnetic vector potential. Electric and magnetic fields and forces are gauge invariant and therefore directly observable, and unlike potentials, appear in the Lorentz force formula.

From the work of this book the scalar potential is sufficient to explain all electromagnetic effects including radiation and radiation reaction. It includes all relativistic effects also. Thus there is an incompatibility between proper electrodynamics and vector potentials. The vector potential appears to be a physically unrelated superfluous idea of vector calculus.

15.2.2 Global Action vs. Local Forces

One interpretation of the Aharonov-Bohm effect claims that the Lagrangian approach to dynamics based on energies is not just a computational aid to the Newtonian approach based on forces. Forces are an incomplete way to formulate physics, and vector potentials must be used instead.

This interpretation is built upon Maxwellian electrodynamics which is an incomplete version of electrodynamics which in turn is based upon many idealizations such as

1. Point particle idealization
2. Linear field superposition principle
3. Failure to conserve energy and momentum
4. Failure to satisfy Mach's Principle
5. Failure to note that the fields of a charge remain attached and have tensile strength

When these idealizations are removed the resulting version of electrodynamics is more complete and more eloquent replacing Special Relativity Theory due to electrical feedback effects on finite size charged particle structures and quantum mechanics due to the resonant properties of the internal charge structures of elementary particles, atoms and molecules.

From this superior version of electrodynamics the interpretation of the Aharonov-Bohm effect is different. The fields of the electrons travel with the electrons and touch the magnetic field of the solenoid. The tensile strength of the electron's fields causes the solenoid to produce interference effects on the observation screen. The fact that the magnetic field is concentrated within the solenoid does not negate the fact that every magnetic field line also exists outside of the solenoid. See Figure 15-3 below. This superior interpretation restores the notions of causality and local contact forces.

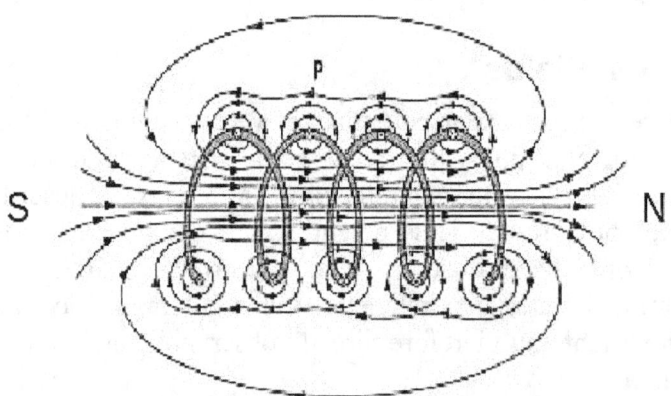

Figure 15-3 Magnetic Field of a Solenoid

15.2.3   Universal wave function vs. Fields

The common interpretation of the Aharonov-Bohm effect has the vector potential causing a phase shift in the quantum mechanical universal wave function. Like the vector potential the quantum mechanical wave function is not directly observable. Like Maxwell's electrodynamics quantum mechanics is built upon many idealizations such as

1. **Point particle idealization**
2. **Energy and momentum not conserved all the time**
   (Heisenberg Uncertainty Principle)
3. **Unobservable unphysical universal wave function ψ**
4. **Failure to satisfy Mach's Principle**
5. **Failure to support principle of locality**
6. **Failure to support Principle of Cause and Effect**

The Aharonov-Bohm effect is claimed to be produced by an unphysical unobserved construct called the Vector Potential, A, interacting with an unphysical unobserved construct called the universal wave function ψ. **What has happened to physical reality? The universal electrodynamic force of this book gives a real physical way to explain the Aharonov-Bohm effect and it is not based on idealizations and unobservable entities!**

15.3 Unipolar Induction

A homopolar or unipolar generator is a DC electrical generator consisting of an electrically conductive disc or cylinder rotating in a plane perpendicular to a uniform static magnetic field. A potential difference is created between the center of the disc and the rim (or the ends of the cylinder. It is also known as a Faraday disc. See Figure 15-4 below.

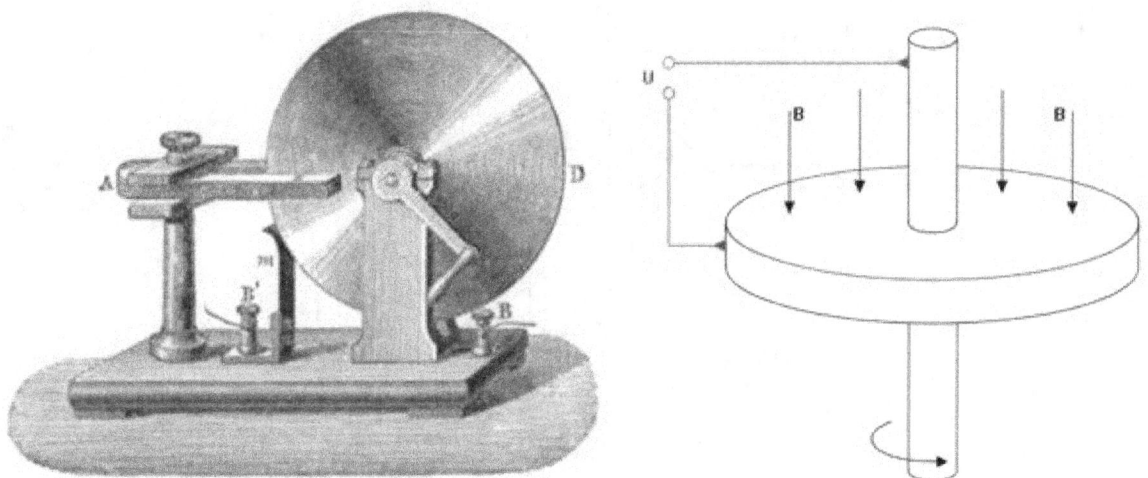

Figure 15-4 Faraday Disc Generator and Schematic

Unipolar inductors occur in astrophysics where a conductor rotates through a magnetic field, such as the movement of the highly conductive plasma in a cosmic body's ionosphere through its magnetic field. For example the motion of the magnetized interplanetary plasma produces electric fields that are essential for the production of aurora and magnetic storms.

The Faraday disc converts kinetic energy to electrical energy. The unipolar inductors or generators can be analyzed using Faraday's law of electromagnetic induction. This law states that an electric current is induced in a closed electrical circuit when the magnetic flux enclosed by the circuit changes in either magnitude or direction. For the Faraday disc it is necessary to consider that the circuits consist of radial "spokes" of the disk connected to the rim and the center and then through the external circuit.

The Lorentz force law can also be used to explain the Faraday disc. This law, formulated 30 years after Faraday's death, states that the force on an electron is proportional to the cross product of its velocity and the magnetic flux vector. In geometrical terms, this means that the force is at right-angles to both the velocity (azimuthal) and the magnetic flux (axial) and therefore in a radial direction. The radial movement of the electrons in the disc produces a charge separation between the center of the disc and its rim. If the circuit is completed an electric current will be produced. [3]

15.4 Marinov Motor

The Marinov motor appears to refute the widely-held belief of physicists that the Lorentz force law suffices to describe all observable electromagnetic force manifestations. The Marinov motor consists of a permanent magnet of roughly toroidal shape (enclosing most of its own magnetic B-field flux) placed inside a conducting ring. If the magnet is held stationary in the laboratory and the ring is supported in bearings, then the ring will rotate continuously in the laboratory, provided direct current is brought into it through sliding contacts (brushes) situated in the vertical plane of the toroid. See Figure 15-5 below.

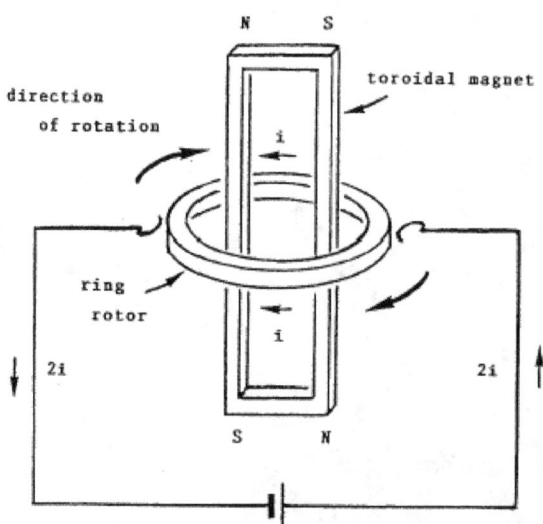

Figure 15-5 Marinov Motor

Note that the 2i current from the battery splits into i flowing clockwise in the front half of the conducting ring and i flowing counterclockwise in the back half of the ring. The magnetic B-flux is largely contained within the toroid and circulates clockwise around the toroidal magnet constructed of 4 bar magnets. The B-flux lines lie in a plane normal to the ring plane. The question is "Why does the ring always turn one way?"

Following Ampere's original description of a permanent magnet in terms of tiny "current whorls", which in modern parlance would be aligned electron spins, one can argue for a certain handedness. If we slice horizontally across a vertical arm of the toroidal magnet at the level of the ring, in thought we expose a planar assemblage of these current whorls, which are considered to cancel each other internally and leave uncompensated only a surface polarization or "magnetization current," which is treated in Maxwell's equations in the same way as a real current. The important thing to note is that this virtual surface current is unidirectional and has a definite sense determined by the sense of the flux internal to the magnet. If the flux vector points up, as in the left side of the toroid shown in the figure, then by the right-hand rule the sense of circulation of surface current in that is counterclockwise, as seen from above. (Note that we are treating the surface current as a conventional plus current, as if it were a real current in a solenoid.)

The "magnetization current" circulating around the surface of the permanent magnet nearest the conductor ring would presumably be treated in the same way as real charge in motion. It would induce an action on the real currents in the adjacent ring portions that would lead to exertion of a ponderomotive force on the ring. An alternative but essentially equivalent approach employs the original ampere force law between current elements. This approach furnishes a simple rule of thumb by which the sense of rotation of the Marinov motor can be inferred: *Adjacent (side by side) parallel current elements mutually attract and anti-parallel elements repel*. This rule also works for the Lorentz force law, under the same notion that surface magnetization (polarization) currents are real ones. Applying this to components of virtual surface current and to components of real ring current in closest proximity gives the correct sense of rotation. Thus, considering the left toroid magnet member of Figure 15-5 with B-flux up, the virtual surface current on the toroid adjacent to the ring is counterclockwise; so it is parallel to the near ring current and antiparallel to the far ring current. The near part of the ring is attracted and the far part is repelled. Both of these actions produce a clockwise rotation of the ring, as observed.

Wesley [4] has analyzed the Marinov motor in terms of the vector potential **A**, which is related to the B-field by **B = ∇ x A.** Because of its normalization or gauge problems this potential has been supposed, since the days of Heaviside, to not be physically "real". Attempts have been made to use the Aharonov-Bohm effect to establish its reality, but they were shown above to be falsely based on many idealizations and approximations. Both Marinov and Wesley noted that a slight formal modification of the vector calculus definition of the vector potential from a partial derivative to a total time derivative could allow a description of the Marinov motor, i.e.

$$E = -\nabla\Phi - \frac{1}{c}\frac{\partial A}{\partial t} \rightarrow -\nabla\Phi - \frac{1}{c}\frac{dA}{dt} \qquad (15-1)$$

This is an illogical and improper way to do theory according to the Axiomatic, Empirical and Structural scientific methods. Note that the Existential and Post-Modern scientific methods, which do not believe in truth, accept these illogical procedures.

15.5 References

1. James Clerk Maxwell, **A Treatise on Electricity & Magnetism, Vol. 1**, unabridged 3rd edition of 1891 (Dover, New York, 1954) **Preface**, p. vi.

2. Batelaan, A. & Tonomura, A. "The Aharonov–Bohm effects: Variations on a Subtle Theme". **Physics Today Vol. 9**, pp. 38–43 (2009)

3. **Electromagnetic Field Theory, 2nd edition** edited by Bo Thidé, Department of Physics and Astronomy, Uppsala University, Sweden
   http://www.plasma.uu.se/CED/Book/EMFT_Book.pdf

4. J. P. Wesley, "The Marinov Motor, Motional Induction without a Magnetic B Field," **Apeiron, Vol. 5**, p. 219 (1998).

About the Author

Charles W. Lucas, Jr.

Education – degrees and awards

Graduated Salutatorian in high school graduating class of 350 at Fairfax High School, Fairfax, VA in 1960

Selected to be a participant in the 10 year long NSF program for gifted high school students. Given an entire undergraduate science and mathematics education the summer of 1960 at Emory and Henry College in Emory, VA by the top college science and math professors in the United States

Received a full expenses four year scholarship to the College of William and Mary consisting of the William and Mary competitive examination scholarship in physics, the National Merit Scholarship, and the Virginia State Merit Scholarship among others

Taught freshman, sophomore and junior level physics courses starting in the fall of 1960 as a result of taking advance placement to the senior level in physics at the college of William and Mary

NSF research fellow in the William and Mary Plasma Physics Laboratory from 1962-64

Elected to Phi Beta Kappa National Honor Society in 1963

Elected to Sigma Pi Sigma National Physics Honor Society in 1964

Graduated from the College of William and Mary with Honors in Physics in June 1964

Received a graduate teaching assistantship in physics at the University of Maryland 1964

Received the University of Maryland's cash prize for outstanding performance as a graduate instructor in September 1965

Graduated from the University of Maryland in 1967 with M.S. in Solid State Physics

Awarded the William and Mary Graduate Fellowship for 1969-70

Graduated with a Ph.D. in Theoretical Intermediate Energy Physics in 1972

Received a two-year postdoctoral fellowship at the Catholic University of America for 1972-74

Officially proclaimed as **Physicist of the year for 2013** by **Who'sWho Worldwide** for the publication of the book **The Universal force Volume 1 – Derived From a More Perfect Union of the Axiomatic and Empirical Scientific Methods**

Overview of Scientific Work

Discovered in 1978 that the union of the ancient Greek Axiomatic method (commonly employed in Euclidean geometry proofs of theorems) with Newton's Empirical Scientific

Method had not been completed in a proper manner. Maxwell used 4 of the 6 empirical laws of electrodynamics to develop his wave equations for electrodynamics. This version of electrodynamics was incomplete and incorporated many idealizations. The theories of Quantum Mechanics and Special Relativity Theory were invented to supplement Maxwell's equations. However these supplementary theories did not accurately describe all data, had problems with their foundational assumptions, and also incorporated unreal idealizations.

Improved Version of Electrodynamics

Developed, starting in 1978, a more perfect union of the Greek Axiomatic method with Newton's Empirical Scientific Method by using the complete set of six empirical laws of electrodynamics as axioms and rigorously deriving an improved version of electrodynamics. Also the point particle and displacement current idealizations were removed from electrodynamics. The resulting improved electrodynamic force law was capable of describing all relativistic effects as a feedback effect due to motion causing a compression and change in shape of finite-size charged particles. Also the discovery was made that standing waves in the finite-size structures of elementary charged particles were responsible for all quantum effects just as in radio transmission from antennas.

New Electrodynamic Theory of Gravity and Inertia

Following the method of Assis, developed a derivation of the force of gravity from the improved electrodynamic force law between vibrating neutral electric dipoles. In normal matter the vibrating neutral electric dipoles consist of atomic electrons and nuclear protons. The same method was extended to derive the force of inertia as the force between an electric charge and vibrating neutral electric dipoles. This work enabled mass to be defined for the first time in the history of science in terms of the charge, amplitude, and frequency of the vibrating neutral electric dipoles. The ratios of the inertial and gravitational mass for the same two bodies were rigorously derived and shown to be equal. The derivation for the force of gravity revealed a previously unknown R x (R x V) second term. This term was shown to cause the force of gravity to be quantized giving rise to the first explanation of the modern version of Bode's Law describing the quantization of the orbits of the planets about the sun and the orbits of the moons about the planets. Due to gravity and inertial forces being based on vibrating neutral electric dipoles, they must decay over time, because all vibrating electrical systems must radiate. This radiation was calculated and found to be the observed cosmic microwave background radiation. Inertial mass was found to depend on distance from the center of the universe in accordance with Mach's Principle. Mach's Principle can explain the motions of the spiral arms of spiral galaxies without resorting to claiming that 95% of the universe is made of "dark matter" as general relativity theory must in order to be in agreement with empirical data. Also the strength of the force of gravity was much higher in the past allowing the gravitational red shift of light to explain the Hubble red shift law without relativity theory. There was also a similar second term for the force of inertia. It explained for the first time the unusual gyroscope experiments of Eric Laithwaite the inventor of the high speed electric railroad in Great Britain.

Declaration of Universal Force Law

Since the improved electrodynamic force law gave rise to a better explanation of electrodynamics, gravity, and inertia, it was declared following the example of Newton to be the

Universal Force Law. The only known force laws not yet explained by the improved electrodynamic force law were the strong interaction force between elementary particles and the nuclear weak interaction force responsible for beta decay. Both of these forces are short-ranged on the order of the size of an elementary particle.

New Electrodynamic Model of Elementary Particles

The improved electrodynamic force plus combinatorial Euclidean geometry was then used to determine the finite-size structures comprising the complete set of elementary particles. Once this was finished each structure was identified with one of the observed elementary particles. This new classical electrodynamic model of elementary particles was able to predict the mass and excited states of all the observed elementary particles. It was able to predict all the decay modes of each elementary particle. Furthermore it was able from symmetry to identify the conserved quantum numbers, previously identified in the Standard Model of elementary particles, and to correlate them with specific structures in elementary particles. The quantum electrodynamic Standard Model of elementary particles requires 19 or more adjustable parameters and is still not able to predict all the experimental data that this classical electrodynamic model predicts with no adjustable parameters representing incomplete physics.

New Electrodynamic Model of the Atom

Once the structures for the finite-size electron, proton, and neutron had been determined, a new model for the atom was developed based on electrodynamics and combinatorial geometry. This new model of the atom was able to predict all the observed emission spectral lines of hydrogen and helium for n = 1/2, 1/3, 1/4, etc. states observed by NASA space experiments in the extreme ultraviolet as well as the emission spectra for n = 1, 2, 3, 4, etc. states previously explained by quantum theories of the atom.

New Electrodynamic Model of the Nucleus

A new model for the nucleus was developed based on electrodynamics and combinatorial geometry. Combinatorial geometry was able to explain why the nuclear shells formed by protons and neutrons had different magic numbers than the atomic electron shells and to predict all the nuclide spin states which the quantum shell model of the nucleus was unable to do. Also the electrodynamic nuclear model was able to predict the decay half-lives of the various nuclides and to predict the observed islands of nuclear stability at high nucleon number. Finally the electrodynamic nuclear model was able to explain the basis for the liquid drop model of the nucleus.

New Electrodynamic Model of Molecules

The improved electrodynamic force law was able to describe the mechanism that holds atoms together to form molecules in terms of electric and magnetic forces. The combination of spherical and chiral symmetry in the electrodynamic force law describes the internal chiral symmetry of all observed molecules, especially the large organic molecules such as proteins and DNA.

New Electrodynamic Explanation of Life Energy

The chiral symmetry of organic molecules causes a spiral spring-like shape and allows a longitudinal vibration which has been identified as the life energy of molecules. All living organisms have this longitudinal vibration at the molecular level and slowly radiate away their life energy over time. An experimental life energy meter has been developed and offered for sale on the internet that can detect the amount of life energy in any plant, animal or living organism. It can also detect if that living organism has a life-threatening disease. Some doctors have experimented with radiating animals and humans with longitudinal lights. They find that most animals and humans can recover naturally from disease when the molecules in their cells are "recharged" with longitudinal radiation without the use of any medicines or drugs.

Scientific Publications and Professional Talks

1. Charles W. Lucas, Jr., "A Microwave Polarimeter" presented to the Virginia Academy of Sciences annual meeting in Richmond, Virginia in 1963.

2. Charles W. Lucas, Jr., "Detecting of Partial Polarization of Microwaves" Honors thesis presented to the College of William and Mary in Williamsburg, Virginia in June 1964.

3. C. W. Lucas, Jr., "Temperature Dependence of the Saturation Magnetization of Nickel", **University of Maryland Department of Physics and Astronomy Technical Report No. 696** July 1967.

4. S. M. Bhagat and C. W. Lucas, Jr., "New Technique for Measurement of the Temperature Dependence of the Saturation Magnetization - Nickel", **Review of Scientific Instruments Vol. 39** p. 255 (1968).

5. C. W. Lucas, Jr. and C. W. Terrill, "Algorithm 404 Complex Gamma Function", **Communications of the ACM Vol. 14**, p. 48 (1971).

6. H. Uberall, B. A. Lamers, C. W. Lucas, Jr., and A. Nagl, "Charged Pion Photo-production from ^{12}C with Excitation of Analog States", **Physics Letters Vol. 44B** p. 324 (1973).

7. V. Devanathan, B. A. Lamers, C. W. Lucas, Jr., A. Nagl, J. B. Seaborn, H. Uberall, and C. Werntz, "Comparison Between Shell Model and Helm Model Predictions for Pion Photo-production in ^{12}C" - contributed to the International Conference on Nuclear Physics, Munich, Germany August 27 - September 1, 1973.

8. B. A. Lamers, G. B. Lamers, C. W. Lucas, Jr. A. Nagl, H. Uberall, and C. Werntz, "Pion Photo-production and Spin Flip States in Self-Conjugate Nuclei" - contribution to the Fifth International Conference in High Energy Physics and Nuclear Structure, Uppsala, Sweden June 1973.

9. B. A. Lamers, C. W. Lucas, Jr., A. Nagl, and H. Uberall, "Charged Pion Photo-production from ^{12}C with Excitation of Analog States", **Bulletin of the American Physical Society II Vol. 18** p. 675 (1973).

10. C. W. Lucas, Jr. and Carl Werntz, "Coherent Pion Photo-production on ^{12}C and ^{40}Ca at 250 MeV", **Bulletin of the American Physical Society II Vol. 18** p 1409 (1973).

11. F. Cannata, B. A. Lamers, C. W. Lucas, Jr., A. Nagl, H. Uberall, and C. Werntz, "Radiative Pion Capture in Flight by ^{12}C", **Bulletin of the American Physical Society II Vol. 18** p. 1584 (1973).

12. F. Cannata, B. A. Lamers, C. W. Lucas, Jr., A. Nagl, H. Uberall, and C. Werntz, "Radiative Pion Capture in Flight and Charged Pion Photo-production on ^{12}C", **Canadian Journal of Physics Vol. 52** p. 1405 (1974).

13. C. Werntz, F. Cannata, and C. W. Lucas, Jr., "Final State Interactions and the Pion Optical Wave Function", **Bulletin of the American Physical Society II Vol. 19** p 575 (1974).

14. F. Cannata, B. A. Lamers, C. W. Lucas, Jr., A. Nagl, H. Uberall, and C. Werntz, "Breakdown of DWIA and Lorentz-Lorenz Effect in Pion Reactions", **Bulletin of the American Physical Society II Vol. 19** p 575 (1974).

15. F. Cannata, C. W. Lucas, Jr., and C. Werntz, "Contributions of Nucleon Momenta to the Pion Optical Potential", **Physical Review C Vol. 10** p 2093 (1974).

16. F. Cannata, C. W. Lucas, Jr., and C. Werntz, "Threshold Production of Pions on ^{6}Li", **Physical Review Letters Vol. 33** p 1316 (1974).

17. F. Cannata, F. J. Kelly, C. W. Lucas, Jr., A. Nagl, H. Uberall, and C. Werntz, "Giant Resonance Excitation in Pion Photo-Production and in Radiative Pion Capture in Flight" - contribution to the Amsterdam Conference on Nuclear Structure, Amsterdam, Netherlands, September 1974.

18. C. Werntz and C. W. Lucas, Jr., "Energy-Dependent Pion Mean Free Path Length for Star Formation", **Bulletin of the American Physical Society II Vol. 20** p 155 (1975).

19. C. W. Lucas, Jr. and Carl Werntz, "Fermi Motion Corrections to the Pion-Nucleus Optical Potential", **Bulletin of the American Physical Society II Vol. 20** p 690 (1975).

20. Carl Werntz and C. W. Lucas, Jr., "Absorption of Pions through Collisional Broadening of the 3, 3 Resonance", **Bulletin of the American Physical Society II Vol. 20** p 1194 (1975).

21. Carl Werntz and C. W. Lucas, Jr., "Theoretical Negative Pion Absorption Cross Sections of Nuclei of Biomedical Significance", **Catholic University Technical Report** February 1976.

22. C. Werntz and C. W. Lucas, Jr., "Energy-Dependent Pion Mean Free Path Length for Star Formation", **Proceedings of the Washington Conference on Nuclear Cross Sections and Technology** March 3-7, 1975 NBS FP425 volume 1 and 2 edited by R. A. Shrack and C. D. Bowman.

23. C. W. Lucas, Jr., "Is Relativity Necessary for Electrodynamics?" **Bulletin of the American Physical Society II Vol. 23** p 544 (1978).

24. C. W. Lucas, Jr., "Is the Mathematics of Relativity Theory Based Upon Reality?" **Bulletin of the American Physical Society II Vol. 23** p 545 (1978).

25. C. W. Lucas, Jr., "Soli Deo Gloria" - contribution to the Thirty-Third Annual Meeting of the American Scientific Affiliation, Holland, Michigan, August 11-14, 1978.

26. C. W. Lucas, Jr., "Lorentz Invariance is Physically Invalid", **Bulletin of the American Physical Society II Vol. 24** (1979).

27. C. W. Lucas, Jr., "No Universal Scientific Theories of Evolution" - contribution to the Third Annual Baltimore Creation Convention, June 7-9, 1984 in Baltimore, Maryland.

28. C. W. Lucas, Jr., "The First Scientific Model for Creation" - contribution to the Third Annual Baltimore Creation Convention, June 7-9, 1984 in Baltimore, Maryland.

29. C. W. Lucas, Jr., "The New Reformation and the Work of Francis Schaeffer" - contribution to the Fourth Annual Baltimore Creation Convention, May 30 - June 1, 1985 in Baltimore, Maryland.

30. C. W. Lucas, Jr., "A Call for Reformation in Modern Science" - contributed paper to the Bible Science Association National Creation Conference August 14-16, 1985 in Cleveland, Ohio.

31. C. W. Lucas, Jr., "A Creationist Call for Reformation in the Selection of Foods and Drugs" - contributed paper to the Fifth Annual Baltimore Creation Convention June 5-7, 1986 in Baltimore, Maryland.

32. C. W. Lucas, Jr., "A Call for Reformation in Modern Science", **Proceedings of the First International Conference on Creationism** August 4-9, 1986 in Pittsburgh, Pennsylvania (Creation Science Fellowship, 362 Ashland Avenue, Pittsburgh, PA 15228) vol. I, pp. 83-87.

33. C. W. Lucas, Jr., "A New Unified Theory of Modern Science", **Proceedings of the First International Conference on Creationism** August 4-9, 1986 in Pittsburgh, Pennsylvania (Creation Science Fellowship, 362 Ashland Avenue, Pittsburgh, PA 15228) vol. II, pp. 127-36.

34. C. W. Lucas, Jr., "A Call for Reformation in Education" - contributed to the Sixth Annual Baltimore Creation Convention June 4-6, 1987 in Baltimore, Maryland.

35. C. W. Lucas, Jr., "A Call for Reformation in Christian Views of Science and Mathematics", presented to Mid-Atlantic Christian Schools Association annual convention November 9-11, 1988 in Lancaster, Pennsylvania.

36. C. W. Lucas, Jr., "A Call for Reformation in Christian Views about Food", presented to Mid-Atlantic Christian Schools Association annual convention November 9-11, 1988 in Lancaster, Pennsylvania.

37. C. W. Lucas, Jr., "The Sanctification of the Christian School" - presented to the Mid-Atlantic Christian Schools Association annual convention November 9-11, 1988 in Lancaster, Pennsylvania.

38. C. W. Lucas, Jr., "A Physical Mechanism by which God Created According to the Scriptures and Science", **Proceedings of the Second International Conference on Creationism** July 30 - August 4, 1990 in Pittsburgh, Pennsylvania (Creation Science Fellowship, 362 Ashland Avenue, Pittsburgh, PA 15228) Vol. I, pp. 127-36.

39. C. W. Lucas, Jr. and Joseph C. Lucas, "Electrodynamics of Real Particles vs. Maxwell's Equations, Relativity Theory, and Quantum Mechanics", **Proceedings of the Twin-**

Cities Creation Conference, 1992, The Twin Cities Creation Science Association, Minneapolis, MN, pp. 243-248.

40. C. W. Lucas, Jr., "Overview of the Electrodynamic Approach to God's Creation and Daily Sustaining of the Universe", **Proceedings of the Twin-Cities Creation Conference**, 1992, The Twin Cities Creation Science Association, Minneapolis, MN, pp. 142-146.

41. C. W. Lucas, Jr. and Joseph W. Lucas, "The Origin of Atomic Structure", **Proceedings of the Third International Conference on Creationism** R. E. Walsh Editor, 1994, July 18-23, 1994 at Duquesne University in Pittsburgh, PA (Creation Science Fellowship, 362 Ashland Avenue, Pittsburgh, PA 15228) Vol. I pp. 305-315.

42. C. W. Lucas, Jr., "A Physical Model for Atoms and Nuclei", **Galilean Electrodynamics Vol. 7**, pp. 3-12 (1996).

43. C. W. Lucas, Jr., "A Physical Model for Atoms and Nuclei", **Proceedings of the Physics Workshop**, held August 24-31, 1997 in Cologne, Germany.

44. C. W. Lucas, Jr., "A New Foundation for Modern Physics", **Proceedings of the Fifth International Conference "Problems of Space, Time and Motion"**, June 22-27, 1998 in St. Petersburg, Russia.

45. Lucas Jr., Charles W. And David L. Bergman, "Physical Models of Matter" **Physics As a Science**. G. Galeczki, P. Marquardt, and J. P. Wesley editors (Hadronic Press, Palm Harbor, FL 34682-1577) pp. 45-68 (1998).

46. C. W. Lucas, Jr. and Joseph W. Lucas, "A New foundation for Modern Science", **Proceedings of the Fourth International Conference on Creationism** R. E. Walsh Editor, 1998, August 3-8, 1998 Geneva, Pennsylvania (Creation Science Fellowship, 362 Ashland Avenue, Pittsburgh, PA 15228) Vol. I pp. 379-394.

47. Lucas Jr., Charles W. and Joseph C. Lucas, "Weber's Force Law for Realistic Finite-Size Elastic Particles" **Journal of New Energy, Vol. 5, No. 3, Proceedings of the 2nd Cologne Workshop "Physics as a Science" 2000**, pp. 70-89 (2001).

48. Lucas Jr., Charles W., "A Classical Electromagnetic Theory of Elementary Particles" **Journal of New Energy, Vol. 6, No. 4, Proceedings of "Physics as a Science" International Workshop**, Arrecife, Lanzarote, Canary Islands July 1-5, 2002, pp. 81-109 (2002).

49. Lucas Jr., Charles W. And Joseph Lucas, "A Physical Model for Atoms and Nuclei Part 1, 2, 3, 4" **Foundations of Science Vol. 5, No. 1**, pp. 1-7 (2002), **Vol. 5. No. 2**, pp. 1-8 (2002), **Vol. 6, No. 1**, pp. 1-10 (2003), **Vol. 6, No. 3**, pp. 1-8 (2003).

50. Lucas Jr., Charles W. and David L. Bergman, "Credibility of Common Sense Science" **Foundations of Science, Vol. 6, No. 2**, pp. 1-10 (2003).

51. Lucas Jr., Charles W., "Radiohalos - Key Evidence for Origin/Age of the Earth" **Proceedings of the Cosmology Conference 2003** held by the Creation Research and Science Education Foundation Columbus, OH, October 31 and November 1, 2003.

52. Lucas Jr., Charles W. and David L. Bergman, "A Classical Electromagnetic Theory of Elementary Particles Part 1 and 2" **Foundations of Science, Vol. 7, No. 4**, pp. 1-8 (2004), **Vol. 8, No. 2**, pp. 1-16 (2005).

53. Lucas Jr., Charles W., "Derivation of a Universal Electromagnetic Force Law for Finite-Size Elastic Charged Particles" **12th Annual Conference of the Natural Philosophy Alliance (NPA) "Foundations of Natural Philosophy", Vol. 2, No. 1**, pp. 85-108, University of Connecticut, at Storrs, CT May 23-25, 2005.

54. Lucas Jr., Charles W., "A Classical Electromagnetic Theory of Everything" **13th Annual Conference of the Natural Philosophy Alliance (NPA) "Science for the Next Generation", Vol. 3, No. 1**, pp. 142-158, University of Connecticut, at Storrs, CT May 23-25, 2005. (Could not fit in **12th Annual Proceedings**)

55. Lucas Jr., Charles W., "Derivation of the Classical Universal Electrodynamic Force Law", "The Electrodynamic Origin of the Force of Inertia", "The Electrodynamic Origin of the Force of Gravity", "A Classical Electrodynamic Theory of the Atom", "A Classical Electrodynamic Theory of the Nucleus", "A Classical Electrodynamic String Theory of Elementary Particles", "The Electrodynamic Origin of Life in Organic Molecules Such as DNA and Proteins" **13th Annual Conference of the Natural Philosophy Alliance (NPA) "Science for the Next Generation",** University of Tulsa, at Tulsa, OK April 3-7, 2006.

56. Lucas Jr., Charles W., "Derivation of the Universal Force Law - Part 1, 2, 3, 4" **Foundations of Science, Vol. 9, No. 2**, pp 1-10 (2006), **Vol. 9, No. 3**, pp. 1-6 (2006), **Vol. 9, No. 4**, pp. 1-13 (2006), **Vol. 10, No. 1**, pp. 1-6 (2007).

57. Lucas Jr., Charles W., "The Electrodynamic Origin of the Force of Inertia Part 1, 2, 3" **Foundations of Science, Vol. 10, No. 4**, pp. 1-9 (2007), **Vol. 11, No. 1**, pp. 1-5 (2008), **Vol. 11, No. 2**, pp. 1-6 (2008).

58. Lucas Jr., Charles W., "The Electrodynamic Origin of the Force of Gravity Part 1, 2, 3" **Foundations of Science, Vol. 11, No. 4**, pp. 1-10 (2008), **Vol. 12, No. 1**, pp. 1-11 (2009), **Vol. 12, No. 2**, pp. 1-12 (2009).

59. Charles W. Lucas, Jr., Roger A. Rydin, "Letter to the Editor" **Nuclear Science and Engineering Vol. 161**, pp. 1-2 (2009).

60. C. W. Lucas, Jr., R. A. Rydin, "Electrodynamic Model of the Nucleus", **Nuclear Science and Engineering Vol. 161**, pp. 255-256 (2009).

61. Lucas Jr., Charles W., "Electrodynamic Origin of Gravitational Forces" **18th Annual Conference of the Natural Philosophy Alliance (NPA) "Science for the Next Generation", Vol. 8,** pp. 375-386, University of Maryland, at College Park, MD July 6-9, 2011.

62. Lucas Jr., Charles W., "The Universal Electrodynamic Force" **18th Annual Conference of the Natural Philosophy Alliance (NPA) "Science for the Next Generation", Vol. 8,** pp. 387-396, University of Maryland, at College Park, MD July 6-9, 2011.

63. Lucas Jr., Charles W., "It's an Electric Universe After All" **1st Annual Electric Universe "The Human Story",** Rio Hotel Las Vegas, NV January 6-9, 2012.

64. Lucas Jr., Charles W., "The Electrodynamic Pulse of Life" **1st Annual Electric Universe "The Human Story",** Rio Hotel Las Vegas, NV January 6-9, 2012.

65. Lucas Jr., Charles W., "Symmetry of Nature Confirms Universal Electrodynamic Force" **19th Annual Conference of the Natural Philosophy Alliance (NPA) "A New Vision for the Universe", Vol. 9,** Albuquerque, NM July 25-28, 2012.

66. Lucas Jr., Charles W., "The Need for Reformation in Modern Science Based on Universal Truth, Structuralism and Euclidean Geometry" **Proceedings of the Natural Philosophy Alliance 19th Annual Conference of the NPA, "A New Vision for the Universe", Vol. 9,** Albuquerque, NM July 25-28, p. 326, 2012.

67. Lucas Jr., Charles W., "The Structure and Symmetry of the Universe" **Proceedings of the Natural Philosophy Alliance 20th Annual Conference of the NPA "A New Vision for the Universe", Vol. 10,** College Park, MD July 10-13, 2013 pp. 174-179.

68. Lucas Jr., Charles W., "Mach's Principle and the Concept of Mass" **Foundations of Science, Vol. 16, No. 3,** pp. 1-7 (2013).

69. Charles W. Lucas, Jr., Eric C. Baxter, Edward A. Boudreaux, and Roger A. Rydin, "A Classical Electro-Dynamic Theory of the Nucleus" **Physics Essays Vol. 26, No. 3,** pp. 392-400 (2013).

TV Network Shows

Produced a series of thirteen 30 minute TV shows on (Scientific) **Origins** for the Cornerstone TV Public Education Network based in Wall, PA in 1987 and 1993.

1. **Atomism & Evolution vs. Christianity OR-2887 10-15-87**
2. **Is there any Truth in Science? OR-2887A 10-15-87**
3. **Universe is Electromagnetic in Nature OR-2887B 10-15-87**
4. **An Electromagnetic Model of Matter OR-2887C 10-15-87**
5. **Atomism vs. Christian Definition of Food OR-2887D 10-15-87**
6. **Atomistic vs. Christian Food Processing OR-2887E 10-15-87**
7. **Atomistic vs. Christian Education OR-2887F 10-15-87**
8. **Getting Together with Dr. Lucas GT-2887 10-15-87**
9. **Polonium Radiohalos OR-9326 7-28-93**
10. **Radiometric Age of the Earth OR-9327 7-28-93**
11. **Theoretical Foundation of Evolution Theory OR-9329 7-29-93**
12. **The Geometry of the Atom OR-9330 7-29-93**
13. **The Electron's Magnetic Properties OR-9331 7-29-93**

Postlude

This is the sixth revision of an important scientific work. It will probably go through many revisions. If you see any errors or things to improve in the scope of this book, the author would appreciate hearing from you. You may contact the author as follows:

Charles W. Lucas, Jr.

29045 Livingston Drive

Mechanicsville, MD 20659-3271

USA

Bill.Lucas001@gmail.com (email)

Any corrections or additions that you suggest that are accepted by the author will be acknowledged in future editions of the book.

Additional Books in the Series

Due to the economics of book publishing this work on the universal force as a foundation for all of science or natural philosophy has been divided up into a series of books. At the time of the publication of Volume 1 the tentative list of the other titles with a brief overview of the content is as follows:

The Universal Force Volume 2 – An Electrodynamic Model of Elementary Particles

An electrodynamic model of elementary particles is developed in terms of finite-size three dimensional structures consisting of closed charge loops. Combinatorial geometry is used to define the complete set of structures. These structures are identified with the known elementary particles. The reactions and decays of the these elementary particles are then described in terms of charge string structures. Finally the rest masses and excited states of the elementary particles are predicted. The claim is made that this theory of elementary particles is more complete than the Standard Model of Elementary Particles based on simplicity (no 19 adjustable constants), superior logical foundation, and better agreement with experimental data.

The Universal Force Volume 3 – An Electrodynamic Model of the Atom and the Nucleus

An electrodynamic model of the atom and nucleus is developed in terms of finite-size three dimensional electrons, protons, and neutrons consisting of closed charge loop structures. Combinatorial geometry is used to define the nuclear and atomic shell structures. The process of nuclear decay and fission is explained in terms of the binding energy of various shell structures. The claim is made that these electrodynamic models of the atom and nucleus are more complete than the Dirac quantum theory of the atom and both the quantum and liquid drop theories of the nucleus based on simplicity (one force law), superior logical foundation, and better agreement with experimental data. The islands of nuclear stability are simply predicted. The atomic mechanism for cold fusion is explained. The emission spectra of hydrogen in the extreme ultraviolet for $n = 1/2, 1/3, 1/4$, etc. states is explained.

The Universal Force Volume 4 – An Electrodynamic Model of Molecules and the Origin of Life

An electrodynamic model of molecules is developed consisting of finite-size three dimensional electrons with electric and magnetic fields to produce binding of atoms. Combinatorial geometry is used to determine the geometrical arrangement of atoms in complex organic molecules and crystal lattices. The pulse of life is identified in organic molecules as a longitudinal vibration of the molecules. The role of this longitudinal vibration in organizing and regulating processes in living cells is explained. Also the role of electrodynamics of the sun in the solar system in support of living systems is explained. The basis for the observed rule that "Life Comes from Life" is explained in terms of longitudinal vibrations. Finally the claim is made that this theory of molecules and the origin of life is more complete than previous approaches.

Appendix A: Derivation of the Biot-Savart and Grassmann Form of Ampere's Law

Maxwell's equations and the covariant formulation of electrodynamics do not apply to point particles, as generally assumed, but only to charge elements of complete circuits or loops. Hermann Grassmann [5]

The magnetic effect of an electric current was discovered in 1819 by the Danish physicist and chemist, Hans Christian Oersted (1777-1851). [1] He found that a compass needle placed under an electric current takes up a direction perpendicular to that of the current.

This observation, which was the first to link electricity and magnetism, stimulated the French physicist and mathematician Andre Marie Ampere (1775-1836) to perform many experiments exploring the implications of this observation. Ampere observed the action of an electric current upon a magnet, the action of a magnet upon an electric current, and the action of one electric current upon another electric current. These actions and reactions were found to be consistent with Newton's third law ("to every action there is always opposed an equal reaction")

Ampere [2] performed a series of four experiments and found that the force between two different current loops obeys the following laws:

1. **The effect of a current is reversed when the direction of the current is reversed.**
2. **The effect of a current flowing in a circuit twisted into small sinuosities is the same as if the circuit were smoothed out.**
3. **The force exerted by a closed circuit on an element of another circuit is at right angles to the latter.**
4. **The force between two elements of circuits is unaffected when all linear dimensions are increased proportionately and the current strengths remain unaltered.**

From these four laws and the assumption that the force between two elements of circuits act along the line joining them, Ampere obtained a mathematical expression for the force. The deduction of the force law was made as follows:

1. From law (2) the effect of **ds** on **ds'** is the vector sum of the effects on **ds'**. From law (1) which supports Newton's third law of action and reaction plus the assumption that the force is linear and homogeneous in ds and ds', the simplest general formula must be

$$\vec{F}_{ij} = ii'\left[\vec{r}(d\vec{s}_i \cdot d\vec{s}_{j'})\varphi(r) + d\vec{s}_i(d\vec{s}_{j'} \cdot \vec{r})\varphi_2(r) + d\vec{s}_{j'}(d\vec{s}_i \cdot \vec{r})\varphi_3(r) \right.$$
$$\left. + \vec{r}(d\vec{s}_i \cdot \vec{r})(d\vec{s}_{j'} \cdot \vec{r})\psi(r)\right] \qquad (A-1)$$

where $\varphi_1(r)$, $\varphi_2(r)$, $\varphi_3(r)$, $\psi(r)$ are functions of r still to be determined.

2. The assumption that the force between two elements of the circuits acts along the line joining them implies that the $\varphi_2(r)$ and $\varphi_3(r)$ terms should be dropped since they are not proportional to the vector **r** giving

$$\vec{F}_{ij} = ii'\vec{r}\left[(d\vec{s}_i \cdot d\vec{s}_{j'})\varphi(r) + (d\vec{s}_i \cdot \vec{r})(d\vec{s}_{j'} \cdot \vec{r})\psi(r)\right] \quad (A-2)$$

3. From law (4) the force $\mathbf{F}_{ij}$ is unaffected when $\mathbf{ds}_i$, $\mathbf{ds}_j$, **r** are all changed by the same factor. The simplest forms for $\varphi(r)$ and $\psi(r)$ for this to be true are giving

$$\varphi(r) = A/r^3 \text{ and } \psi(r) = B/r^5 \quad (A-3)$$

where A and B denote constants still to be determined.

4. From law (3) the projection of $\mathbf{F}_{ij}$ along **ds'** must vanish when integrated around the circuit s, i.e.

$$\vec{F}_{ij} = ii'\vec{r}\left[\frac{A}{r^3}(d\vec{s}_i \cdot d\vec{s}_{j'}) + \frac{B}{r^5}(d\vec{s}_i \cdot \vec{r})(d\vec{s}_{j'} \cdot \vec{r})\right] \quad (A-4)$$

5. For the limiting case $ds_j = -dr$ where ds_j is fixed, one can obtain a relationship between A and B, i.e.

$$\oint(\vec{F}_{ij} \cdot d\vec{s}_{j'})\, d\vec{s}_i = 0$$

$$= ii'\oint\left[\frac{A}{r^3}(\vec{r} \cdot d\vec{s}_{j'})(\vec{r} \cdot d\vec{s}_{j'}) + \frac{B}{r^5}(d\vec{s}_i \cdot \vec{r})(d\vec{s}_{j'} \cdot \vec{r})\right]d\vec{s}_i \quad (A-5)$$

$$0 = -ii'\oint\left[\frac{A}{2r^3}d(d\vec{s}_{j'} \cdot \vec{r}) + \frac{B}{r^4}(d\vec{s}_{j'} \cdot \vec{r})d\vec{r}\right]d\vec{s}_i \quad (A-6)$$

6. In order for this to be true the integrand must be a complete differential such that

$$d\left(\frac{A}{2r^3}\right) = -\frac{3}{2}\frac{A}{r^4}dr = \frac{B}{r^4}dr \quad (A-7)$$

giving

$$B = -\frac{3}{2}A \quad (A-8)$$

Thus the Ampere law for the force between two current loops is

$$\vec{F}_{ij} = -\frac{A}{2}ii'\vec{r}\left[\frac{2}{r^3}(d\vec{s}_i \cdot d\vec{s}_{j'}) - \frac{3}{r^5}(d\vec{s}_i \cdot \vec{r})(d\vec{s}_{j'} \cdot \vec{r})\right] \quad (A-9)$$

As Whittaker [3] points out there is a weakness in Ampere's law. The weakness is the assumption that the force between two circuit elements acts only along the line joining them. In the analogous case of the action between magnetic molecules, the experimentally observed force is not entirely directed along the line joining the molecules. Also Helmholtz [4] assumes that the interaction between two current elements is derivable from a potential, like Weber, and this entails the existence of a couple in addition to a force along the line joining the elements. Thus in order to obtain the more general force law between current elements, one must include all of the terms linear and homogeneous in **ds** and **ds'** in equation (A-1).

Now Ampere has already obtained one set of terms that satisfies Newton's third law. In order to get an additional set of terms that satisfy Newton's third law for the forces between current elements that are not directed along the line between them, a second full set of terms need to be added. Noting that these terms must also form a complete differential, the more general equation for **F**$_{ij}$ is

$$\vec{F}_{ij} = -\frac{ii'\vec{r}}{c}\left[\frac{2}{r^3}(d\vec{s}_i \cdot d\vec{s}_{j'}) - \frac{3}{r^5}(d\vec{s}_i \cdot \vec{r})(d\vec{s}_{j'} \cdot \vec{r})\right] \qquad (A-10)$$
$$+ \chi(r)(d\vec{s}_{j'} \cdot \vec{r})d\vec{s}_i + \chi(r)(d\vec{s}_i \cdot \vec{r})d\vec{s}_{j'} + \chi(r)(d\vec{s}_i \cdot d\vec{s}_{j'})\vec{r}$$
$$+ \frac{1}{r}\chi'(r)(d\vec{s}_i \cdot \vec{r})(d\vec{s}_{j'} \cdot \vec{r})\vec{r}$$

where for CGS units the constant A/2= 1/c.

The simplest solution for $\chi(r)$ that satisfies Ampere's four laws for current elements is

$$\chi(r) = \frac{ii'}{cr^3} \qquad (A-11)$$

Substituting in equation (A10), one notes some cancellation of terms to obtain

$$\vec{F}_{ij} = \frac{ii'}{cr^3}\left[(d\vec{s}_i \cdot \vec{r})d\vec{s}_{j'} + (d\vec{s}_{j'} \cdot \vec{r})d\vec{s}_i - (d\vec{s}_i \cdot d\vec{s}_{j'})\vec{r}\right] \qquad (A-12)$$

For forces between current loops, if one calculates the force **F**$_i$ on the current element ds$_{j'}$ due to the entire j' current loop,

$$\vec{F}_i = \frac{ii'}{cr^3}\left[(d\vec{s}_i \cdot \vec{r})\oint d\vec{s}_{j'} + \oint(d\vec{s}_{j'} \cdot \vec{r})d\vec{s}_i - \oint(d\vec{s}_i \cdot d\vec{s}_{j'})\vec{r}\right] \qquad (A-13)$$

then the first term evaluates to zero, because the j' loop is a closed loop. Thus this term does not make any effective contribution to the force. In other words Ampere's general force law for forces between current loops is not mathematically unique. Note that Grassmann [5] omits this term by making the assumption that two current elements in the same straight line have no mutual action or longitudinal force between them. However it is necessary to see this term in order to understand that Newton's third law is satisfied by both Grassmann and Ampere force laws. Assis [6, 7] has shown that Ampere's and Grassmann's expressions for the force always yield the same results when considering the net force on any current element of a closed circuit of arbitrary shape.

The general form of the force law of equation (A-12) can be written in terms of a triple vector product

$$\vec{F}_{ij} = i \, d\vec{s}_i \times \frac{i' d\vec{s}_{j'} \times \vec{r}_{ij}}{c r_{ij}{}^3} = i \, d\vec{s}_i \times d\vec{B}_{ij} \qquad (A-14)$$

where

$$d\vec{B}_{ij} = i' \frac{d\vec{s}_{j'} \times \vec{r}_{ij}}{c r_{ij}{}^3} \qquad (A-15)$$

Heaviside [8, 9] noted in 1888 that the original Ampere law in the form of equation (A-9) was not actually being used in practice. He suggested that the Father of Electrodynamics have his name attached to the more general force law of equation (A-14) that is based on his experimental observations. Unfortunately Grassmann and to some extent Biot and Savart also had their name associated with various forms of this equation. There has been confusion ever since. Maxwell attributed his differential equation which results from the more general form of the force law to Ampere.

Equations (A-14) and (A-15) only have meaning as one element of a sum over a continuous set where the sum represents the magnetic induction of a current loop or circuit. The continuity equation

$$\nabla \cdot \vec{J} = 0 \qquad (A-16)$$

is not satisfied for the current element id**s** by itself, because the current comes from nowhere and disappears after traversing the length d**s**.

The German mathematician Herman Gunther Grassmann (1809-1877) rewrote equations (A-14) and (A-15) by replacing id**s** by -q**v** where q is the charge in the segment d**s** and **v** is its velocity. Integrating both sides of the equation obtain

$$\vec{F}_{ij} = q_i \left(\vec{v}_i \times \vec{B}_{ij} \right) \qquad (A-17)$$

where

$$\vec{B}_{ij} = \frac{q_j \left(\vec{v}_j \times \vec{r}_{ij} \right)}{c r_{ij}{}^3} \qquad (A-18)$$

Equation (A-17) is known as the Grassmann force law. Grassmann's substitution of -q**v** for id**s** is based upon the following reasoning. If we let Q be the total charge in the circuit, and assume that the total charge of a closed system or circuit remains constant then

$$\frac{dQ}{dt} = 0 \ \ with \ \ Q = \sum_{i=1}^{n} q_i ds_i \qquad (A-19)$$

Taking the derivative explicitly of Q one obtains

$$\frac{dQ}{dt} = \sum_{i=1}^{n} \left(ds_i \frac{dq_i}{dt} + q_i \frac{ds_i}{dt} \right) = 0 \qquad (A-20)$$

Requiring each term in the series i = 1 to n to individually be equal to 0 gives the relation

$$ds_i \frac{dq_i}{dt} = -q_i \frac{ds_i}{dt} \qquad (A-21)$$

or in general

$$I_i d\vec{s}_i = -\vec{v}_i q_i \qquad (A-22)$$

Thus it appears that on the most elementary basis Grassmann's force law of equations (A-17) and (A-18) applies to the forces between elements of charges in complete loops or circuits and not arbitrary point charges. If one does not include all the charge elements of the complete circuit, one is not using the Grassmann law correctly. This implies that Maxwell's equations and the covariant formulation of electrodynamics do not apply to point particles, as generally assumed, but only to charge elements of complete circuits or loops.

A: References

1. Hans Christian Oersted, **Experimenta circa effectum conflictus electrici in acum magneticam** (Copenhagen, 1820); English translation in Thompson's **Annals of Philosophy, xvi**, p. 273 (1820).
2. Ampere, **Mem. de l'Acad. VI,** p. 175 (1825).
3. Sir Edmund Whittaker, **A History of the Theories of Aether & Electricity** (Dover Publications, Inc., New York, 1989) p. 86.
4. Helmholtz, **Berl. Akad. Monatsber.**, pp. 91-107 (1873).
5. Hermann Grassmann, **Annalen der Physik und Chemie LXIV**, pp. 1-18 (1845).
6. A. K. T. Assis and Marcelo A. Bueno, "Equivalence between Ampere and Grassmann's Forces", **IEEE Transactions on Magnetics, vol. 32**, pp. 431-436 (1996).
7. Marcelo Bueno and A. K. T. Assis, "Proof of the Identity between Ampere and Grassmann's Forces", **Physica Scripta, vol. 56**, pp. 554-559 (1997).
8. Heaviside, **Electrician**, pp. 229-230 (Dec. 28, 1888).
9. **Heaviside's Electrical Papers, ii**, (Cambridge University Press), pp. 500-502 (1891).

Appendix B: Helmholtz Decomposition Theorem

Helmholtz's theorem, also known as the fundamental theorem of vector calculus, states that any sufficiently smooth, rapidly decaying vector field in three dimensions can be resolved into the sum of an <u>irrotational</u> (curl-free) vector field and a <u>solenoidal</u> (divergence-free) vector field; this is known as the **Helmholtz decomposition**. [1] It is named after Hermann von Helmholtz.

Helmholtz's theorem implies that any such vector field **F** can be considered to be generated by a pair of potentials: a scalar potential φ and a vector potential **A**. Let **F** be a vector field on a bounded domain V in $\mathbf{R}^3$, which is twice continuously differentiable. Then **F** can be decomposed into a curl-free component and a divergence-free component

$$\vec{F} = -\nabla\varphi + \nabla \times \vec{A} \qquad (B-1)$$

where

$$\varphi(r) = \frac{1}{4\pi}\oint \frac{\nabla' \cdot \vec{F}(\vec{r}')}{|\vec{r} - \vec{r}'|}\, dV' - \frac{1}{4\pi}\oint \frac{\vec{F}(\vec{r}') \cdot d\vec{S}'}{|\vec{r} - \vec{r}'|} \qquad (B-2)$$

$$A(r) = \frac{1}{4\pi}\oint \frac{\nabla' \times \vec{F}(\vec{r}')}{|\vec{r} - \vec{r}'|}\, dV' + \frac{1}{4\pi}\oint \frac{\vec{F}(\vec{r}') \times d\vec{S}'}{|\vec{r} - \vec{r}'|}$$

If V is $\mathbf{R}^3$ itself (unbounded), and **F** vanishes sufficiently fast at infinity, then the second component of both scalar and vector potential are zero. That is

$$\varphi(r) = \frac{1}{4\pi}\oint \frac{\nabla' \cdot \vec{F}(\vec{r}')}{|\vec{r} - \vec{r}'|}\, dV' \qquad (B-3)$$

$$A(r) = \frac{1}{4\pi}\oint \frac{\nabla' \times \vec{F}(\vec{r}')}{|\vec{r} - \vec{r}'|}\, dV'$$

In electrodynamics F(r') is associated with the electrical current j(r'). The second terms in the definition of the scalar and vector potentials are missing in Maxwellian electrodynamics, but the experimental evidence is that the fields are an extension of the charges and have tensile strength up to the finite bounded size of the universe!

237

Also it is important to note that the scalar and vector potentials above are not the same kind of potential that Newton and others used in classical physics. The classical potential is based upon conservation of energy and the continuity equation. It is interesting that Maxwellian electrodynamics uses the term potentials but does not conserve magnetic energy for dynamic systems due to failure to incorporate Lenz's Law. Thus this is a different kind of potential that comes from vector calculus. Also note that the Copenhagen version of quantum mechanics for electrodynamics in the atom and elementary particles does not satisfy the continuity equation, because the Heisenberg uncertainty principle allows instantaneous exceptions, i.e. instantaneous creation and annihilation of charge particle pairs.

B: References

1. http://en.wikipedia.org/wiki/Helmholtz_decomposition

Appendix C: Derivation of Blackbody Radiation Formula

In 1901 Max Planck [14] was able to find a mathematical expression that fit the blackbody radiation data. His attempts to work backwards to find the correct physical theory resulted in the birth of Quantum Physics. However, this theory was never fully satisfactory. It was based on the notion that point-charges undergoing simple harmonic motion in the blackbody were absorbing and emitting radiation. This picture led to oscillations of point-electron charges *that were too big to remain in the lattice of the solid.* Also, the empirical laws of electrodynamics were violated by Planck's theory. Both Ampère's Law and Faraday's Law require continuous emission and absorption of radiation for simple harmonic motion of point-electron charges. Finally, the Quantum Theory of blackbody radiation was not compatible with optical reflection, refraction, and diffraction phenomena due to its emission of radiation that is *discontinuous* in time.

The problem with Planck's work — which was to develop a proper scientific theory to predict his mathematical expressions that described blackbody radiation — was that he had an inadequate model for charged elementary particles in nature. He had the notion that elementary particles could be approximated as point-particles. This notion is still found today in Quantum Theories and in Relativity Theory.

Bergman's toroidal "ring" model of elementary particles behaves quite differently with absorption and emission of radiation than is the case for point-particles. See Figure C-1. Radiation may be continuously absorbed by the ring structure. Since it is a continuous ring structure, the laws of electrodynamics do not require it to immediately re-radiate the energy absorbed. [18] Also the flow of current in a ring physically explains the magnetic moment of the electron due to the current flow in a loop.

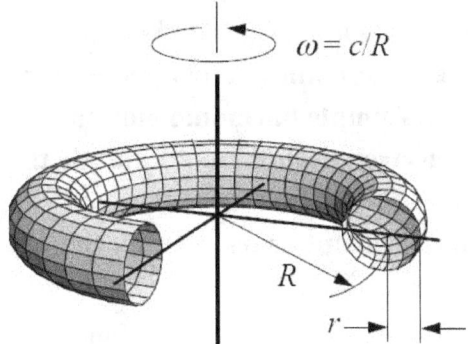

$$\omega = c/R$$

Figure C-1 Bergman Toroidal Ring Electron [2]

When electromagnetic energy or light is absorbed by the ring, there is a disturbance of the flow of charge around the ring, resulting in oscillations of the electric charge distribution flowing around the ring at the speed of light. These oscillations reflect the wavelength of the light being absorbed. The flow of charge around the ring may be thought of as the superposition of the original continuous flow plus the oscillations of charge resulting from the absorption of various light waves.

The original state of the ring, *i.e.* the continuous flow of charge around the ring, is known as a *stationary state*. No change can be detected over time. Additional stationary states of the ring structure will occur when the oscillations of the charge produced by the absorption of radiation produce standing-waves, *i.e.* the wavelength is exactly an integral number of circumferences of the ring, *i.e.* $n(2\pi r) = \lambda$ $n = 1,2,3,...$ or the circumference of the ring is exactly an integral number of wavelengths, *i.e.* $(2\pi r) = n\lambda$ $n = 1,2,3,...$ When a ring is in a stationary state, the distribution of charge is stable, and the surrounding electromagnetic fields form a standing-wave with an integer number of nodes.

The ring may retain the radiation energy indefinitely. The laws of electrodynamics do not require it to emit any radiation. However, if the ring has a collision or significant interaction with another moving ring, an additional oscillation of the charge density may result making the ring unstable. The laws of electrodynamics now require the ring to radiate. At this point one makes the reasonable assumption that radiation from ring structures may only occur from one slightly excited stationary state to another stationary state.

From Bergman's paper [9, equation 7 and equation 35]

$$E_n = n\frac{e^2}{2\pi\varepsilon_0 cR}log_e\frac{8R}{r} = n\frac{hc}{\lambda} \qquad (C-1)$$

and $(2\pi R) = n\lambda$, the energy of the stationary states is

$$E_n = \frac{nhc}{2\pi R} = \frac{nhc}{2\pi\dfrac{\lambda}{2\pi}} = nh\nu \ where \ \lambda\nu = c \qquad (C-2)$$

Although this result is mathematically identical to Planck's result, it is fundamentally different in the following ways:

1. **It does not violate any known law of electrodynamics.**
2. **It does not use an unrealistic point-particle model for the electron with a magnetic moment, which undergoes simple harmonic motion.**
3. **It does not require an electron oscillation amplitude that is too large for the electron to remain in the atom.**
4. **Simple harmonic motion of point charges is *not* the physical mechanism involved in blackbody radiation.**

Let us calculate the energy density $\rho_T(\lambda)$ as a function of wavelength λ for a specific temperature T, under the assumption above that the radiation from the ring electrons only occurs during a transition from one stationary state to another stationary state. The first step is the evaluation of the average energy contained in each standing wave of wavelength λ or frequency $\nu = c/\lambda$. According to Classical Physics, the particular energy of some wave can have any value from zero to infinity. The actual value is proportional to the square of its average amplitude, *i.e.*

$$\rho(\lambda) \propto \frac{E^2}{4\pi} \qquad (C-3)$$

However, if we have a system containing a large number of identical ring-electrons which are in thermal equilibrium with each other at a temperature T, the classical theory of statistical mechanics requires that the energies of the standing-waves be distributed according to a definite probability distribution whose form is specified by T.

From the **law of equipartition of energy** the average kinetic energy ε of the standing wave in the rings is

$$\overline{\varepsilon}_{KE} = \frac{kT}{2} \qquad (C-4)$$

where $k = 1.38 \times 10^{-16}$ erg/deg is Boltzmann's constant. For an electromagnetic wave where only the amplitude of the wave executes simple harmonic oscillations, the total average energy is just twice the average kinetic energy, $i.e.$

$$\overline{\varepsilon} = kT = \frac{\sum_{n=0}^{\infty} nh\nu e^{-nh\nu/kT}}{\sum_{n=0}^{\infty} e^{-nh\nu/kT}} = \frac{h\nu}{e^{-h\nu/kT} - 1} \qquad (C-5)$$

The Boltzmann probability of finding the wave in an energy state between ε and $\varepsilon + d\varepsilon$ for a system containing a large number of ring-electrons with waves is

$$P(\varepsilon) = Ae^{-\varepsilon/kT} \qquad (C-6)$$

The average energy of a wave is given by

$$\overline{\varepsilon} = \frac{\int_0^\infty \varepsilon P(\varepsilon) d\varepsilon}{\int_0^\infty P(\varepsilon) d\varepsilon} \qquad (C-7)$$

Now

$$\rho_T(\nu)d\nu = \frac{\overline{\varepsilon}N(\nu)d\nu}{V_{Ring}} \quad where \ V_{Ring} = (2\pi R)\pi r^2 \qquad (C-8)$$

where N (ν) is the number of allowed frequencies in the frequency interval ν to $\nu + d\nu$. It can be shown that N (ν) $d\nu$ is independent of the shape of the ring and depends only on its volume $V = 2\pi R (\pi r^2)$, [15, p. 57] $i.e.$

$$N(\nu)d\nu = \frac{8\pi(2\pi R)(\pi r^2)\nu^2 d\nu}{c^3} \qquad (C-9)$$

Thus

$$\rho_T(\nu)d\nu = \frac{h\nu}{e^{-h\nu/kT} - 1} \frac{8\pi(2\pi R)(\pi r^2)\nu^2 d\nu}{(2\pi R)(\pi r^2)c^3} = \frac{8\pi\nu^2}{c^3} \frac{h\nu}{e^{-h\nu/kT} - 1} \qquad (C-10)$$

Transforming to the variable λ where $\nu = c/\lambda$, $d\nu = -(c/\lambda^2) \, d\lambda$, and $\rho_T (\lambda)d\lambda = \rho_T (\nu)d\nu$

$$\rho_T(\lambda)d\lambda = \frac{8\pi hc}{\lambda^5} \frac{hv}{e^{-hc/k\lambda T} - 1} \qquad (C-11)$$

This is mathematically the same as the blackbody spectral distribution derived by Planck (see Figure C-2). However, it has a very different physical interpretation, and it does not violate the laws of electrodynamics.

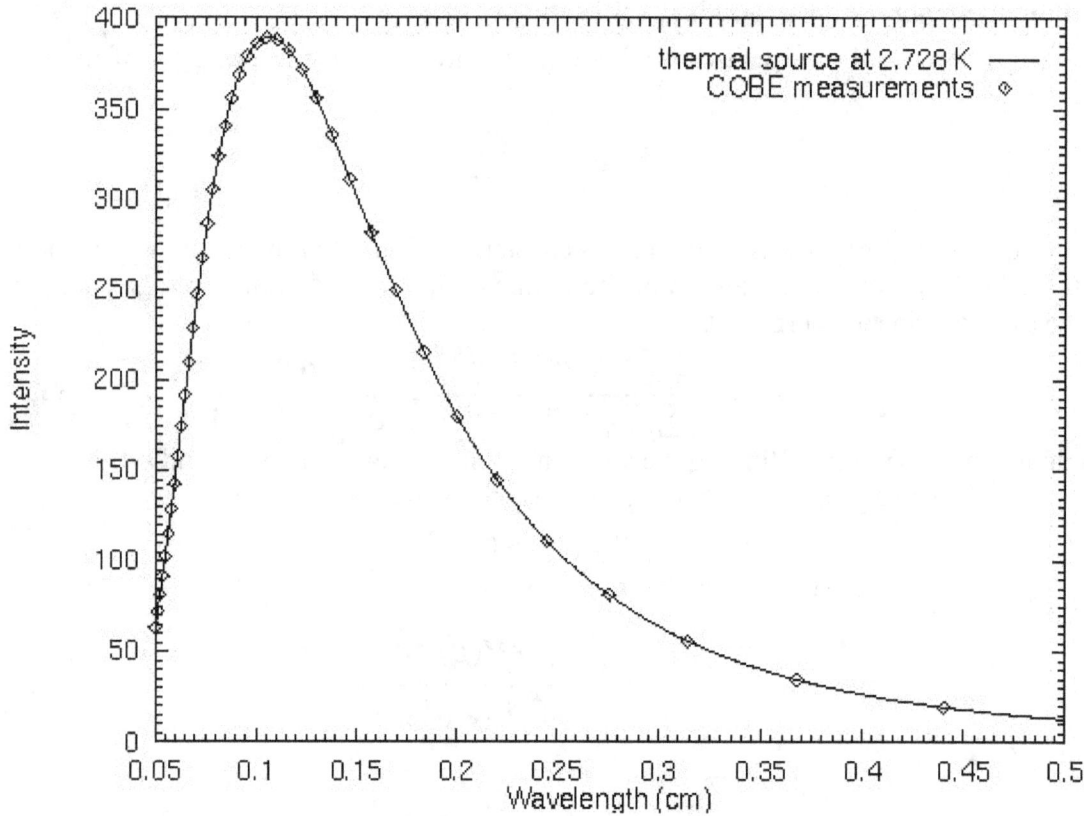

Figure C-2 COBE Blackbody Spectral Distribution

C: References

1. Planck, Max, "Uber das Gesetz der Energieverteilung im Normalspektrum", **Annals de Physik,** Vol. 4, *p.* 553 (1901).

2. Panofsky, W. K. and M. Phillips, **Classical Electricity and Magnetism** (Addison-Wesley, Reading, Massachusetts, 1962), p. 370.

3. Bergman, David L. and J. Paul Wesley, "Spinning Charged Ring Model of the Electron Yielding Anomalous Magnetic Moment", **Galilean Electrodynamics,** Vol. 1, No. 5, *p.* 63 (1990).

4. Eisberg, Robert Martin, **Fundamentals of Modern Physics** (John Wiley and Sons, Inc., New York, 1961), *pp.* 77-81.

Appendix D: Velocity Dependent Generalized Potentials

Lagrange's equations can be put in the form

$$\frac{d}{dt}\left(\frac{\partial L}{\partial \dot{q}_j}\right) - \frac{\partial L}{\partial q_j} = 0 \qquad (D-1)$$

even if the system is not conservative in the usual sense, provided the generalized forces can be obtained from a function $U(q_j, dq_j/dt)$ by the prescription

$$Q_j = -\frac{\partial U}{\partial q_j} + \frac{d}{dt}\left(\frac{\partial U}{\partial \dot{q}_j}\right) \qquad (D-2)$$

In such cases equation (D-1) still follows from equation (D-3)

$$\frac{d}{dt}\left(\frac{\partial T}{\partial \dot{q}_j}\right) - \frac{\partial T}{\partial q_j} = Q_j \qquad (D-3)$$

with the Lagrangian given by

$$L = T - U \qquad (D-4)$$

U is called the "generalized potential" or "velocity-dependent potential".

Velocity dependent potentials were introduced in Lagrangian mechanics by the German mathematician Ernst Christian Julius Schering (1833-1897) in 1873 [1] as a way of dealing with the pre-Maxwellian velocity dependent electrodynamic theory of Wilhelm Eduard Weber (1804-1891). [2] It was coined as the Schering potential by Edmund Taylor Whitaker (1873-1956) in the first edition of his **Analytical Dynamics** [3], but he dropped this attribution in later editions.

Equation (D-3) is often referred to as the Lagrange's equations, but this designation is normally reserved for the form of equation (D-3) when the system is conservative, i.e. when the forces are derivable from a scalar potential function, V, which is the potential energy of the system.

$$F_i = -\nabla_i V \qquad (D-5)$$

In the past the most important application for the use of the generalized potential was the electromagnetic forces on moving charges. In Gaussian units the Maxwell equations are

$$\nabla \times \vec{E} + \frac{1}{c}\frac{\partial \vec{B}}{\partial t} = 0 \qquad \nabla \cdot \vec{D} = 4\pi\rho$$

$$\nabla \times \vec{H} - \frac{1}{c}\frac{\partial \vec{D}}{\partial t} = \frac{4\pi}{c}\vec{j} \qquad \nabla \cdot B = 0 \qquad (D-6)$$

The force on a charge q is not given entirely by the electric force

$$\vec{F} = q\vec{E} = -q\nabla\varphi \qquad (D-7)$$

so that the system is not conservative in this sense. Instead, the complete force is given by the Lorentz force

$$\vec{F} = q\left\{\vec{E} + \frac{1}{c}(\vec{v} \times \vec{B})\right\} \qquad (D-8)$$

Here E is not the gradient of a scalar function, but from the divergence of B = 0, it follows that B can be represented by the curl of a vector

$$\vec{B} = \nabla \times \vec{A} \qquad (D-9)$$

where A is called the magnetic vector potential in vector calculus. Using the magnetic vector potential the curl E equation in equation (D-6) becomes

$$\nabla \times \vec{E} + \frac{1}{c}\frac{\partial}{\partial t}(\nabla \times \vec{A}) = \nabla \times \left(\vec{E} + \frac{1}{c}\frac{\partial\vec{A}}{\partial t}\right) = 0 \qquad (D-10)$$

Hence we can set

$$\vec{E} + \frac{1}{c}\frac{\partial\vec{A}}{\partial t} = -\nabla\varphi \ or \ \vec{E} = -\nabla\varphi - \frac{1}{c}\frac{\partial\vec{A}}{\partial t} \qquad (D-11)$$

In terms of the potentials φ and A, the Lorentz force becomes

$$\vec{F} = q\left\{-\nabla\varphi - \frac{1}{c}\frac{\partial\vec{A}}{\partial t} + \frac{1}{c}(\vec{v} \times \nabla \times \vec{A}))\right\} \qquad (D-12)$$

This approach based on the Maxwell equation (D-6) that the divergence of **B** = 0, which says that there are no magnetic monopoles is now being questioned. Magnetic monopole quasi-particles appear to exist in some condensed matter systems. [4, 5, 6]

Figure D-1 Dirac Strings and Magnetic Monopoles in Spin Ice Dy2Ti2O7 [7, 8]

The approach of the original form of the generalized potential U developed for Weber's electrodynamics can handle magnetic monopoles. **The vector potential approach cannot!**

D: References

1. E. Schering, "**Gött. Abh.**", **18**, 3 (1873).
2. W. Weber, "**Annalen d. Phys**", **LXXIII**, 193 (1848).
3. E. T. Whittaker, **A Treatise on the Analytical Dynamics of Particles and Rigid Bodies** (Cambridge University Press, Cambridge, 1904) 1st. edition.
4. http://en.wikipedia.org/wiki/Magnetic_monopole
5. http//www.popsci.com/technology/article/2009-10/newly-dicovered-monopole-particles-flow-electric-currents
6. http://www.sciencedaily.com/releases/2009/09/090903163725.htm
7. D.J.P. Morris, D.A. Tennant, S.A. Grigera, B. Klemke, C. Castelnovo, R. Moessner, C. Czter-nasty, M. Meissner, K.C. Rule, J.-U. Hoffmann, K. Kiefer, S. Gerischer, D. Slobinsky, and R.S. Perry. "Dirac Strings and Magnetic Monopoles in Spin Ice $Dy_2Ti_2O_7$." **Science**, 2009
8. http://physicsworld.com/cws/article/news/2009/sep/03/magnetic-monopoles-spotted-in-spin-ices

Appendix E: Poincaré's Views on Relativity, Gravity, Lorentz Transformation and Electrodynamics

In the introduction to his famous paper on the "Dynamics of the Electron" [1] in which Poincaré talks about relativity theory and the Lorentz transformation, he says (my translation from the original French to English)

> **We cannot be content with a simple juxtaposition of formulas that agree with each other by good fortune alone; these formulas must, in a manner of speaking, interpenetrate. The mind will be satisfied only when it believes it has perceived the reason for this agreement, and the belief is strong enough to entertain the illusion that it could have been predicted.**

> **But the question may be viewed from a different perspective, better shown via an analogy. Let us imagine a pre-Copernican astronomer who reflects on Ptolemy's system; he will notice that for all the planets, one of two circles—epicycle or deferent—is traversed in the same time. This fact cannot be due to chance, and consequently between all the planets there is a mysterious link we can only guess at.**

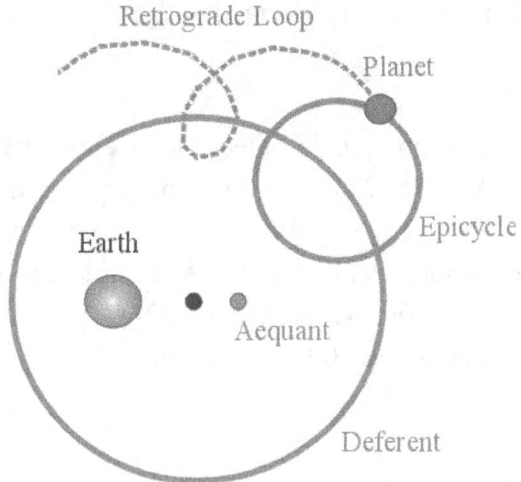

Figure E-1 Ptolemy's Epicycle and Deferent Orbits

> **Copernicus, however, destroys this apparent link by a simple change in the coordinate axes that were considered fixed. Each planet now describes a single circle (about the sun), and orbital periods become independent.**

> **It is possible that something analogous is taking place here. If we were to admit the postulate of relativity, we would find the same number in the law of gravitation and the laws of electromagnetism—the speed of light—and**

we would find it again in all other forces of any origin whatsoever. This state of affairs may be explained in one of two ways: either everything in the universe would be of electromagnetic origin, or this aspect—shared, as it were, by all physical phenomena—would be a mere epiphenomenon, something due to our methods of measurement. How do we go about measuring? The first response will be: we transport objects considered to be invariable solids, one on top of the other. But that is no longer true in the current theory, if we admit the Lorentzian contraction. In this theory, two lengths are equal, by definition, if they are traversed by light in equal times.

Perhaps if we were to abandon this definition Lorentz's theory would be as fully overthrown as was Ptolemy's system by Copernicus's intervention. Should that happen someday, it would not prove that Lorentz's efforts were in vain, because regardless of what one may think, Ptolemy was useful to Copernicus.

I, too, have not hesitated to publish these few partial results, even if at this very moment the discovery of magneto-cathode rays seems to threaten the entire Lorentzian theory.

Poincaré also used arguments from metatheory, like those above, to say that no two fundamental theories should use the same fundamental constants such as c and that no two fundamental theories should have the same form such as $1/R^2$ such as the electrodynamic force and the force of gravity. Poincaré believed that the simplest explanation was that "there is nothing in the world that is not of electromagnetic origin".

E: References

1. Henri Poincaré, "La dynamique de l'électron," **Rendiconti del Circolo matematico di Palermo 21**, pp. 129-176 (1906); reprinted in **Oeuvres de Henri Poincaré, vol. 9** (Paris: Gauthier-Villars, 1954), pp. 494-550. A brief summary of the main results had previously been given in the **Comptes rendus of the Académie des Sciences, Paris**, **140**, 1504-8 (5 June 1905); Poincaré, **Oeuvres, vol. 9**, pp. 489-93. See http://www.phys.lsu.edu/mog/100/poincare.pdf

INDEX

Action-at-a-distance, 26, 45, 68, 69, 84

Ampere, 27, 77, 81, 82, 83, 84, 99, 102, 231, 232, 233, 234, 235

Analysis and Synthesis, 24, 180

Anthropology, 76

Aristotle, 78

Assis, 16, 233, 235

Atomism, 67

Axiomatic, 1, 2, 6, 9, 13, 16, 17, 18, 19, 26, 29, 33, 61, 68, 69, 76, 77, 181, 202, 205, 206, 211

Axioms, 81

Balfour Stewart, 50

Big Bang, 11

Biot-Savart law, 83

Black Body Radiation, 50

Bode's Law, 4, 5, 36, 142, 150, 164, 191, 193, 198, 220

Bourbaki, 75, 76, 77, 78

Causality, 67

Centripetal Forces, 22

Combinatorial Geometry, 10, 15

Common Sense Science, 15

Constructive Theories, 10

Contact Forces, 22, 45, 67, 81, 92

Coulomb Gauge, 84

Cullwick, 44, 46, 79, 84, 93

Dark Energy, 11, 24

Dark Matter, 10, 11, 24

Dave Bergman, 15

Democritus, 67

Descartes, 22, 23, 31

Determinism, 78, 92

Discrete Quanta, 41

Economics, 9, 11, 76, 230

Einstein, 10, 16, 24, 41, 69, 78, 89

Electrodynamic Feedback Effects, 15

Empirical Scientific Methods, 13, 16

Euclid, 1, 9, 11, 18, 31, 68, 206

Existentialism, 26, 69, 76

Experimental Philosophy, 24, 180

Falsifiability, 77

Faraday, 27, 77, 81, 82, 84, 85, 86, 87, 88, 93

Feedback Effects, 93

Finite-size, 10, 15, 81, 83, 84, 88, 91, 93, 105

Fizeau, 39

Force of Inertia, 20, 108

Galilean Invariance, 81, 87, 88

Galilean Relativity, 29, 30, 39, 84

Galilean Transformation, 15, 82, 84, 86, 92, 93, 96

Galileo, 22, 180

Gauss, 27, 77, 81, 82, 91

General Relativity, 9, 10, 24

Grassmann, 83, 233, 234, 235

Gravitational Mass, 21

Heaviside, 234, 235

Helmholtz, 81, 233, 235, 237

Helmholtz's Theorem, 81

Hertz and Hallwachs, 41

Hierarchy of Electrodynamic Interactions, 113, 151, 164

Hooper, 44, 45, 46, 78, 79, 84, 85, 93, 101, 102

Huygens, 21, 22, 23, 31

Hypotheses, 9, 11, 21, 22, 24, 26, 69, 180, 181

Inertial Mass, 21, 26, 68

Isaac Newton, 1, 9, 15, 16, 19, 31, 34, 68, 119, 151, 155, 165, 167, 177, 194, 204

Iteration, 89, 90

Karl Popper, 77, 79

Kirchoff, 50

Larmor Formula, 101

Law of Cause and Effect, 67, 69, 77

Lee Smolin, 9, 13

Lenz, 27, 30, 44, 45, 77, 81, 82, 89, 96, 238

Leucippus, 67

Lienard, 100, 102

Lienard-Wichert, 100

Linguistics, 11, 75, 76

Literary Theory, 76

Longitudinal Forces, 84

Lorentz, 29, 30, 82, 88, 89, 92, 103

Mach, 10, 26, 44, 45, 62, 81, 89, 92, 93, 120, 156, 160, 164, 181, 195, 196, 198, 202, 203, 204, 205, 208, 214, 215

Maxwell, 2, 28, 29, 30, 33, 34, 35, 45, 60, 61, 66, 81, 83, 84, 85, 87, 89, 93, 96, 99, 104, 107, 121, 150, 179, 205, 207, 211, 213, 215, 217, 218, 220, 224, 231, 234, 235, 243, 244

Metatheory, 77, 109, 110, 248

Michelson and Morley, 39

MOND, 145, 146

Moon and Spencer, 85

Natural Philosophy, 24, 31, 67, 68, 78, 108, 151, 165, 177

Newton's 3rd Law, 30

Oersted, 84, 93, 231, 235

Photoelectric Effect, 41, 43

Photon, 34, 78

Pillars of Modern Science, 10, 11

Poincaré, 12, 13, 33, 39, 58, 61, 66, 69, 70, 71, 72, 73, 74, 75, 77, 78, 111, 120, 154, 205, 206, 207, 247, 248

Postmodern Philosophy, 11, 76

Poynting vector, 101

Principia, 1, 19, 21, 22, 23, 31, 78, 151, 165, 177, 180, 196, 204

Principles of Natural Philosophy, 19

Psychology, 9, 11, 76

Quantum Mechanics, 9

Radiation Reaction, 103, 104, 105

Realism, 78, 92

Reformation, 11

Relativity, 9, 10, 11, 16, 34, 69, 78, 82, 93

Resistive Forces, 23

Retardation Effects, 45, 84

Self-Field Effects, 84

Self-Fields, 103

Semiotics, 76

Standard Model of Cosmology, 10, 11

Standard Model of Elementary Particles, 10

String Theory, 9, 11, 101

Structuralism, 75, 76, 77

Superposition Principle, 85, 100

Symmetry of the Universal Force, 11

Theory of Principle, 10

Thermodynamics, 69

Universal Force Law, 6, 7, 9, 10, 11, 151, 208

Universal Law of Gravitation, 20

Universal Wave Function, 69

Weber, 16, 93, 100, 233

Whittaker, 233, 235